Lecture Notes in Applied and Computational Mechanics

Volume 25

Series Editors

Prof. Dr.-Ing. Friedrich Pfeiffer
Prof. Dr.-Ing. Peter Wriggers

Lecture Notes in Applied and Computational Mechanics

Edited by F. Pfeiffer and P. Wriggers

Vol. 25: Chen C.-N.
Discrete Element Analysis Methods of Generic Differential Quadratures
294 p. 2006 [3-540-28947-X]

Vol. 24: Schenk C., Schuëller G. (Eds.)
Uncertainty Assessment of Large Finite Element Systems
400 p. 2005 [3-540-25343-2]

Vol. 23: Frémond M., Maceri F. (Eds.)
Mechanical Modelling and Computational Issues in Civil Engineering
400 p. 2005 [3-540-25567-3]

Vol. 22: Chang C.H.
Mechanics of Elastic Structures with Inclined Members: Analysis of Vibration, Buckling and Bending of X-Braced Frames and Conical Shells
190 p. 2004 [3-540-24384-4]

Vol. 21: Hinkelmann R.
Efficient Numerical Methods and Information-Processing Techniques for Modeling Hydro- and Environmental Systems
305 p. 2005 [3-540-24146-9]

Vol. 20: Zohdi T.I., Wriggers P.
Introduction to Computational Micromechanics
196 p. 2005 [3-540-22820-9]

Vol. 19: McCallen R., Browand F., Ross J. (Eds.)
The Aerodynamics of Heavy Vehicles: Trucks, Buses, and Trains
567 p. 2004 [3-540-22088-7]

Vol. 18: Leine, R.I., Nijmeijer, H.
Dynamics and Bifurcations of Non-Smooth Mechanical Systems
236 p. 2004 [3-540-21987-0]

Vol. 17: Hurtado, J.E.
Structural Reliability: Statistical Learning Perspectives
257 p. 2004 [3-540-21963-3]

Vol. 16: Kienzler R., Altenbach H., Ott I. (Eds.)
Theories of Plates and Shells: Critical Review and New Applications
238 p. 2004 [3-540-20997-2]

Vol. 15: Dyszlewicz, J.
Micropolar Theory of Elasticity
356 p. 2004 [3-540-41835-0]

Vol. 14: Frémond M., Maceri F. (Eds.)
Novel Approaches in Civil Engineering
400 p. 2003 [3-540-41836-9]

Vol. 13: Kolymbas D. (Eds.)
Advanced Mathematical and Computational Geomechanics
315 p. 2003 [3-540-40547-X]

Vol. 12: Wendland W., Efendiev M. (Eds.)
Analysis and Simulation of Multifield Problems
381 p. 2003 [3-540-00696-6]

Vol. 11: Hutter K., Kirchner N. (Eds.)
Dynamic Response of Granular and Porous Materials under Large and Catastrophic Deformations
426 p. 2003 [3-540-00849-7]

Vol. 10: Hutter K., Baaser H. (Eds.)
Deformation and Failure in Metallic Materials
409 p. 2003 [3-540-00848-9]

Vol. 9: Skrzypek J., Ganczarski A.W. (Eds.)
Anisotropic Behaviour of Damaged Materials
366 p. 2003 [3-540-00437-8]

Vol. 8: Kowalski, S.J.
Thermomechanics of Drying Processes
365 p. 2003 [3-540-00412-2]

Vol. 7: Shlyannikov, V.N.
Elastic-Plastic Mixed-Mode Fracture Criteria and Parameters
246 p. 2002 [3-540-44316-9]

Vol. 6: Popp K., Schiehlen W. (Eds.)
System Dynamics and Long-Term Behaviour of Railway Vehicles, Track and Subgrade
488 p. 2002 [3-540-43892-0]

Vol. 5: Duddeck, F.M.E.
Fourier BEM: Generalization of Boundary Element Method by Fourier Transform
181 p. 2002 [3-540-43138-1]

Vol. 4: Yuan, H.
Numerical Assessments of Cracks in Elastic-Plastic Materials
311 p. 2002 [3-540-43336-8]

Vol. 3: Sextro, W.
Dynamical Contact Problems with Friction: Models, Experiments and Applications
159 p. 2002 [3-540-43023-7]

Vol. 2: Schanz, M.
Wave Propagation in Viscoelastic and Poroelastic Continua
170 p. 2001 [3-540-41632-3]

Vol. 1: Glocker, C.
Set-Valued Force Laws: Dynamics of Non-Smooth Systems
222 p. 2001 [3-540-41436-3]

Discrete Element Analysis Methods of Generic Differential Quadratures

Chang-New Chen

Springer

Professor Chang-New Chen
Department of Systems and
Naval Mechatronic Engineering
National Cheng Kung University
Tainan, Taiwan
Republic of China
E-mail: cchen@mail.ncku.edu.tw

With 86 Figures

ISSN print edition: 1613-7736
ISSN electronic edition: 1860-0816

ISBN-10 3-540-28947-X Springer Berlin Heidelberg New York
ISBN-13 978-3-540-28947-0 Springer Berlin Heidelberg New York

Library of Congress Control Number: 2005933610

Springer is a part of Springer Science+Business Media

springeronline.com

Typesetting: by the author and TechBooks using a Springer LaTeX macro package
Cover design: *design & production* GmbH, Heidelberg

Printed on acid-free paper SPIN: 10816924 89/TechBooks 5 4 3 2 1 0

Preface

Following the advance in computer technology, the numerical technique has made significant progress in the past decades. Among the major techniques available for numerically analyzing continuum mechanics problems, finite difference method is most early developed. It is difficult to deal with continuum mechanics problems showing complex curvilinear geometries by using this method. The other method that can consistently discretize continuum mechanics problems showing arbitrarily complex geometries is finite element method. In addition, boundary element method is also a useful numerical method.

In the past decade, the differential quadrature and generic differential quadratures based discrete element analysis methods have been developed and used to solve various continuum mechanics problems. These methods have the same advantage as finite element method of consistently discretizing continuum mechanics problems having arbitrarily complex geometries. This book includes my research results obtained in developing the related novel discrete element analysis methods using both of the extended differential quadrature based spacial and temporal elements. It is attempted to introduce the developed numerical techniques as applied to the solution of various continuum mechanics problems, systematically.

This book is divided into sixteen chapters. Chapter 1 gives an overview to the developed numerical techniques. Chapter 2 introduces the generalization of differential quadrature – extended differential quadrature, with various approximation techniques. Chapter 3 uses the one-dimensional elasticity problem as example to present the discretization procedures, and assemblage and solution techniques for the resulting overall algebraic systems, of differential quadrature element method as applied to the solutions of static deformation and free vibration problems. The application of differential quadrature element method to various structural mechanics problems like Euler-Bernoulli beams, trusses, frames, Timoshenko beams and curved beams is dealt with in Chapters 4 to 9. Chapter 10 uses the two-dimensional problem as example to introduce the development of irregular elements for the discrete element analyses of continuum mechanics problems having arbitrarily curved domain configurations. The application of differential quadrature element method to general field problems with two-dimensional domains is presented in Chapter 11. Chapter 12 deals with the application of differential quadrature element method to the analysis of two-dimensional elasticity problems with composite materials. The application of differential quadrature element method to Kirchhoff-Love plate problems is introduced in Chapter 13. The application of differential quadrature to the derivation of novel finite difference operators and the use of the related analysis method, differential quadrature finite difference method, to the solutions of various continuum mechanics problems are presented in Chapter 14. The application of a generalized coordinate differential quadrature element method to some continuum mechanics problems is introduced in Chapter 15. Finally, the application of using extended differential quadrature to the development of direct time integration schemes for solving general discrete transient equation systems is presented in Chapter 16.

Chang-New Chen

Contents

1

Introduction

The method of differential quadrature (DQ), proposed by Bellman, is an effective technique for solution of differential or partial differential equations [1–11]. The DQ approximates a derivative or partial derivative of a variable function ϕ with respect to a coordinate variable ξ at a node as a weighted linear sum of the function values at all nodes along that coordinate direction. The variable function is a function of the coordinate variable. Thus, the DQ discretization for a partial derivative of order m can be expressed by

$$\frac{\partial^m \phi_\alpha}{\partial \xi^m} = D^{\xi^m}_{\alpha\bar{\alpha}} \Phi_{\bar{\alpha}}, \quad \alpha, \bar{\alpha} = 1, 2, ..., N_\xi \tag{1.1}$$

where N_ξ is the number of nodes in ξ direction, α is a free index taking the values 1,2,...,N_ξ, $\bar{\alpha}$ is a dummy index regarded as being summed from 1 to N_ξ, $\Phi_{\bar{\alpha}}$ are function values at the nodes and $D^{\xi^m}_{\alpha\bar{\alpha}}$ are weighting coefficients attached to the function values at the N_ξ node points.

1.1 Computation of Weighting Coefficients

Various methods can be used to calculate the weighting coefficients $D^{\xi^m}_{\alpha\bar{\alpha}}$.

1.1.1 Explicit Computation

Shifted Legendre polynomials can be used to explicitly calculate the weighting coefficients. Assume that the range of ξ is $0 \le \xi \le 1$. Legendre polynomials can be generated from the following recurrent relation

$$(n+1)P_{n+1}(\xi) = \xi(2n+1)P_n(\xi) - nP_{n-1}(\xi) \tag{1.2}$$

with the initial member $P_0(\xi) = 0$. The following Legendre polynomials can also be defined

$$\tilde{P}_n(\xi) = \frac{2^n n! \, , n!}{(2n)!} \sum_{k=0}^{m} (-1)^k \frac{(2n-2k)! \, , \xi^{(n-2k)}}{2^n k! \, , (n-k)! \, , (n-2k)!} \tag{1.3}$$

Shifted Legendre polynomials can be defined by using the Legendre polynomials $\tilde{P}_n(\xi)$ and expressed as

$$\hat{P}_n(\xi) = \tilde{P}_n(\xi)(1 - 2\xi) \tag{1.4}$$

The following test functions are defined by using the shifted Legendre polynomials

$$p_\beta(\xi) = \frac{\hat{P}_N(\xi)}{(\xi - \xi_\beta)\hat{P}_N^{(1)}(\xi_\beta)} \tag{1.5}$$

where ξ_β are roots of $\hat{P}_N(\xi)$ and $\hat{P}^{(1)}(\xi_\beta)$ the first derivative of $\hat{P}_N(\xi)$ with respect to ξ at β. The test function is a polynomial of degree $(N-1)$ holding the relation $p_\beta(\xi_\alpha) = \delta_{\alpha\beta}$. The introduction of $p_\beta(\xi)$ into the definition equation of the first order DQ discretization equation at the discrete node point ξ_α of a root of $\hat{P}_N(\xi)$ leads to obtaining the following weighting coefficients

$$D^\xi_{\alpha\beta} = \frac{\hat{P}_N^{(1)}(\xi_\alpha)}{(\xi_\alpha - \xi_\beta)\hat{P}_N^{(1)}(\xi_\beta)}, \quad \text{for} \quad \alpha \neq \beta \tag{1.6}$$

Consider the following transformed Legendre differential equation

$$\xi(1-\xi)\hat{P}_N^{(2)}(\xi) + (1-2\xi)\hat{P}_N^{(1)}(\xi) + N(N+1)\hat{P}_N(\xi) = 0 \tag{1.7}$$

By using the L'Hospital's rule and the above equation, the weighting coefficients for $\alpha = \beta$ can be obtained

$$D^\xi_{\alpha\alpha} = \frac{1 - 2\xi_\alpha}{2\xi_\alpha(\xi_\alpha - 1)} \tag{1.8}$$

1.1.2 Implicit Computation

The variable function can be approximated by an appropriate analytical function such as the polynomial,

$$\phi(\xi) = \xi^{p-1}, \quad p = 1, 2, ..., N_\xi \tag{1.9}$$

Substituting Eq. (1.9) in Eq. (1.1) leads to N_ξ sets of linear algebraic system with one row of $D^{\xi^m}_{\alpha\bar{\alpha}}$ as the unknowns of one linear algebraic set. $D^{\xi^m}_{\alpha\bar{\alpha}}$ can be obtained by solving the N_ξ sets of linear algebraic system. The coefficient matrix of each set of linear algebraic system is a Vandermonde matrix. A unique solution can thus be obtained.

DQ has been used to solve various continuum mechanics problems with the boundary conditions only involving prescribed variable functions [7–17].

For solving flexural deformation problems of structures with the kinematic boundary conditions involving deflection slopes, a δ-grid arrangement is used to define the DQ discretization. The δ-grid is designed to approximately define certain boundary conditions at a point close to the boundary [18–19]. Consequently, the definition of boundary conditions is inconsistent.

The DQ can be only used to discretize problems having regular domain configurations. Consequently, its application is very limited.

A discrete element analysis method – differential quadrature element method (DQEM) has been developed for solving generic scientific and engineering problems having arbitrarily irregular domain configurations and external environments [20]. The DQEM has been used to develop solution algorithms for solving various problems [21–27].

Like finite element method, for the DQEM the domain of problems is separated into many subdomains or elements. Then the differential quadrature discretization is carried out on an element-basis. The governing differential or partial differential equation defined on the elements, the transition conditions on inter-element boundaries and the boundary conditions on the boundary of problem domains are in computable algebraic forms after the differential quadrature discretization. In order to solve the problem, all discretized governing equations, transition conditions and boundary conditions have to be assembled to obtain a global algebraic system. Since all relations governing a continuous problem are satisfied, the essence of this method is to find a rigorous solution numerically. For solving problems having a curved domain boundary, the interior elements can be regular. However, in order to solve the problem having an arbitrary analysis domain configuration, elements connected to or near the analysis domain boundary might need to be irregular. The mapping technique can be used to develop irregular elements. Therefore this method has the same advantage as the finite element method of geometric flexibility. Hence a generic engineering or scientific problem can be converted into a numerical differential quadrature element algorithm. And the related computer code can be systematically developed.

The gradient of a response function in the problem domain will depend on the domain configuration and distribution of external causes. The adaptive concept can be used to efficiently solve a generic engineering or scientific problem. The DQEM is suitable for adaptively discretize a continuous problem by using various elements with variously assumed variable functions, simultaneously.

In treating a concentrated external cause existing in the problem domain, two approaches are available. One of which is to generate the mesh by locating the concentrated external cause on some inter-element boundaries and including it into the natural transition conditions. The second one is to locate the concentrated external cause in an element domain and use certain continuous function defined over the element domain to approximate it based on the rule of force equivalence.

There is also a discrete element analysis technique QEM which also adopts the DQ. The original QEM was proposed to solve truss and frame structures [28]. In this method, the truss element is limited to a three-node second-order approximation, while a δ-grid arrangement is used to define the DQ discretization of the flexural deformation [18]. Consequently, the definition of boundary conditions and inter-element transition conditions, is inconsistent. When developing the plane stress and plate bending QEM models, Striz et al adopted a hybrid technique to incorporate the DQ discretization into a Galerkin finite element formulation and define a discrete element analysis procedure [29].

In contrast to the QEM, the DQEM which can be used to develop solution algorithms for the analysis of flexural problem of structures with the inter-element transition conditions and boundary conditions numerically, exactly satisfied [20–24]. Some numerical results involving the use of DQ to the solutions of problems having a specifically nonrectangular domain configuration can also be found in existing articles [30–33].

Generalization of DQ has been carried out. An extended differential quadrature (EDQ) was developed. For the EDQ, a certain order derivative or partial derivative of the variable function with respect to the coordinate variables at a discrete point is expressed as the weighted linear sum of the values of variable function and/or its possible derivatives or partial derivatives at all nodes [34–35]. The node and discrete point can be different. Consequently, more analytical functions can be used to define the EDQ discretization. By using certain analytical functions such as Hermite polynomials, not only derivatives can be considered as independent variables but also only simple algebraic operations are necessary for computing the weighting coefficients. The weighting coefficients for a grid model defined by a coordinate system having arbitrary dimensions can also be generated. The configuration of a grid model can be arbitrary. Consequently, problems with an arbitrary domain configuration can be solved. Discrete element analysis method for solving problems having an irregular domain without adopting the mapping technique can also be constructed using this method to the element basis discretization [37].

Instead of using the function variables at nodes to define the differential quadrature, a generalized coordinate differential quadrature (GCDQ) adopts the generalized coordinates to define the differential quadrature discretization and the related discrete element analysis method – generalized coordinate differential quadrature element method (GCDQEM) [38]. The computation of GCDQ weighting coefficients is always explicit.

A differential quadrature finite difference method (DQFDM) has also been developed [39–41]. DQ or EDQ which adopts the values of variable function at discrete points, only, to define the EDQ discretization, is used to derive the finite difference operators. For a specified DQFDM grid with a specified number of grid points, the order of DQ or GDQ approximation for deriving the finite difference operators defined at a discrete point in a discrete element or a subdomain can be flexible. Different orders of derivatives existing in the fundamental equations can have different orders of DQ or EDQ approximations. It is different from the DQEM in which the orders of DQ approximations for all orders of derivatives defined in an element are unique. The approximate analytical functions in a DQEM element always span over the whole element. Higher order DQFDM analysis models can be easily developed. Irregular DQFDM analysis models can be developed by using the mapping technique or EDQ. DQ can also be used to the integral statement of the finite element method. A discrete element analysis method - differential quadrature finite element method (DQFEM) has also been developed [42–43]. This method uses DQ or EDQ to the element basis integral statement and carries out the weak formulation.

The various discrete element analysis methods also have the same advantage as the finite element method of general geometry and systematic boundary treatment. These methods need less computer memory requirements than the FEM. For solving free vibration problems, the mass matrix is diagonal which requires a little storage space. The mass matrix is simpler to form and cheaper to use as compared to the consistent mass matrix used in the FEM analysis.

DQ and EDQ can also be used to develop the time-element by time-element and stages by stages direct time integration methods for solving discrete transient problems existing in various areas of science and engineering [44–46].

2

Generalization of DQ – Extended Differential Quadrature

For the EDQ, a derivative or partial derivative of the variable function with respect to the coordinate variables at an arbitrary discrete point is expressed as a weighted linear sum of the values of function and/or its possible derivatives at all grid nodes. The grid pattern can be fixed while the selection of discrete points for defining discrete fundamental relations is flexible.

The configuration of EDQ grid model can be irregular. The design of grid model is flexible. Weighting coefficients for general multi-coordinate grid models having arbitrary configurations can also be calculated. Three methods for generating the weighting coefficients can be used. Typical procedures of EDQ discretizations for grid models defined by one, two, three, four and arbitrary finite-coordinate grid models are summarized.

2.1 One-Coordinate Grid Model

The grid configuration of the one-coordinate grid model can be a straight or curved line. The EDQ discretization for a derivative of order m of the variable function ϕ at discrete point α can be expressed by

$$\frac{d^m \phi_\alpha}{d\xi^m} = D_{\alpha i}^{\xi^m} \tilde{\Phi}_i, \quad i = 1, 2, ..., \bar{N} \tag{2.1}$$

where $\bar{N}$ is the number of degrees of freedom and $\tilde{\Phi}_i$ the values of variable function and/or its possible derivatives at the N nodes.

Three possible approaches can be used to calculate the weighting coefficients. The first approach is to assume that the variable function can be a set of appropriate analytical functions denoted by $\Upsilon_p(\xi)$. The substitution of $\Upsilon_p(\xi)$ in Eq. (2.1) leads to a linear algebraic system for implicitly determining the weighting coefficients $D_{\alpha i}^{\xi^m}$.

Assume that the variable function can be approximated by using interpolation functions

$$\phi(\xi) = \Psi_p(\xi)\tilde{\Phi}_p, \quad p = 1, 2, ..., \bar{N} \tag{2.2}$$

where $\Psi_p(\xi)$ are the corresponding interpolation functions of $\tilde{\Phi}_p$. Adopting $\Psi_p(\xi)$ as the variable function $\phi(\xi)$ then substitute it in Eq. (2.1), a linear algebraic system for determining $D_{\alpha i}^{\xi^m}$ can be obtained. And the mth order differentiation of Eq. (2.2) at discrete point α also leads to the EDQ discretization equation (2.1) in which $D_{\alpha i}^{\xi^m}$ is expressed by

$$D_{\alpha i}^{\xi^m} = \frac{d^m \Psi_i}{d\xi^m} |_\alpha \tag{2.3}$$

Using the above equation, the weighting coefficients can be easily obtained by simple algebraic calculations. This second approach is an explicit method.

The third approach is to approximate the variable function by the following equation

$$\phi(\xi) = \Upsilon_p(\xi)c_p, \quad p = 1, 2, ..., \bar{N} \tag{2.4}$$

where $\Upsilon_p(\xi)$ are appropriate analytical functions and c_p are unknown coefficients. The constraint conditions at all nodes can be expressed as

$$\tilde{\Phi}_p = \chi_{p\bar{p}} c_{\bar{p}} \tag{2.5}$$

where $\chi_{p\bar{p}}$ are composed of the values of $\Upsilon_p(\xi)$ and/or their possible derivatives at all nodes. Solving Eq. (2.5) for $c_{\bar{p}}$ and substituting them back into Eq. (2.4), the variable function can be rewritten as

$$\phi(\xi) = \Upsilon_p(\xi)\chi_{\bar{p}p}^{-1}\tilde{\Phi}_{\bar{p}} \tag{2.6}$$

Using Eq. (2.6), the weighting coefficients can also be obtained.

$$D_{\alpha i}^{\xi^m} = \frac{\partial^m \Upsilon_{\bar{p}}}{\partial \xi^m} |_\alpha \chi_{i\bar{p}}^{-1} \tag{2.7}$$

This third method calculates the weighting coefficients implicitly.

2.2 Two-Coordinate Grid Model

The grid configuration of a two-coordinate grid model can be a triangle, a quadrilateral or a certain other configuration. The dimensions for defining the discrete point and node can be different. By adopting a one-dimensional node identification method to express both the discrete point and node, the EDQ discretization for a partial derivative of order $m+n$ at discrete point α can be expressed by

$$\frac{\partial^{(m+n)}\Phi_\alpha}{\partial \xi^m \partial \eta^n} = D_{\alpha i}^{\xi^m \eta^n} \tilde{\Phi}_i, \quad i = 1, 2, ..., \bar{N} \tag{2.8}$$

The variable function can be a set of appropriate analytical functions denoted by $\Upsilon_j(\xi, \eta)$. The substitution of $\Upsilon_j(\xi, \eta)$ in equation (2.8) leads to a linear algebraic system for determining $D_{\alpha i}^{\xi^m \eta^n}$. The set of analytical functions can also be expressed by a tensor having an order other than one. The variable function can also be approximated by using interpolation functions

$$\phi(\xi,\eta) = \Psi_j(\xi,\eta)\tilde{\Phi}_j, \quad j = 1, 2, ..., \bar{N} \tag{2.9}$$

where $\tilde{\Phi}_j$ are the values of variable function and/or its possible partial derivatives at the N nodes, and $\Psi_j(\xi,\eta)$ are their corresponding interpolation functions. Adopting the set of $\Psi_j(\xi,\eta)$ as the variable function $\phi(\xi,\eta)$, the same procedure can also be used to find $D_{\alpha i}^{\xi^m \eta^n}$. And the $(m+n)$th order partial differentiation of Eq. (2.9) at discrete point α also leads to the EDQ discretization equation (2.8) in which $D_{\alpha i}^{\xi^m \eta^n}$ is expressed by

$$D_{\alpha i}^{\xi^m \eta^n} = \frac{\partial^{(m+n)} \Psi_i}{\partial \xi^m \partial \eta^n} |_\alpha \tag{2.10}$$

The variable function can also be approximated by

$$\phi(\xi,\eta) = \Upsilon_j(\xi,\eta)c_j, \quad j = 1, 2, ..., \bar{N} \tag{2.11}$$

Then the weighting coefficients can also be obtained by

$$D_{\alpha i}^{\xi^m \eta^n} = \frac{\partial^{(m+n)} \Upsilon_{\bar{j}}}{\partial \xi^m \partial \eta^n} |_\alpha \, X_{i\bar{j}}^{-1} \tag{2.12}$$

In Eq. (2.11), the unknown coefficients and appropriate analytical functions can also be expressed by certain other tensors having orders other than one.

By adopting a two-dimensional node identification method to express both discrete point and node, the EDQ discretization for a partial derivative of order $m+n$ at discrete point (α,β) can be expressed by

$$\frac{\partial^{(m+n)} \phi_{\alpha\beta}}{\partial \xi^m \partial \eta^n} = D_{\alpha\beta ij}^{\xi^m \eta^n} \tilde{\Phi}_{ij} \tag{2.13}$$

The variable function can be a set of appropriate analytical functions denoted by $\Upsilon_{pq}(\xi,\eta)$. The substitution of $\Upsilon_{pq}(\xi,\eta)$ in Eq. (2.13) leads to a linear algebraic system for determining $D_{\alpha\beta ij}^{\xi^m \eta^n}$. The set of analytical functions can also be expressed by a tensor having an order other than two. The variable function can also be approximated by

$$\phi(\xi,\eta) = \Psi_{pq}(\xi,\eta)\tilde{\Phi}_{pq} \tag{2.14}$$

where $\tilde{\Phi}_{pq}$ are the values of variable function and/or its possible partial derivatives at the nodes, and $\Psi_{pq}(\xi,\eta)$ are their corresponding interpolation functions. Adopting the set of $\Psi_{pq}(\xi,\eta)$ as the variable function $\phi(\xi,\eta)$, the same procedure can also be used to determine $D_{\alpha\beta ij}^{\xi^m \eta^n}$. And the $(m+n)$th order partial differentiation of Eq. (2.14) at discrete point (α, β) also leads to the EDQ discretization equation (2.13) in which $D^{\xi^m \eta^n}_{\alpha\beta ij}$ is expressed by

$$D^{\xi^m \eta^n}_{\alpha\beta ij} = \frac{\partial^{(m+n)} \Psi_{ij}}{\partial \xi^m \partial \eta^n} |_{\alpha\beta} \tag{2.15}$$

If the variable function $\phi(\xi, \eta)$ can be expressed by a different form

$$\phi(\xi, \eta) = \Psi_{\bar{p}}(\xi) \Psi_{\bar{q}}(\eta) \tilde{\Phi}_{\bar{p}\bar{q}}, \quad \bar{p} = 1, 2, ..., N_{\bar{\xi}}, \quad \bar{q} = 1, 2, ..., N_{\bar{\eta}} \tag{2.16}$$

where $N_{\bar{\xi}}$ and $N_{\bar{\eta}}$ are the numbers of degrees of freedom attached to the nodes in ξ and η directions, respectively, $D^{\xi^m \eta^n}_{\alpha\beta ij}$ can be obtained by the following equation

$$D^{\xi^m \eta^n}_{\alpha\beta ij} = D^{\xi^m}_{\alpha i} D^{\eta^n}_{\beta j} \tag{2.17}$$

There are three types of triangular grid:
(a) Pascal triangular grid,
(b) triangular grid having no interior node,
(c) triangular grid having interior nodes but not the Pascal triangular grid.
And there are also three types of quadrilateral grid:
(a) Lagrange family grid,
(b) quadrilateral grid having no interior node,
(c) quadrilateral grid having interior nodes but not the Lagrange family grid.
The discrete points based grid types can be similarly discussed.

An example is the use of complete polynomials for defining the 2-D triangular EDQ model with Pascal triangular grid and 2-D node identification, and calculating the related weighting coefficients. Let n denote the order of the complete polynomials. These complete polynomials are expressed as

$$\Upsilon_{pk}(x, y) = x^{p+1-k} y^{k-1}, \quad 0 \le p \le n, \quad 1 \le k \le n+1$$

With the above analytical functions, the weighting coefficients can be calculated by using one of the two implicit methods with which a solution of a linear algebraic system is necessary.

2.3 Three-Coordinate Grid Model

The grid configuration of a three-coordinate grid model can be a triangle with the variable function defined by area coordinates, a tetrahedron, a triangular prism, a hexahedron or a certain other configuration. The dimensions for defining the discrete point and node can be different. By adopting a one-dimensional node identification method to express both the discrete point and the node, the EDQ discretization for a partial derivative of order $m + n + o$ at discrete point α can be expressed by

$$\frac{\partial^{(m+n+o)}\phi_\alpha}{\partial\xi^m\eta^n\zeta^o} = D_{\alpha i}^{\xi^m\eta^n\zeta^o}\tilde{\Phi}_i, \quad i = 1, 2, ..., \bar{N} \tag{2.18}$$

The variable function can be a set of appropriate analytical functions denoted by $\Upsilon_j(\xi,\eta,\zeta)$. The substitution of $\Upsilon_j(\xi,\eta,\zeta)$ in Eq. (2.18) leads to a linear algebraic system for determining weighting coefficients $D_{\alpha i}^{\xi^m\eta^n\zeta^o}$. The set of analytical functions can also be expressed by a tensor having an order other than one. The variable function can also be approximated by

$$\phi(\xi,\eta,\zeta) = \Psi_j(\xi,\eta,\zeta)\Phi_j, \quad j = 1, 2, ..., \bar{N} \tag{2.19}$$

where $\Psi_j(\xi,\eta,\zeta)$ are the corresponding interpolation functions of $\tilde{\Phi}_j$. Adopting the set of $\Psi_j(\xi,\eta,\zeta)$ as the variable function $\phi(\xi,\eta,\zeta)$, the same procedure can also be used to determine $D_{\alpha i}^{\xi^m\eta^n\zeta^o}$. And the $(m+n+o)$th order partial differentiation of Eq. (2.19) at discrete point α also leads to the EDQ discretization equation (2.18) in which $D_{\alpha i}^{\xi^m\eta^n\zeta^o}$ is expressed by

$$D_{\alpha i}^{\xi^m\eta^n\zeta^o} = \frac{\partial^{(m+n+o)}\Psi_i}{\partial\xi^m\partial\eta^n\partial\zeta^o}|_\alpha \tag{2.20}$$

The variable function can also be approximated by

$$\phi(\xi,\eta,\zeta) = \Upsilon_j(\xi,\eta,\zeta)c_j, \quad j = 1, 2, ..., \bar{N} \tag{2.21}$$

Then the weighting coefficients can also be obtained by

$$D_{\alpha i}^{\xi^m\eta^n\zeta^o} = \frac{\partial^{(m+n+o)}\Upsilon_{\bar{j}}}{\partial\xi^m\partial\eta^n\partial\zeta^o}|_\alpha \, \chi_{i\bar{j}}^{-1} \tag{2.22}$$

In Eq. (2.21), the unknown coefficients and appropriate analytical functions can also be expressed by certain other tensors having orders other than one.

By adopting a two-dimensional node identification method to express both the discrete point and the node, the EDQ discretization for a partial derivative of order $m+n+o$ at discrete point (α,γ) can be expressed by

$$\frac{\partial^{(m+n+o)}\phi_{\alpha\gamma}}{\partial\xi^m\partial\eta^n\partial\zeta^o} = D_{\alpha\gamma ik}^{\xi^m\eta^n\zeta^o}\tilde{\Phi}_{ik} \tag{2.23}$$

The variable function can be a set of appropriate analytical functions $\Upsilon_{jr}(\xi,\eta,\zeta)$. The substitution of $\Upsilon_{jr}(\xi,\eta,\zeta)$ in Eq. (2.23) leads to a linear algebraic system for determining $D_{\alpha\gamma ik}^{\xi^m\eta^n\zeta^o}$. The set of analytical functions can also be expressed by a tensor having an order other than two. The variable function can also be approximated by

$$\phi(\xi,\eta,\zeta) = \Psi_{jr}(\xi,\eta,\zeta)\tilde{\Phi}_{jr} \tag{2.24}$$

where $\tilde{\Phi}_{jr}$ are values of variable function and/or its possible partial derivatives at the nodes, and $\Psi_{jr}(\xi,\eta,\zeta)$ are their corresponding interpolation functions.

Adopting the set of $\Psi_{jr}(\xi, \eta, \zeta)$ as the variable function $\phi(\xi, \eta, \zeta)$, the same procedure can also be used to determine $D_{\alpha\gamma ik}^{\xi^m\eta^n\zeta^o}$. And the $(m+n+o)$th order partial differentiation of Eq. (2.24) at discrete point (α, γ) also leads to the EDQ discretization equation (2.23) in which $D_{\alpha\gamma ik}^{\xi^m\eta^n\zeta^o}$ is expressed by

$$D_{\alpha\gamma ik}^{\xi^m\eta^n\zeta^o} = \frac{\partial^{(m+n+o)}\Psi_{ik}}{\partial\xi^m\partial\eta^n\partial\zeta^o}\Big|_{\alpha\gamma} \tag{2.25}$$

Consider that the interpolation functions can be expressed by forming products of two sets of functions with one set defined by two coordinate variables while the other set defined by the remaining coordinate variable. Then a representative of this type of interpolation functions can be $\Psi_j(\xi, \eta)\Psi_r(\zeta)$. By using this type of interpolation functions, the weighting coefficients show to have the following form:

$$D_{\alpha\gamma ik}^{\xi^m\eta^n\zeta^o} = D_{\alpha i}^{\xi^m\eta^n} D_{\gamma k}^{\zeta^o} \tag{2.26}$$

By adopting a three-dimensional node identification method to express both the discrete point and node, the EDQ discretization for a partial derivative of order $m+n+o$ at discrete point (α, β, γ) can be expressed by

$$\frac{\partial^{(m+n+o)}\phi_{\alpha\beta\gamma}}{\partial\xi^m\partial\eta^n\zeta^o} = D_{\alpha\beta\gamma ijk}^{\xi^m\eta^n\zeta^o}\tilde{\Phi}_{ijk} \tag{2.27}$$

The variable function can be a set of appropriate analytical functions denoted by $\Upsilon_{pqr}(\xi, \eta, \zeta)$. The substitution of $\Upsilon_{pqr}(\xi, \eta, \zeta)$ in equation (2.27) leads to a linear algebraic system for determining $D_{\alpha\beta\gamma ijk}^{\xi^m\eta^n\zeta^o}$. The set of analytical functions can also be expressed by a tensor having an order other than three. The variable function can also be approximated by

$$\phi(\xi, \eta, \zeta) = \Psi_{pqr}(\xi, \eta, \zeta)\tilde{\Phi}_{pqr} \tag{2.28}$$

where $\tilde{\Phi}_{pqr}$ are the values of variable function and/or its possible partial derivatives at the nodes, and $\Psi_{pqr}(\xi, \eta, \zeta)$ their corresponding interpolation functions. Adopting the set of $\Psi_{pqr}(\xi, \eta, \zeta)$ as the variable function $\Phi(\xi, \eta, \zeta)$, the same procedure can also be used to determine $D_{\alpha\beta\gamma ijk}^{\xi^m\eta^n\zeta^o}$. And the $(m+n+o)$th order partial differentiation of Eq. (2.28) at discrete point (α, β, γ) also leads to the EDQ discretization equation (2.27) in which $D_{\alpha\beta\gamma ijk}^{\xi^m\eta^n\zeta^o}$ is expressed by

$$D_{\alpha\beta\gamma ijk}^{\xi^m\eta^n\zeta^o} = \frac{\partial^{(m+n+o)}\Psi_{ijk}}{\partial\xi^m\partial\eta^n\partial\zeta^o}\Big|_{\alpha\beta\gamma} \tag{2.29}$$

Consider that the interpolation functions can be expressed by the product of two sets of functions with one set defined by two coordinate variables while the other set defined by the remaining coordinate variable. Then a representative of this type of interpolation functions can be $\bar{\Psi}_{pq}(\xi, \eta)\tilde{\Psi}_r(\zeta)$. By using this type of interpolation functions, the weighting coefficients show to have the following form:

$$D^{\xi^m \eta^n \zeta^o}_{\alpha\beta\gamma ijk} = D^{\xi^m \eta^n}_{\alpha\beta ij} D^{\zeta^o}_{\gamma k} \tag{2.30}$$

If the variable function $\phi(\xi, \eta, \zeta)$ can be approximated by the following equation

$$\begin{aligned} &\phi(\xi, \eta, \zeta) = \hat{\Psi}_{\bar{p}}(\xi)\bar{\Psi}_{\bar{q}}(\eta)\tilde{\Psi}_{\bar{r}}(\zeta)\tilde{\Phi}_{\bar{p}\bar{q}\bar{r}}, \\ &\bar{p} = 1, 2, ..., N_{\bar{\xi}}, \quad \bar{q} = 1, 2, ..., N_{\bar{\eta}}, \quad \bar{r} = 1, 2, ..., N_{\bar{\zeta}} \end{aligned} \tag{2.31}$$

where $N_{\bar{\xi}}$, $N_{\bar{\eta}}$ and $N_{\bar{\zeta}}$ are the numbers of degrees of freedom attached to the nodes in ξ, η and ζ directions, respectively, $D^{\xi^m \eta^n \zeta^o}_{\alpha\beta\gamma ijk}$ can be obtained by the following equation

$$D^{\xi^m \eta^n \zeta^o}_{\alpha\beta\gamma ijk} = D^{\xi^m}_{\alpha i} D^{\eta^n}_{\beta j} D^{\zeta^o}_{\gamma k} \tag{2.32}$$

For the triangular grid with the coordinate variables ξ, η and ζ being the area coordinates L_1, L_2 and L_3, the formulations are the same as the above procedures.

Consider the Pascal triangular grid with the coordinate variables being the area coordinates L_1, L_2 and L_3. Let $\bar{p}$ denote the level number of a node in L_1 direction with $\bar{p} = 1$ the base level $L_1 = 0$, $\bar{q}$ denote the level number of the node in L_2 direction and $\bar{r}$ denote the level number of the node in L_3 direction. By adopting the three-dimensional node identification method, the variable function can be approximated by the following equation

$$\phi(L_1, L_2, L_3) = \Psi_{\bar{p}}(L_1)\Psi_{\bar{q}}(L_2)\Psi_{\bar{r}}(L_3)\Phi_{\bar{p}\bar{q}\bar{r}} \tag{2.33}$$

where $\Psi_{\bar{p}}(L_1)$, $\Psi_{\bar{q}}(L_2)$ and $\Psi_{\bar{r}}(L_3)$ are Lagrange interpolation functions defined in the range $0 \leq L_k \leq 1$ and expressed by

$$\Psi_{\bar{\alpha}}(L_k) = \frac{M(L_k)}{M^{(1)}(L_{k\bar{\alpha}})} \tag{2.34}$$

where

$$\begin{aligned} M(L_k) &= \prod_{\gamma=1,\gamma\neq\bar{\alpha}}^{n+1} (L_k - L_{k\gamma}), \\ M^{(1)}(L_{k\bar{\alpha}}) &= \frac{dM(L_{k\bar{\alpha}})}{dL_k} = \prod_{\gamma=1,\gamma\neq\bar{\alpha}}^{n+1} (L_{k\bar{\alpha}} - L_{k\gamma}) \end{aligned}$$

with n the order of approximation. At a node, the order of approximation and level numbers have the relation $\bar{p} + \bar{q} + \bar{r} = n + 3$. Then, the weighting coefficients for a partial derivative of order $m+n+o$ at discrete point (α, β, γ) can be obtained by the following equation

$$D^{L_1^m L_2^n L_3^o}_{\alpha\beta\gamma ijk} = D^{L_1^m}_{\alpha i} D^{L_2^n}_{\beta j} D^{L_3^o}_{\gamma k} \tag{2.35}$$

The types of triangular grid with the coordinate variables being the area coordinates are the same as those listed in Section 2.2. There are three types of tetrahedral grid:
(a) Pascal tetrahedral grid,
(b) tetrahedral grid having no interior node,
(c) tetrahedral grid having interior nodes but not being the Pascal tetrahedral grid.
There are also three types of triangular prism grid:
(a) Lagrange family grid having the Pascal triangular cross sections,
(b) triangular prism grid having no interior node,
(c) triangular prism grid having interior nodes but not being the Lagrange family grid.
And there are also three types of hexahedral grid:
(a) Lagrange family grid,
(b) hexahedral grid having no interior node,
(c) hexahedral grid having interior nodes but not being the Lagrange family grid.
The discrete points based grid types can be similarly discussed.

2.4 Four-Coordinate Grid Model

The grid configuration of a four-coordinate grid model can be a triangular prism with three of the four coordinate variables being the three area coordinates for the triangular cross sections, the tetrahedron with the variable function defined by the volume coordinates or a certain other configuration. By adopting a one-dimensional node identification method, the EDQ discretization for a partial derivative of order $m+n+o+p$ can be expressed by

$$\frac{\partial^{(m+n+o+p)}\phi_i}{\partial\xi^m\partial\eta^n\partial\zeta^o\partial\vartheta^p} = D_{i\bar{i}}^{\xi^m\eta^n\zeta^o\vartheta^p}\tilde{\Phi}_{\bar{i}}, \quad i=1,2,...,N, \quad \bar{i}=1,2,...,\bar{N} \tag{2.36}$$

The variable function can be a set of appropriate analytical functions denoted by $\Upsilon_j(\xi,\eta,\zeta,\vartheta)$. The substitution of $\Upsilon_j(\xi,\eta,\zeta,\vartheta)$ in Eq. (2.36) leads to N sets of linear algebraic system for determining $D_{i\bar{i}}^{\xi^m\eta^n\zeta^o\vartheta^p}$. The variable function can be approximated by

$$\phi(\xi,\eta,\zeta,\vartheta) = \Psi_j(\xi,\eta,\zeta,\vartheta)\tilde{\Phi}_j, \quad j=1,2,...,\bar{N} \tag{2.37}$$

where $\tilde{\Phi}_j$ are the values of variable function and its possible partial derivatives at the N nodes, and $\Psi_j(\xi,\eta,\zeta,\vartheta)$ are their corresponding interpolation functions. Adopting the set of $\Psi_j(\xi,\eta,\zeta,\vartheta)$ as the variable function $\Phi(\xi,\eta,\zeta,\vartheta)$, then the same procedure can also be used to find $D_{i\bar{i}}^{\xi^m\eta^n\zeta^o\vartheta^p}$. And the $(m+n+o+p)$th order partial differentiations of Eq. (2.37) at all discrete points also lead to the EDQ discretization equation (2.36) in which $D_{i\bar{i}}^{\xi^m\eta^n\zeta^o\vartheta^p}$ is expressed by

$$D_{i\bar{i}}^{\xi^m\eta^n\zeta^o\vartheta^p} = \frac{\partial^{(m+n+o+p)}\Psi_{\bar{i}}}{\partial\xi^m\partial\eta^n\partial\zeta^o\partial\vartheta^p}|_i \tag{2.38}$$

The variable function can also be approximated by

$$\phi(\xi,\eta,\zeta,\vartheta) = \Upsilon_j(\xi,\eta,\zeta,\vartheta)c_j, \quad j=1,2,...,\bar{N} \tag{2.39}$$

Then the weighting coefficients can also be obtained by

$$D_{i\bar{i}}^{\xi^m\eta^n\zeta^o\vartheta^p} = \frac{\partial^{(m+n+o+p)}\Upsilon_{\bar{j}}}{\partial\xi^m\partial\eta^n\partial\zeta^o\partial\vartheta^p}|_i \, \chi_{\bar{i}\bar{j}}^{-1} \tag{2.40}$$

By adopting a two-dimensional node identification method, the EDQ discretization for a partial derivative of order $m+n+o+p$ can be expressed by

$$\frac{\partial^{(m+n+o+p)}\phi_{i\delta}}{\partial\xi^m\partial\eta^n\partial\zeta^o\partial\vartheta^p} = D_{i\delta\bar{i}\bar{\delta}}^{\xi^m\eta^n\zeta^o\vartheta^p}\tilde{\Phi}_{\bar{i}\bar{\delta}} \tag{2.41}$$

The variable function can be a set of analytical functions denoted by $\Upsilon_{js}(\xi,\eta,\zeta,\vartheta)$. Then $D_{i\delta\bar{i}\bar{\delta}}^{\xi^m\eta^n\zeta^o\vartheta^p}$ can be obtained by using $\Upsilon_{js}(\xi,\eta,\zeta,\vartheta)$ in Eq. (2.41). The variable function can also be approximated by

$$\phi(\xi,\eta,\zeta,\vartheta) = \Psi_{js}(\xi,\eta,\zeta,\vartheta)\tilde{\Phi}_{js} \tag{2.42}$$

where $\tilde{\Phi}_{js}$ are values of variable function and its possible partial derivatives at nodes, and $\Psi_{js}(\xi,\eta,\zeta,\vartheta)$ are their corresponding interpolation functions. Adopting the set of $\Psi_{js}(\xi,\eta,\zeta,\vartheta)$ as the variable function $\Phi(\xi,\eta,\zeta,\vartheta)$, $D_{i\delta\bar{i}\bar{\delta}}^{\xi^m\eta^n\zeta^o\vartheta^p}$ can also be obtained by using $\Psi_{js}(\xi,\eta,\zeta,\vartheta)$ in Eq. (2.41). And the $(m+n+o+p)$th order partial differentiations of Eq. (2.42) at all discrete points also lead to the EDQ discretization equation (2.41) in which $D_{i\delta\bar{i}\bar{\delta}}^{\xi^m\eta^n\zeta^o\vartheta^p}$ is expressed by

$$D_{i\delta\bar{i}\bar{\delta}}^{\xi^m\eta^n\zeta^o\vartheta^p} = \frac{\partial^{(m+n+o+p)}\Psi_{\bar{i}\bar{\delta}}}{\partial\xi^m\partial\eta^n\partial\zeta^o\partial\vartheta^p}|_{i\delta} \tag{2.43}$$

Consider that the interpolation functions can be expressed by forming products of two sets of functions with one set defined by three coordinate variables while the other set defined by the remaining coordinate variable. Then a representative of this type of interpolation functions can be $\bar{\Psi}_j(\xi,\eta,\zeta)\tilde{\Psi}_s(\vartheta)$. By using this type of interpolation functions, the weighting coefficients show to have the following form:

$$D_{i\delta\bar{i}\bar{\delta}}^{\xi^m\eta^n\zeta^o\vartheta^p} = D_{i\bar{i}}^{\xi^m\eta^n\zeta^o}D_{\delta\bar{\delta}}^{\vartheta^p} \tag{2.44}$$

By adopting a three-dimensional node identification method, the EDQ discretization for a partial derivative of order $m+n+o+p$ can be expressed by

$$\frac{\partial^{(m+n+o+p)}\phi_{i\gamma\delta}}{\partial\xi^m\partial\eta^n\partial\zeta^o\partial\vartheta^p} = D_{i\gamma\delta\bar{i}\bar{\gamma}\bar{\delta}}^{\xi^m\eta^n\zeta^o\vartheta^p}\tilde{\Phi}_{\bar{i}\bar{\gamma}\bar{\delta}} \tag{2.45}$$

The variable function can be a set of analytical functions denoted by $\Upsilon_{jrs}(\xi,\eta,\zeta,\vartheta)$. Then $D^{\xi^m\eta^n\zeta^o\vartheta^p}_{i\gamma\delta\bar{i}\bar{\gamma}\bar{\delta}}$ can be obtained by using $\Upsilon_{jrs}(\xi,\eta,\zeta,\vartheta)$ in Eq. (2.45). The variable function can also be approximated by

$$\phi(\xi,\eta,\zeta,\vartheta) = \Psi_{jrs}(\xi,\eta,\zeta,\vartheta)\tilde{\Phi}_{jrs} \tag{2.46}$$

where $\tilde{\Phi}_{jrs}$ are values of variable function and its possible partial derivatives at the nodes, and $\Psi_{jrs}(\xi,\eta,\zeta,\vartheta)$ are their corresponding interpolation functions. Adopting the set of $\Psi_{jrs}(\xi,\eta,\zeta,\vartheta)$ as the variable function $\Phi(\xi,\eta,\zeta,\vartheta)$, $D^{\xi^m\eta^n\zeta^o\vartheta^p}_{i\gamma\delta\bar{i}\bar{\gamma}\bar{\delta}}$ can also be obtained by using $\Psi_{jrs}(\xi,\eta,\zeta,\vartheta)$ in Eq. (2.45). And the $(m+n+o+p)$th order partial differentiations of Eq. (2.46) at all discrete points also lead to the EDQ discretization equation (2.45) in which $D^{\xi^m\eta^n\zeta^o\vartheta^p}_{i\gamma\delta\bar{i}\bar{\gamma}\bar{\delta}}$ is expressed by

$$D^{\xi^m\eta^n\zeta^o\vartheta^p}_{i\gamma\delta\bar{i}\bar{\gamma}\bar{\delta}} = \frac{\partial^{(m+n+o+p)}\Psi_{\bar{i}\bar{\gamma}\bar{\delta}}}{\partial\xi^m\partial\eta^n\partial\zeta^o\partial\vartheta^p}|_{i\gamma\delta} \tag{2.47}$$

Consider that the interpolation functions can be expressed by forming products of two sets of functions with one set defined by three coordinate variables while the other set defined by the remaining coordinate variable. Then a representative of this type of interpolation functions can be $\bar{\Psi}_{jr}(\xi,\eta,\zeta)\tilde{\Psi}_s(\vartheta)$. By using this type of interpolation functions, the weighting coefficients show to have the following form:

$$D^{\xi^m\eta^n\zeta^o\vartheta^p}_{i\gamma\delta\bar{i}\bar{\gamma}\bar{\delta}} = D^{\xi^m\eta^n\zeta^o}_{i\gamma\bar{i}\bar{\gamma}} D^{\vartheta^p}_{\delta\bar{\delta}} \tag{2.48}$$

By adopting a four-dimensional node identification method, the EDQ discretization for a partial derivative of order $m+n+o+p$ can be expressed by

$$\frac{\partial^{(m+n+o+p)}\phi_{\alpha\beta\gamma\delta}}{\partial\xi^m\partial\eta^n\partial\zeta^o\partial\vartheta^p} = D^{\xi^m\eta^n\zeta^o\vartheta^p}_{\alpha\beta\gamma\delta\bar{\alpha}\bar{\beta}\bar{\gamma}\bar{\delta}}\tilde{\Phi}_{\bar{\alpha}\bar{\beta}\bar{\gamma}\bar{\delta}} \tag{2.49}$$

The variable function can be a set of appropriate analytical functions denoted by $\Upsilon_{pqrs}(\xi,\eta,\zeta,\vartheta)$. Then $D^{\xi^m\eta^n\zeta^o\vartheta^p}_{\alpha\beta\gamma\delta\bar{\alpha}\bar{\beta}\bar{\gamma}\bar{\delta}}$ can be obtained by using $\Upsilon_{pqrs}(\xi,\eta,\zeta,\vartheta)$ in Eq. (2.49). The variable function can also be approximated by

$$\phi(\xi,\eta,\zeta,\vartheta) = \Psi_{pqrs}(\xi,\eta,\zeta,\vartheta)\tilde{\Phi}_{pqrs} \tag{2.50}$$

where $\tilde{\Phi}_{pqrs}$ are values of variable function and its possible partial derivatives at the nodes, and $\Psi_{pqrs}(\xi,\eta,\zeta,\vartheta)$ are their corresponding interpolation functions. Adopting the set of $\Psi_{pqrs}(\xi,\eta,\zeta,\vartheta)$ as the variable function $\Phi(\xi,\eta,\zeta,\vartheta)$, $D^{\xi^m\eta^n\zeta^o\vartheta^p}_{\alpha\beta\gamma\delta\bar{\alpha}\bar{\beta}\bar{\gamma}\bar{\delta}}$ can also be obtaining by using $\Psi_{pqrs}(\xi,\eta,\zeta,\vartheta)$ in Eq. (2.49). And the $(m+n+o+p)$th order partial differentiations of Eq. (2.50) at all discrete points also lead to the EDQ discretization equation (2.48) in which $D^{\xi^m\eta^n\zeta^o\vartheta^p}_{\alpha\beta\gamma\delta\bar{\alpha}\bar{\beta}\bar{\zeta}\bar{\delta}}$ is expressed by

$$D^{\xi^m\eta^n\zeta^o\vartheta^p}_{\alpha\beta\gamma\delta\bar{\alpha}\bar{\beta}\bar{\gamma}\bar{\delta}} = \frac{\partial^{(m+n+o+p)}\Psi_{\bar{\alpha}\bar{\beta}\bar{\gamma}\bar{\delta}}}{\partial\xi^m\partial\eta^n\partial\zeta^o\partial\vartheta^p}|_{\alpha\beta\gamma\delta} \tag{2.51}$$

Consider that the interpolation functions can be expressed by forming products of two sets of functions with one set defined by three coordinate variables while the other set defined by the remaining coordinate variable. Then a representative of this type of interpolation functions can be $\bar{\Psi}_{pqr}(\xi,\eta,\zeta)\tilde{\Psi}_s(\vartheta)$. By using this type of interpolation functions, the weighting coefficients show to have the following form:

$$D^{\xi^m\eta^n\zeta^o\vartheta^p}_{\alpha\beta\gamma\delta\bar{\alpha}\bar{\beta}\bar{\gamma}\bar{\delta}} = D^{\xi^m\eta^n\zeta^o}_{\alpha\beta\gamma\bar{\alpha}\bar{\beta}\bar{\gamma}} D^{\vartheta^p}_{\delta\bar{\delta}} \tag{2.52}$$

For a triangular prism of Lagrange family grid having the Pascal triangular cross sections, the variable function can be approximated by

$$\phi(L_1,L_2,L_3,\zeta) = \Psi_{\bar{p}}(L_1)\Psi_{\bar{q}}(L_2)\Psi_{\bar{r}}(L_3)\Psi_{\bar{s}}(\zeta)\tilde{\Phi}_{\bar{p}\bar{q}\bar{r}\bar{s}} \tag{2.53}$$

Then, the weighting coefficients for a partial derivative of order $m+n+o+p$ can be obtained by the following equation

$$D^{L_1^m L_2^n L_3^o \zeta^p}_{\alpha\beta\gamma\delta\bar{\alpha}\bar{\beta}\bar{\gamma}\bar{\delta}} = D^{L_1^m}_{\alpha\bar{\alpha}} D^{L_2^n}_{\beta\bar{\beta}} D^{L_3^o}_{\gamma\bar{\gamma}} D^{\zeta^p}_{\delta\bar{\delta}} \tag{2.54}$$

And for the Pascal tetrahedral grid with the coordinate variables being the volume coordinates $\bar{L}_1$, $\bar{L}_2$, $\bar{L}_3$ and $\bar{L}_4$, the variable function can be approximated by the following equation which uses the four-dimensional node identification method

$$\phi(\bar{L}_1,\bar{L}_2,\bar{L}_3,\bar{L}_4) = \Psi_{\bar{p}}(\bar{L}_1)\Psi_{\bar{q}}(\bar{L}_2)\Psi_{\bar{r}}(\bar{L}_3)\Psi_{\bar{s}}(\bar{L}_4)\tilde{\Phi}_{\bar{p}\bar{q}\bar{r}\bar{s}} \tag{2.55}$$

where $\bar{p}$, $\bar{q}$, $\bar{r}$ and $\bar{s}$ are numbers of levels of a specified discrete point with respect to $\bar{L}_1$, $\bar{L}_2$, $\bar{L}_3$ and $\bar{L}_4$, respectively, and $\Psi_{\bar{p}}(\bar{L}_1)$, $\Psi_{\bar{q}}(\bar{L}_2)$, $\Psi_{\bar{r}}(\bar{L}_3)$ and $\Psi_{\bar{s}}(\bar{L}_4)$ can be defined by using the Lagrange interpolation functions expressed by Eq. (2.34). Then, the weighting coefficients for a partial derivative of order $m+n+o+p$ can be obtained by the following equation

$$D^{\bar{L}_1^m \bar{L}_2^n \bar{L}_3^o \bar{L}_4^p}_{\alpha\beta\gamma\delta\bar{\alpha}\bar{\beta}\bar{\gamma}\bar{\delta}} = D^{\bar{L}_1^m}_{\alpha\bar{\alpha}} D^{\bar{L}_2^n}_{\beta\bar{\beta}} D^{\bar{L}_3^o}_{\gamma\bar{\gamma}} D^{\bar{L}_4^p}_{\delta\bar{\delta}} \tag{2.56}$$

2.5 Arbitrary Finite-Coordinate Model

Consider an arbitrary M-coordinate EDQ model with the number of coordinate variables larger than two. The dimension and grid configuration for defining the discrete point and node can be different. By adopting an one-dimensional node identification method to express both the discrete point and node, the EDQ discretization for a partial derivative of order $m+...+p$ at discrete point α can be expressed by

$$\frac{\partial^{(m+...+p)}\phi_\alpha}{\partial\xi^m...\partial\vartheta^p} = D^{\xi^m...\vartheta^p}_{\alpha i}\tilde{\Phi}_i, \quad i=1,2,...,\bar{N} \tag{2.57}$$

The variable function can be a set of appropriate analytical functions denoted by $\Upsilon_j(\xi,...,\vartheta)$. The substitution of $\Upsilon_j(\xi,...,\vartheta)$ in (2.18) leads to a linear algebraic system for determining $D_{\alpha i}^{\xi^m...\vartheta^p}$. The set of analytical functions can also be expressed by a tensor having an order other than one. The variable function can also be approximated by

$$\phi(\xi,...,\vartheta) = \Psi_j(\xi,...,\vartheta)\tilde{\Phi}_j, \quad j = 1,2,...,\bar{N} \tag{2.58}$$

where $\tilde{\Phi}_j$ are values of variable function and/or its possible partial derivatives at the N nodes, and $\Psi_j(\xi,...,\vartheta)$ are their corresponding interpolation functions. Adopting the set of $\Psi_j(\xi,...,\vartheta)$ as the variable function $\phi(\xi,...,\vartheta)$, the same procedure can also be used to determine $D_{\alpha i}^{\xi^m...\vartheta^p}$. And the $(m+...+p)$th order partial differentiation of Eq. (2.58) at discrete point α also leads to the EDQ discretization equation (2.57) in which $D_{\alpha i}^{\xi^m...\vartheta^p}$ is expressed by

$$D_{\alpha i}^{\xi^m...\vartheta^p} = \frac{\partial^{(m+...+p)}\Psi_i}{\partial\xi^m...\partial\vartheta^p}|_\alpha \tag{2.59}$$

The variable function can also be approximated by

$$\phi(\xi,...,\vartheta) = \Upsilon_j(\xi,...,\vartheta)c_j, \quad j = 1,2,...,\bar{N} \tag{2.60}$$

Then the weighting coefficients can also be obtained by

$$D_{\alpha i}^{\xi^m...\vartheta^p} = \frac{\partial^{(m+...+p)}\Upsilon_{\bar{j}}}{\partial\xi^m...\partial\vartheta^p}|_\alpha \, \chi_{i\bar{j}}^{-1} \tag{2.61}$$

In Eq. (2.60), the unknown coefficients and appropriate analytical functions can also be expressed by certain other tensors having orders other than one.

By adopting a two-dimensional node identification method to express both the discrete point and node, the EDQ discretization for a partial derivative of order m+...+p at discrete point (α,δ) can be expressed by

$$\frac{\partial^{(m+...+p)}\phi_{\alpha\delta}}{\partial\xi^m...\partial\vartheta^p} = D_{\alpha\delta il}^{\xi^m...\vartheta^p}\tilde{\Phi}_{il} \tag{2.62}$$

The variable function can be a set of appropriate analytical functions denoted by $\Upsilon_{js}(\xi,...,\vartheta)$. The substitution of $\Upsilon_{js}(\xi,...,\vartheta)$ in Eq. (2.62) leads to a linear algebraic system for determining $D_{\alpha\delta il}^{\xi^m...\vartheta^p}$ The set of analytical functions can also be expressed by a tensor having an order other than two. The variable function can also be approximated by

$$\phi(\xi,...,\vartheta) = \Psi_{js}(\xi,...,\vartheta)\tilde{\Phi}_{js} \tag{2.63}$$

where $\tilde{\Phi}_{js}$ are values of variable function and/or its possible partial derivatives at the nodes, and $\Psi_{js}(\xi,...,\vartheta)$ are their corresponding interpolation functions. Adopting the set of $\Psi_{js}(\xi,...,\vartheta)$ as the variable function $\phi(\xi,...,\vartheta)$, the same procedure can also be used to determine $D^{\xi^m...\vartheta^p}_{\alpha\delta il}$. And the $(m+...+p)$th order partial differentiation of Eq. (2.63) at discrete point (α, δ) also leads to the EDQ discretization equation (2.62) in which $D^{\xi^m...\vartheta^p}_{\alpha\delta il}$ is expressed by

$$D^{\xi^m...\vartheta^p}_{\alpha\delta il} = \frac{\partial^{(m+...+p)}\Psi_{il}}{\partial\xi^m...\partial\vartheta^p}|_{\alpha\delta} \tag{2.64}$$

Consider that the interpolation functions can be expressed by forming products of two sets of functions with one set defined by three coordinate variables while the other set defined by the remaining coordinate variable. Then a representative of this type of interpolation functions can be $\bar{\Psi}_j(\xi, ...)\tilde{\Psi}_s(\vartheta)$. By using this type of interpolation functions, the weighting coefficients show to have the following form:

$$D^{\xi^m...\vartheta^p}_{\alpha\delta il} = D^{\xi^m}_{\alpha i}\cdots D^{\vartheta^p}_{\delta l} \tag{2.65}$$

If the dimension of node identification and the number of coordinate variables are the same in expressing both the discrete point and node, the EDQ discretization for a partial derivative of order $m+...+p$ at discrete point $(\alpha, ..., \delta)$ can be expressed by

$$\frac{\partial^{(m+...+p)}\phi_{\alpha...\delta}}{\partial\xi^m...\partial\vartheta^p} = D^{\xi^m...\vartheta^p}_{\alpha...\delta i...l}\tilde{\Phi}_{i...l} \tag{2.66}$$

The variable function can be a set of appropriate analytical functions denoted by $\Upsilon_{p...s}(\xi, ..., \vartheta)$. The substitution of $\Upsilon_{p...s}(\xi, ..., \vartheta)$ in Eq. (2.66) leads to a linear algebraic system for determining $D^{\xi^m...\vartheta^p}_{\alpha...\delta i...l}$. The set of analytical functions can also be expressed by a tensor having an order other than M. The variable function can also be approximated by

$$\phi(\xi, ..., \vartheta) = \Psi_{p...s}(\xi, ..., \vartheta)\tilde{\Phi}_{p...s} \tag{2.67}$$

where $\tilde{\Phi}_{p...s}$ are the values of variable function and/or its possible partial derivatives at the nodes, and $\Psi_{p...s}(\xi, ..., \vartheta)$ are their corresponding interpolation functions. Adopting the set of $\Psi_{p...s}(\xi, ..., \vartheta)$ as the variable function $\phi(\xi, ..., \vartheta)$, the same procedure can also be used to determine $D^{\xi^m...\vartheta^p}_{\alpha...\delta i...l}$. And the $(m+...+p)$th order partial differentiation of Eq. (2.67) at discrete point $(\alpha, ..., \delta)$ also leads to the EDQ discretization equation (2.66) in which $D^{\xi^m...\vartheta^p}_{\alpha...\delta i...l}$ is expressed by

$$D^{\xi^m...\vartheta^p}_{\alpha...\delta i...l} = \frac{\partial^{(m+...+p)}\Psi_{i..l}}{\partial\xi^m...\vartheta^p}|_{\alpha...\delta} \tag{2.68}$$

Consider that the interpolation functions can be expressed by the product of two sets of functions with one set defined by $M-1$ coordinate variables while the other set defined by the remaining coordinate variable. Then a representative of this type of interpolation functions can be $\bar{\Psi}_{p...}(\xi, ...)\tilde{\Psi}_s(\vartheta)$. By using this type of interpolation functions, the weighting coefficients show to have the following form:

$$D^{\xi^m ... \vartheta^p}_{\alpha...\delta i...l} = D^{\xi^m ...}_{\alpha...i...} D^{\vartheta^p}_{\delta l} \tag{2.69}$$

If the interpolation functions can be expressed by the product of P sets of functions with P larger than two and a coordinate variable only appearing in one set of functions, similar procedures can be used to determine the weighting coefficients which are components of the outer product of P tensors.

2.6 Sample Applications

Various analytical functions such as sinc functions, Lagrange polynomials, Chebyshev polynomials, Bernoulli polynomials, Euler polynomials, rational functions, ..., etc. can be used to define the weighting coefficients. To solve problems having singularity properties, certain singular functions can be used for the EDQ discretization. The problems having infinite domains can also be treated.

2.6.1 Lagrange DQ Model

General polynomials can be used to implicitly calculate EDQ weighting coefficients. By using Lagrange interpolation functions, the DQ weighting coefficients can explicitly be calculated. Consider the one-dimensional discretization using only one DOF representing the variable function at the node to define the DQ and Lagrange interpolation functions for explicitly express the weighting coefficients. Lagrange interpolation functions $\hat{L}_\beta(\xi)$ can be expressed by

$$\hat{L}_\beta(\xi) = \frac{M(\xi)}{(\xi - \xi_\beta)M^{(1)}(\xi_\beta)} \tag{2.70}$$

where

$$M(\xi) = \prod_{\gamma=1}^{N}(\xi - \xi_\gamma), \quad M^{(1)}(\xi_\beta) = \frac{dM(\xi_\beta)}{d\xi} = \prod_{\gamma=1,\gamma\neq\beta}^{N}(\xi_\beta - \xi_\gamma)$$

The weighting coefficients can be derived

$$\begin{aligned} D^{\xi}_{\alpha\beta} &= \tfrac{d\hat{L}_\beta}{d\xi}\,|_\alpha \\ &= \tfrac{(\xi_\alpha-\xi_\beta)M^{(1)}(\xi_\beta)M^{(1)}(\xi_\alpha) - M(\xi_\alpha)M^{(1)}(\xi_\beta)}{[(\xi_\alpha-\xi_\beta)M^{(1)}(\xi_\beta)]^2} \\ &= \begin{cases} \tfrac{M^{(1)}(\xi_\alpha)}{(\xi_\alpha-\xi_\beta)M^{(1)}(\xi_\beta)}, & \text{for} \quad \alpha \neq \beta \\ -\sum_{\gamma=1,\gamma\neq\alpha}^{N} D_{\alpha\gamma}, & \text{for} \quad \alpha = \beta \end{cases} \end{aligned} \tag{2.71}$$

The truncation error of the first order Lagrange DQ discretization is $\frac{du^N(\zeta(\xi_\alpha))}{N!\, d\xi^N} \times \prod_{\beta=1,\beta\neq\alpha}^{N-1}(\xi_\alpha - \xi_\beta)$ for some number $\zeta(\xi_\alpha)$. The weighting coefficients $D_{\alpha\beta}^{\xi^m}$ for an mth order derivative with m larger than one can be similarly obtained. When the uniform grid is used, Eq. (2.71) is reduced to

$$D_{\alpha\beta}^{\xi} = (-1)^{\alpha+\beta} \frac{(\alpha-1)!\,(N-\alpha)!}{\Delta\xi(\alpha-\beta)(\beta-1)!\,(N-\beta)!} \quad \text{for} \quad \alpha \neq \beta \tag{2.72}$$

where $\Delta\xi = \xi_\alpha - \xi_\beta$. The mth order weighting coefficients $D_{\alpha\beta}^{\xi^m}$ can be similarly calculated. Assume that the nodes are symmetric with respect to the middle point and let $r = N - \alpha + 1$ and $s = N - \beta + 1$. Then $D_{\alpha\beta}^{\xi^m} = D_{rs}^{\xi^m}$ if m is even, while $D_{\alpha\beta}^{\xi^m} = -D_{rs}^{\xi^m}$ if m is odd. The number of arithmetic operations can thus be reduced in calculating the weighting coefficients. Assume that the nodes are symmetric with respect to the middle point and let $r = N - \alpha + 1$ and $s = N - \beta + 1$. Then $D_{\alpha\beta}^{\xi^m} = D_{rs}^{\xi^m}$ if m is even, while $D_{\alpha\beta}^{\xi^m} = -D_{rs}^{\xi^m}$ if m is odd. The number of arithmetic operations can thus be reduced in calculating the weighting coefficients.

Because only the variable functions at nodes are used to define the DQ, the higher order weighting coefficients can also be calculated by the following recurrent procedure by using the first order weighting coefficients $D_{\alpha\beta}^{\xi}$

$$D_{\alpha\beta}^{\xi^2} = \sum_{\gamma=1}^{N} D_{\alpha\gamma}^{\xi} D_{\gamma\beta}^{\xi},$$

$$D_{\alpha\beta}^{\xi^3} = \sum_{\gamma=1}^{N} D_{\alpha\gamma}^{\xi^2} D_{\gamma\beta}^{\xi},$$

$$\vdots$$

$$D_{\alpha\beta}^{\xi^m} = \sum_{\gamma=1}^{N} D_{\alpha\gamma}^{\xi^{m-1}} D_{\gamma\beta}^{\xi} \tag{2.73}$$

The above recurrent computation procedure is equivalent to the procedure developed by Shu and Richard [30,47-48].

Consider a prismatic bar subjected to a quartically distributed force. The governing equilibrium equation is

$$EA\frac{d^2u}{dx^2} = -x^4 \tag{2.74}$$

where u is axial displacement, E is Young's modulus and A is area of cross section. In the analysis, EA is set to be 1 and the length of the bar is equal to 2. The boundary conditions are $u(0) = 0$ and $EA\frac{du(2)}{dx} = 0$. Each node has a deformation parameter of axial displacement. Lagrange DQ model is used to analyze the problem.

Using the DQ discretization, Eq. (2.74) can be discretized

$$\frac{EA}{4}\sum_{i=1}^{N} D_{\alpha i}^{\xi^2} u_i = -x_\alpha^4 \tag{2.75}$$

The discrete natural boundary condition can be expressed by $\frac{EA}{2}\sum_{i=1}^{N} D_{\alpha i}^{\xi} u_i = 0$. By considering the kinematic boundary condition, discrete equilibrium equations and natural boundary condition, a linear equation system can be constructed. The displacements at nodes can thus be found by solving the linear equation system. In the analysis, nodes are equally spaced. Numerical tests are carried out by gradually increasing the discrete points. Numerical results of displacement at the free end and axial force at the fixed end are summarized and presented. They are shown in Table 2.1. It shows that results converge to the exact solutions by increasing the number of nodes up to seven.

Table 2.1 Results of the bar problem

Number of nodes	Displacement at free end	Axial force at fixed end
3	$.20000\times 10^1$	$.20000\times 10^1$
5	$.95000\times 10^1$	$.61667\times 10^1$
7	$.10667\times 10^2$	$.64000\times 10^1$
Exact solution	$.10667\times 10^2$	$.64000\times 10^1$

2.6.2 Hermite EDQ model

In the Hermite EDQ discretization, each node has two deformation parameters of lateral displacement and displacement gradient. Also consider that the Hermite EDQ model has a unit length of the range and using Hermite interpolation, displacement $w(\xi)$ can be approximated by

$$\phi(\xi) = \sum_{\beta=1}^{N} H_\beta(\xi)\varPhi_\beta + \sum_{\beta=1}^{N} \tilde{H}_\beta(\xi)\frac{d\varPhi_\beta}{d\xi} \tag{2.76}$$

where

$$H_\beta(\xi) = \left[1 - 2(\xi - \xi_\beta)\hat{L}_{\beta,\xi}(\xi_\beta)\right]\hat{L}_\beta^2(\xi) \tag{2.77}$$

and

$$\tilde{H}_\beta(\xi) = (\xi - \xi_\beta)\hat{L}_\beta^2(\xi) \tag{2.78}$$

are Hermite interpolation functions, and $\hat{L}_\beta(\xi)$ the Lagrange interpolation functions. Then the following relation can be defined

$$\frac{d\phi}{d\xi}|_\alpha = \sum_{\beta=1}^{N} \frac{dH_\beta}{d\xi}|_\alpha \Phi_\beta + \sum_{\beta=1}^{N} \frac{d\tilde{H}_\beta}{d\xi}|_\alpha \frac{d\Phi}{d\xi} = \sum_{\beta=1}^{N} \bar{D}^\xi_{\alpha\beta}\Phi_\beta + \sum_{\beta=1}^{N} \tilde{D}^\xi_{\alpha\beta}\frac{d\Phi_\beta}{d\xi} \tag{2.79}$$

where

$$\begin{aligned}\bar{D}^\xi_{\alpha\beta} &= 2\hat{L}_\beta(\xi_\alpha)\{[1 - 2(\xi_{(\alpha)} - \xi_{(\beta)})\hat{L}_{(\beta),\xi}(\xi_{(\beta)})]\hat{L}_{(\beta),\xi}(\xi_{(\alpha)}) \\ &\quad -\hat{L}_{(\beta),\xi}(\xi_{(\beta)})\hat{L}_{(\beta)}(\xi_{(\alpha)})\} \\ &= 2\hat{L}_\beta(\xi_\alpha)\{[1 - 2(\xi_{(\alpha)} - \xi_{(\beta)})D^\xi_{(\beta)(\beta)}]D^\xi_{(\alpha)(\beta)} \\ &\quad -D^\xi_{(\beta)(\beta)}\hat{L}_{(\beta)}(\xi_{(\alpha)})\}\end{aligned} \tag{2.80}$$

and

$$\begin{aligned}\tilde{D}^\xi_{\alpha\beta} &= \hat{L}_\beta(\xi_\alpha)[\hat{L}_{(\beta)}(\xi_{(\alpha)}) + 2(\xi_{(\alpha)} - \xi_{(\beta)})\hat{L}_{(\beta),\xi}(\xi_{(\alpha)})] \\ &= \hat{L}_\beta(\xi_\alpha)[\hat{L}_{(\beta)}(\xi_{(\alpha)}) + 2(\xi_{(\alpha)} - \xi_{(\beta)})D^\xi_{(\alpha)(\beta)}]\end{aligned} \tag{2.81}$$

Then the weighting coefficients $D^\xi_{\alpha\beta}$ defined by Hermite interpolation, with the range of α and β being $\bar{N}$ or $2N$, can be explicitly formed by using the elements of $\bar{D}^\xi_{\alpha\beta}$ and $\tilde{D}^\xi_{\alpha\beta}$. The weighting coefficients for higher-order derivatives can be similarly calculated.

A Bernoulli-Euler cantilever beam subjected to a laterally distributed force $q = x^5$ is solved by using the Hermite EDQ model. The governing equilibrium equation is expressed by

$$EI\frac{d^4w}{dx^4} = x^5 \tag{2.82}$$

where w is the lateral displacement and EI is the flexural rigidity. In the analysis, EI is set to be 1. and the length of the beam is also equal to 1. The boundary conditions are $w(0) = 0$, $dw(0)/dx = 0$, $EId^2w(1)/dx^2 = 0$ and $EId^3w(1)/dx^3 = 0$. Using Hermite EDQ model, the equilibrium equation Eq. (2.82) at an interior discrete point α can be discretized

$$EI\sum_{i=1}^{\bar{N}} D^{\xi^4}_{\alpha i}\tilde{w}_i = x^5_\alpha \tag{2.83}$$

where $\tilde{w}_i \equiv \lfloor w_1\ \theta_1\ w_2\ \theta_2\ ...\ w_N\ \theta_N \rfloor$, with $\theta_i = dw_i/dx$. The two natural boundary conditions can also be discretized

$$EI\sum_{i=1}^{\bar{N}} D^{\xi^3}_{(2N-1)i}\tilde{w}_i = 0, \quad EI\sum_{i=1}^{\bar{N}} D^{\xi^2}_{(2N)i}\tilde{w}_i = 0 \tag{2.84}$$

In the analysis, nodes are equally spaced. Considering a $(\bar{N}-1)$th order Hermite EDQ model and defining $\Delta x = 1./(\bar{N}-1)$, the interior discrete points for defining discrete equilibrium equations are located at $x = (p-1)\Delta x$, with $p = 2, ..., \bar{N}-1$. Numerical results of displacement at free end and bending moment at fixed end for the analyses using three different node grids are listed in Table 2.2. The first and last nodes which are also discrete points are used to define the four boundary conditions. Numerical results in Table 2.2 converge to exact solutions by increasing the DOF up to ten.

Table 2.2 Numerical results of the Bernoulli-Euler beam problem

Number of DOF	Displacement at free end	Bending moment at fixed end
6	$.1534667\times10^{-1}$	$.5013334\times10^{-1}$
8	$.4021241\times10^{-1}$	$.1306400\times10^{0}$
10	$.4398148\times10^{-1}$	$.1428571\times10^{0}$
Exact solution	$.4398148\times10^{-1}$	$.1428571\times10^{0}$

2.6.3 Other Approximations

(a) Chebyshev polynomials

With the range of ξ being $-1 \leq \xi \leq 1$, Chebyshev polynomials can be generated from the following recurrent formula:

$$T_{n+1}(\xi) = 2\xi T_n(\xi) - T_{n-1}(\xi) \tag{2.85}$$

with the two initial members $T_0(\xi) = 1$ and $T_1(\xi) = \xi$. Chebyshev polynomials can be used to implicitly calculate the related DQ or EDQ weighting coefficients of the Chebyshev DQ or EDQ models.

(b) Bernoulli polynomials

Bernoulli polynomials can be generated from the following equation

$$B_n(\xi) = \sum_{k=0}^{n} \binom{n}{k} x^{n-k} \bar{B}_k \tag{2.86}$$

where $\bar{B}_k$ are Bernoulli numbers which can be found in a mathematics handbook [49]. Bernoulli polynomials can be used to implicitly calculate the related DQ or EDQ weighting coefficients of the Bernoulli DQ or EDQ models.

(c) Euler polynomials

Euler polynomials can be generated from the following recurrent formula

$$E_n(\xi) = \frac{2}{n+1} \sum_{k=1}^{n+1} \left(1 - 2^k\right) \binom{n+1}{k} x^{n+k-1} \bar{E}_k \tag{2.87}$$

where $\bar{E}_k$ are Euler numbers. Euler polynomials can be used to implicitly calculate the related DQ or EDQ weighting coefficients of the Euler DQ or EDQ models.

(d) Laguerre polynomials

Laguerre polynomials and the first derivatives can be generated from the following recurrent formulas:

$$\begin{aligned} G_{n+1}(\xi) &= (2n + 1 - \xi)G_n(\xi) - n^2 G_{n-1}(\xi) \\ G^{(1)}_{n+1}(\xi) &= \frac{n+1}{\xi} \left[G_{n+1}(\xi) - (n+1)G_n(\xi)\right] \end{aligned} \tag{2.88}$$

with the two initial members $G_0(\xi) = 1$ and $G_1(\xi) = 1 - \xi$. The Laguerre polynomials and the first derivatives can be used to implicitly calculate the related DQ or EDQ weighting coefficients of the Laguerre DQ or EDQ models.

(e) Sinc functions

Sinc functions are suitable for dealing problems with singularity property or infinite domain. Assume that the range of ξ is $0 \le \xi \le 1$ and that the discrete points are equally spaced. Then the following transformation relation can be defined

$$s_\alpha = \frac{\pi}{h} (\xi - \alpha h) \tag{2.89}$$

with $h = \frac{1}{N-1}$. The sinc functions can be defined as [50]

$$\begin{aligned} S_\alpha(\xi, h) &= \frac{\sin s_\alpha}{s_\alpha}, \\ S_\alpha(\xi_\beta) &= \delta_{\alpha\beta} \end{aligned} \tag{2.90}$$

Consider that only one DOF representing the variable function at the discrete point is used to define the DQ. Then sinc functions can be interpolation functions used to define the DQ weighting coefficients. Using Eqs. (2.90) in Eq. (1.1) weighting coefficients in explicit form can be obtained. Following are first order weighting coefficients of the sinc DQ model

$$D^\xi_{\alpha\beta} = \begin{cases} \frac{\pi}{h} \frac{s_{\alpha\beta} \cos s_{\alpha\beta} - \sin s_{\alpha\beta}}{s^2_{\alpha\beta}}, & \text{for} \quad \alpha \ne \beta \\ 0, \quad \text{for} \quad \alpha = \beta \end{cases} \tag{2.91}$$

Higher order weighting coefficients can be calculated by either the differentiation of sinc functions or using the recurrent procedure represented by Eq. (2.73).

(f) Harmonic interpolation

Assume that the range of ξ is $0 \leq \xi \leq 1$. Let $n = (N-2)/2$ for N an even number and $n = (N-1)/2$ for N an odd number. The variable function $\phi(\xi)$ can be approximated by the following harmonic interpolation

$$\phi(\xi) = \frac{1}{2}a_0 + \sum_{k=1}^{n} (a_k \cos 2\pi k\xi + b_k \sin 2\pi k\xi) \tag{2.92}$$

If N is even, the undetermined constants a_0, a_k and b_k are expressed as

$$a_0 = \frac{2}{N-1}\sum_{\beta=1}^{N} \Phi_\beta,$$

$$a_k = \frac{2}{N-1}\sum_{\beta=1}^{N} \Phi_\beta \cos 2\pi k\xi_\beta,$$

$$b_k = \frac{2}{N-1}\sum_{\beta=1}^{N} \Phi_\beta \sin 2\pi k\xi_\beta, \quad k = 1, ..., (N-2)/2 \tag{2.93}$$

The introduction of Eq. (2.93) in Eq. (2.92) leads to the following equation

$$\phi(\xi) = \frac{1}{2}a_0 + \sum_{\beta=1}^{N} \frac{2}{N-1} \sum_{k=1}^{(N-2)/2} (\cos 2\pi k\xi_\beta \cos k\xi + \sin 2\pi k\xi_\beta \sin k\xi)\Phi_\beta \tag{2.94}$$

with Φ_β the values of variable function at nodes. Using the above equation in Eq. (1.1), the harmonic DQ model can be defined with the related weighting coefficients expressed by

$$D^m_{\alpha\beta} = \sum_{\beta=1}^{N} \frac{2}{N-1} \sum_{k=1}^{(N-2)/2} (\cos 2\pi k\xi_\beta \frac{d^m \cos k\xi_\alpha}{d\xi^m} + \sin 2\pi k\xi_\beta \frac{d^m \sin k\xi_\alpha}{d\xi^m}) \tag{2.95}$$

Higher order weighting coefficients can also be calculated by using $D^\xi_{\alpha\beta}$ in the recurrent relation equation (2.73).

If N is odd, the undetermined constants a_0, a_k and b_k are expressed as

$$a_0 = \frac{2}{N-1}\sum_{\beta=1}^{N} \phi_\beta,$$

$$a_k = \frac{2}{N-1}\sum_{\beta=1}^{N} \phi_\beta \cos 2\pi k\xi_\beta,$$

$$b_k = \frac{1}{N-1} \sum_{\beta=1}^{N} \phi_\beta \sin 2\pi k \xi_\beta, \quad k = 1, ..., (N-1)/2 \tag{2.96}$$

Introducing Eq. (2.96) in Eq. (2.92) then using the obtained $\phi(\xi)$ in Eq. (1.1), the related weighting coefficients can also be obtained

$$D_{\alpha\beta}^{m} = \sum_{\beta=1}^{N} \frac{2}{N-1} \sum_{k=1}^{(N-1)/2} \left(\cos 2\pi k \xi_\beta \frac{d^m \cos k\xi_\alpha}{d\xi^m} \right. \\ \left. + \sin 2\pi k \xi_\beta \frac{d^m \sin k\xi_\alpha}{d\xi^m} \right) \tag{2.97}$$

The buckling of a simply supported prismatic beam subjected to an axial load is solved by using the Chebyshev, Bernoulli and Euler Polynomials Based EDQs. Chebyshev polynomials, Bernoulli polynomials and Euler polynomials are used to calculate weighting coefficients and solve the problem, separately. In using Chebyshev polynomials to solve the problem, the beam is mapped onto the region $\Omega = \{-1 \le \xi \le 1\}$. Chebyshev polynomials are 1, ξ, $-1+2\xi^2$, $-3\xi + 4\xi^3$, $1 - 8\xi^2 + 8\xi^4, \ldots$, etc. In using Bernoulli or Euler polynomials to solve the problem, the beam is mapped onto the region $\Omega = \{0 \le \xi \le 1\}$. Bernoulli polynomials are 1, $-\frac{1}{2} + \xi$, $\frac{1}{6} - \xi + \xi^2$, $\frac{\xi}{2} - \frac{3\xi^2}{2} + \xi^3$, $-\frac{1}{30} + \xi^2 - 2\xi^3 + \xi^4, \ldots$, etc. Euler polynomials are 1, $-\frac{1}{2} + \xi$, $-\xi + \xi^2$, $\frac{1}{4} - \frac{3\xi^2}{2} + \xi^3$, $\xi - 2\xi^3 + \xi^4, \ldots$, etc. Among the three methods for calculating weighting coefficientsWeighting coefficients, the two implicit methods can be used.

Consider that the bar is subjected to a compressive end force P and buckles in $x - z$ plane. Let E, L and I denote the Young's modulus, beam length and moment of inertia of the cross section, respectively. The governing equation can be expressed by

$$EI \frac{d^4 w}{dx^4} + P \frac{d^2 w}{dx^2} = 0 \tag{2.98}$$

The kinematic boundary conditions are $w(0) = w(L) = 0$, while the natural boundary conditions are $M(0) = M(L) = 0$ with M the bending moment. Assume that the EDQ model has a unit length. Then by using EDQ to discretize Eq. (2.98) at discrete point α leads to the following equation

$$\frac{4EI}{L^2} \sum_{i=1}^{\bar{N}} D_{\alpha i}^{\xi^4} \tilde{w}_i + P \sum_{i=1}^{\bar{N}} D_{\alpha i}^{\xi^2} \tilde{w}_i = 0 \tag{2.99}$$

In the analysis, three degrees of freedom representing the displacement and the first and second order derivatives of displacement with respect to x are assigned to each of the two boundary nodes, while only one degree of freedom representing the displacement is assigned to each of the interior nodes. The nodes are equally spaced. Refinement procedure is used to analyze the problem. Three different grids representing the arrangement of discrete points are utilized to solve the problem separately. In using Chebyshev polynomials to solve the problem, the type 1 grid is defined by $\xi_1 = -1$; $\xi_i = -1 + \frac{2i}{\bar{N}-1}$, for $i = 2, ..., \bar{N}-3$; $\xi_{\bar{N}-2} = 1$. The type 2 grid adopts the Chebyshev sampling points as the discrete points. These Chebyshev sampling points are defined by $\xi \mid_\alpha = -cos\left[\frac{\pi(\alpha-1)}{\bar{N}-3}\right]$, $\alpha = 1, ..., \bar{N}-2$. The type 3 grid is defined by $\xi_1 = -1$; $\xi_i = -cos\left[\frac{\pi i}{\bar{N}-1}\right]$, for $i = 2, ..., \bar{N}-3$; $\xi_{\bar{N}-2} = 1$. The interior discrete points are used to define discrete buckling equilibrium equations. Since in using Bernoulli and Euler polynomials to solve the problem the bar is mapped onto a different region, the values of coordinates of a specified type of grid are different from those used for the Chebyshev analysis. The corresponding type 1 grid is defined by $\xi_1 = 0$; $\xi_i = \frac{i}{\bar{N}-1}$, for $i = 2, ..., \bar{N}-3$; $\xi_{\bar{N}-2} = 1$. The corresponding type 2 grid is defined by $\frac{1}{2}\left[1 - cos\frac{\pi(i-1)}{\bar{N}-3}\right]$, $\alpha = 1, ..., \bar{N}-2$. The type 3 grid is defined by $\xi_1 = 0$; $\frac{1}{2}\left[1 - cos\frac{\pi i}{\bar{N}-1}\right]$, for $i = 2, ..., \bar{N}-3$; $\xi_{\bar{N}-2} = 1$. The refinement procedure is used to analyze the problem. The relative error e_r of the DQEM results with respect to the exact solution is the convergence indicator. Defining $\bar{C} = \frac{PL^2}{EI}$ as the load factor, numerical results of the first critical load factor are summarized and listed in Table 2.3. It shows that the results always converge well following the increase of the order of Chebyshev polynomials for all of the three different grids. It also shows that the results of type 2 grid converge the best. Consequently, it is known that the EDQ discretization adopting the type 2 grid approximates the best. Non-discrete analyses using Bernoulli and Euler functions can also be found in existing articles [51–53].

Table 2.3. The lowest critical load factor $\bar{C}_1$ of a compressed simply supported prismatic beam

Type of element grid	DOF per element	Chebyshev polynomials	Bernoulli polynomials	Euler polynomials
1	6	$.900000\times10^1$	$.833333\times10^1$	$.833333\times10^1$
	8	$.992318\times10^1$	$.100097\times10^2$	$.100097\times10^2$
	10	$.986833\times10^1$	$.986559\times10^1$	$.986559\times10^1$
2	6	$.106667\times10^2$	$.106667\times10^2$	$.106667\times10^2$
	8	$.986729\times10^1$	$.986729\times10^1$	$.986729\times10^1$
	10	$.986963\times10^1$	$.986963\times10^1$	$.986963\times10^1$
3	6	$.884458\times10^1$	$.884458\times10^1$	$.884458\times10^1$
	8	$.991373\times10^1$	$.991373\times10^1$	$.991373\times10^1$
	10	$.986893\times10^1$	$.986893\times10^1$	$.986893\times10^1$
Exact solution		$.986960\times10^1$		

The static deflection of a clamped square plate with the length of sides being a is also solved by using the Chebyshev EDQ model. The governing equation is expressed by

$$\frac{\partial^4 w}{\partial x^4} + 2\frac{\partial^4 w}{\partial x^2 \partial y^2} + \frac{\partial^4 w}{\partial y^4} = \frac{q(x,y)}{D} \tag{2.100}$$

where w is the lateral displacement, D is the flexural rigidity and $q(x, y)$ is the distributed load. By locating the origin of the coordinate system at the center of the plate, the boundary conditions are

$$w = 0, \quad \frac{\partial w}{\partial x} = 0, \quad \text{for} \quad x = \pm\frac{a}{2}; \quad w = 0, \quad \frac{\partial w}{\partial y} = 0, \quad \text{for} \quad y = \pm\frac{a}{2} \tag{2.101}$$

In the analysis, one element is used to represent the problem domain. Various grids can be used. The two-dimensional node identification method adopting the Lagrange family grid is used. For interior nodes only the degree of freedom representing the lateral displacement is assigned while multiple degrees of freedom are assigned to a node on the boundary to represent the lateral displacement and displacement gradient in the direction normal to the boundary line. Chebyshev polynomials are used to define the approximate displacement function. Thus, the problem domain is mapped onto the region $\Omega = \{-1 \le \xi \le 1, \ -1 \le \eta \le 1\}$. And Eq. (2.100) is transformed

$$\frac{\partial^4 w}{\partial \xi^4} + 2\frac{\partial^4 w}{\partial \xi^2 \partial \eta^2} + \frac{\partial^4 w}{\partial \eta^4} = \frac{a^4 q}{16D} \tag{2.102}$$

where ξ and η are natural coordinates. Let $T_{\tilde{p}}(\xi_i)$ denote the Chebyshev polynomials. The displacement function can be defined by the two-dimensional expansion of the Chebyshev polynomials

$$w(\xi,\eta) = T_p(\xi)T_q(\eta)c_{pq}, \qquad p = 1, ..., N_\xi + 2; \quad q = 1, ..., N_\eta + 2 \tag{2.103}$$

The weighting coefficients $D^{\xi_i^m}_{\alpha\beta}$ for mth order partial derivative in ξ_i direction can be calculated by using the implicit methods 1 or 3. At an interior discrete point (α, β) Eq. (2.102) can be discretized

$$\sum_{i=1}^{N_\xi+2} D^{\xi^4}_{\alpha i}\tilde{w}_{i\beta} + 2\sum_{i=1}^{N_\xi+2}\sum_{j=1}^{N_\eta+2} D^{\xi^2}_{\alpha i}D^{\eta^2}_{\beta j}\tilde{w}_{ij} + \sum_{j=1}^{N_\eta+2} D^{\eta^4}_{\beta j}\tilde{w}_{\alpha j} = \frac{a^4 q_{\alpha\beta}}{16D} \tag{2.104}$$

The discrete boundary conditions are

$$\begin{aligned} w_{1j} = w_{N_\xi j} &= 0, \quad \text{for} \quad j = 2, 3, ..., N_\eta - 1; \\ w_{1j,\xi} = w_{N_\xi j,\xi} &= 0, \quad \text{for} \quad j = 1, 2, ..., N_\eta \\ w_{i1} = w_{iN_\eta} &= 0, \quad \text{for} \quad i = 1, 2, ..., N_\xi; \\ w_{i1,\eta} = w_{iN_\eta,\eta} &= 0, \quad \text{for} \quad i = 1, 2, ..., N_\xi \end{aligned} \tag{2.105}$$

In the above equation, N_ξ and N_η are the numbers of level of the Lagrange family grid in ξ and η directions, respectively.

The Lagrange family grid is designed to have a form that at the discrete points the Chebyshev polynomials in both ξ and η directions are zeros. These Chebyshev sampling points are defined by $\xi_i \mid_{\bar{\alpha}} = -cos\left(\frac{\pi(\bar{\alpha}-1)}{N_{\xi_i}-1}\right)$, $\bar{\alpha} = 1, ..., N_{\xi_i}$. Numerical results of the center displacements for the uniformly loaded plate are listed in Table 2.4. The ordinary finite difference method (FDM) and finite element method (FEM) Adini-Clough-Melosh element solutions are also included in the table. It shows that the current results converge fast by gradually increasing the Chebyshev sampling points. The Chebyshev EDQ needs much less degrees of freedom (DOF) than the FEM and FDM to converge.

Table 2.4. Results of the Chebyshev EDQ analysis of the uniformly loaded plate

Method	Mesh type	DOF	$\pi^6 Dw/(16qa^4)$
EDQ	5×5	9	$.7707\times10^{-1}$
	7×7	25	$.7603\times10^{-1}$
	9×9	49	$.7603\times10^{-1}$
FEM	8×8	108	$.7833\times10^{-1}$
	12×12	300	$.7710\times10^{-1}$
	16×16	588	$.7661\times10^{-1}$
FDM	21×21	361	$.7766\times10^{-1}$
	31×31	841	$.7676\times10^{-1}$
	41×41	1521	$.7642\times10^{-1}$
Exact solution			$.7603\times10^{-1}$

2.7 Generation of EDQ Using DQ

Some EDQ models can be generated by using DQ through the establishment of a transformation relation between the set of discrete function variables π_α of the equivalent DQ element and the set of discrete EDQ parameters $\tilde{\pi}_\alpha$ [54–56]. The weighting coefficients of these EDQ models are calculated by using the related transformation matrices and the weighting coefficients of the equivalent DQ model. For illustration, the $C^1 - C^0 - C^1$ and $C^2 - C^0 - C^2$ EDQ models used to the spacial discretization of DQEM analysis, and the $C^1 - C^0$ EDQ model used to the temporal discretization of EDQ based direct time integration methods for structural dynamic problems are introduced. C^m represents that the node has the DOF of the function variable and its derivatives up to the mth order.

2.7.1 $C^1 - C^0 - C^1$ EDQ Model

Consider the $C^1 - C^0 - C^1$ EDQ model with each of the two boundary nodes having two DOF of the function variable and its first order derivative, and each of the interior nodes having only one DOF of the function variable. This EDQ model is compatible and conformable, and can automatically set the kinematic transition conditions on the inter-element boundary of two adjacent elements and the kinematic boundary conditions for the DQEM analysis of flexural deformation problems of structures.

Let $\bar{D}_{\alpha i}^{\xi^m}$ denote the weighting coefficients for the equivalent DQ model defined on the natural space which is a $C^0 - C^0 - C^0$ DQ model. For the DQEM analysis of flexural deformation problems of structures using the DQ model, the DOF assigned to the first two and last two nodes are used to define either the transition conditions at the inter-element boundary of two adjacent real physical elements or boundary conditions at the real physical domain boundary. One of the two nodes at or next to one boundary node of the equivalent DQ element is at the inter-element boundary of two real physical elements used to separate the analysis domain or at the real physical domain boundary, while the other one can be either outside or inside the real physical elements, and next to the element boundary of the real physical element. If the two extra nodes are inside the real physical element, the equivalent DQ element coincide with the real physical element. If the two extra nodes are outside the real physical element, the equivalent DQ element extends over the two element boundary points of the real physical element. Consequently, for the DQEM analysis of flexural deformation problems using DQ elements, the two adjacent equivalent DQ elements are overlapped partially and the equivalent DQ element containing the domain boundary is extended over the physical domain boundary.

Let $\hat{N}$ denote the number of the two boundary nodes of the $C^1 - C^0 - C^1$ EDQ element plus the other $\bar{N} - 4$ interior nodes with $\bar{N} = N$. Also let $\bar{l}$ and l denote the physical element lengths of the equivalent physical DQ element and the physical $C^1 - C^0 - C^1$ EDQ element. $\bar{l}$ equals l for the transformation model with the physical EDQ element coinciding the physical DQ element, while $\bar{l}$ is larger than l for the transformation model with the two extra nodes of the equivalent physical DQ element outside the physical EDQ element.

For the DQEM analysis of flexural deformation problems using EDQ elements, the DOF assigned to an auxiliary node of the equivalent DQ element can be transformed to obtain one rotational DOF assigned to the related element boundary node of the EDQ element. Figure 2.1 shows the $C^1 - C^0 - C^1$ EDQ model generated by using the equivalent DQ model with the two auxiliary nodes inside the physical EDQ element. Assume that ξ is a natural coordinate having one unit length for both the $C^1 - C^0 - C^1$ EDQ model and the equivalent DQ model defined on the natural space. The transformation relation between the $C^1 - C^0 - C^1$ EDQ model and the equivalent DQ model can be represented by the following equation

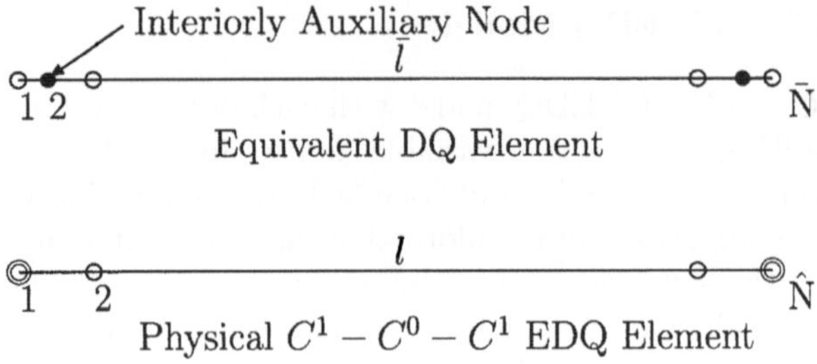

Fig. 2.1. $C^1 - C^0 - C^1$ EDQ model generated by using the equivalent DQ model with two auxiliary nodes inside the physical EDQ model.

$$\{\tilde{\pi}\} = [T]\{\pi\} \tag{2.106}$$

where

$$\{\pi\} = \lfloor \pi_1 \; \pi_2 \; \pi_3 \; ... \; \pi_{N-2} \; \pi_{N-1} \; \pi_N \rfloor^T, \tag{2.107}$$

For this transformation, $\{\tilde{\pi}\}$ and $[T]$ in Eq. (2.106) are expressed by

$$\{\tilde{\pi}\} = \lfloor \pi_1 \; \tfrac{d\pi_1}{d\xi} \; \pi_3 \; ... \; \pi_{N-2} \; \pi_N \; \tfrac{d\pi_N}{d\xi} \rfloor^T, \tag{2.108}$$

and

$$[T] = \begin{bmatrix} 1 & 0 & 0 & \cdots & 0 & 0 & 0 \\ \bar{D}^{\xi}_{11} & \bar{D}^{\xi}_{12} & \bar{D}^{\xi}_{13} & \cdots & \bar{D}^{\xi}_{1(N-2)} & \bar{D}^{\xi}_{1(N-1)} & \bar{D}^{\xi}_{1N} \\ 0 & 0 & 1 & \cdots & 0 & 0 & 0 \\ \cdot & \cdot & \cdot & \cdots & \cdot & \cdot & \cdot \\ \cdot & \cdot & \cdot & \cdots & \cdot & \cdot & \cdot \\ \cdot & \cdot & \cdot & \cdots & \cdot & \cdot & \cdot \\ 0 & 0 & 0 & \cdots & 1 & 0 & 0 \\ 0 & 0 & 0 & \cdots & 0 & 0 & 1 \\ \bar{D}^{\xi}_{N1} & \bar{D}^{\xi}_{N2} & \bar{D}^{\xi}_{N3} & \cdots & \bar{D}^{\xi}_{N(N-2)} & \bar{D}^{\xi}_{N(N-1)} & \bar{D}^{\xi}_{NN} \end{bmatrix} \tag{2.109}$$

Figure 2.2 shows the $C^1 - C^0 - C^1$ EDQ model generated by using the equivalent DQ model with the two auxiliary nodes outside the physical EDQ element. For this transformation, $\{\tilde{\pi}\}$ and $[T]$ in Eq. (2.106) are expressed by

$$\{\tilde{\pi}\} = \lfloor \pi_2 \; \tfrac{d\pi_2}{d\xi} \; \pi_3 \; ... \; \pi_{N-2} \; \pi_{N-1} \; \tfrac{d\pi_{N-1}}{d\xi} \rfloor^T, \tag{2.110}$$

and

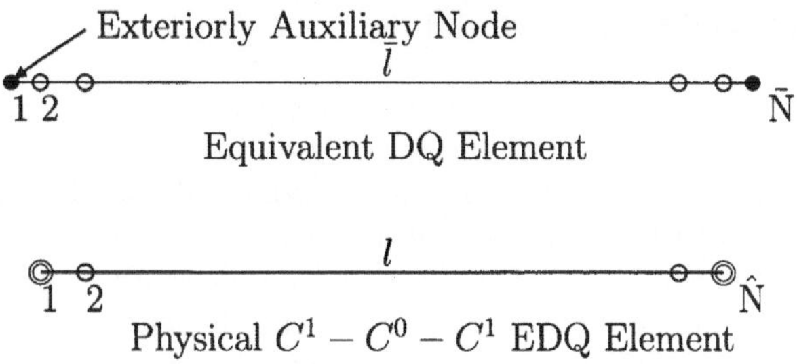

Fig. 2.2. $C^1 - C^0 - C^1$ EDQ model generated by using the equivalent DQ model with two auxiliary nodes outside the physical EDQ model.

$$[T] = \begin{bmatrix} 0 & 1 & 0 & \cdots & 0 & 0 & 0 \\ \bar{D}^{\xi}_{21} & \bar{D}^{\xi}_{22} & \bar{D}^{\xi}_{23} & \cdots & \bar{D}^{\xi}_{2(N-2)} & \bar{D}^{\xi}_{2(N-1)} & \bar{D}^{\xi}_{2N} \\ 0 & 0 & 1 & \cdots & 0 & 0 & 0 \\ \cdot & \cdot & \cdot & \cdots & \cdot & \cdot & \cdot \\ \cdot & \cdot & \cdot & \cdots & \cdot & \cdot & \cdot \\ \cdot & \cdot & \cdot & \cdots & \cdot & \cdot & \cdot \\ 0 & 0 & 0 & \cdots & 1 & 0 & 0 \\ 0 & 0 & 0 & \cdots & 0 & 1 & 0 \\ \bar{D}^{\xi}_{(N-1)1} & \bar{D}^{\xi}_{(N-1)2} & \bar{D}^{\xi}_{(N-1)3} & \cdots & \bar{D}^{\xi}_{(N-1)(N-2)} & \bar{D}^{\xi}_{(N-1)(N-1)} & \bar{D}^{\xi}_{(N-1)N} \end{bmatrix} \tag{2.111}$$

Assume that $D^{\xi^m}_{\alpha i}$ are weighting coefficients for the $C^1 - C^0 - C^1$ EDQ model defined on the natural space. By using the fact that the derivatives with respect to the local element based physical coordinates at the $\hat{N}$ grid nodes are the same for both the $C^1 - C^0 - C^1$ EDQ model and the DQ model, the following relation can be obtained

$$\bar{D}^{\xi^m}_{\alpha i}\,\pi_i = D^{\xi^m}_{\alpha i}\,\tilde{\pi}_i, \quad \alpha = 1, 2, ..., \hat{N}; i = 1, 2, ..., \bar{N} \tag{2.112}$$

The introduction of Eq. (2.106) into the above equation leads to the following transformation equation for calculating the weighting coefficients for the $C^1 - C^0 - C^1$ EDQ model

$$D^{\xi^m}_{\alpha i} = \bar{D}^{\xi^m}_{\alpha j}\,T^{-1}_{ji}, \quad \alpha = 1, 2, ..., \hat{N}; i = 1, 2, ..., \bar{N} \tag{2.113}$$

$D^{\xi^m}_{\alpha i}$ is a $\hat{N} \times \bar{N}$ matrix with $\hat{N} = \bar{N} - 2$. For applying the $C^1 - C^0 - C^1$ EDQ model to the DQEM analysis, the physical first order derivatives at the two EDQ element boundary nodes are discrete EDQ parameters. Consequently, the elements of the second and last columns of the related EDQ weighting coefficient matrix $D^{\xi^m}_{\alpha i}$ need to be multiplied by l.

2.7.2 $C^2 - C^0 - C^2$ EDQ Model

Consider the $C^2 - C^0 - C^2$ EDQ model with each of the two boundary nodes having three DOF of the function variable and its first and second order derivatives, and each of the interior nodes having only one DOF of the function variable. This EDQ model can automatically set the kinematic transition conditions and the continuity of the second order derivatives of displacements on the inter-element boundary of two adjacent elements and the kinematic boundary conditions and the bending moment related natural boundary condition for the DQEM analysis of flexural deformation problems of structures.

To obtain the $C^2 - C^0 - C^2$ EDQ model, two DOF of two nodes of the equivalent DQ model close to a boundary node of the physical $C^2 - C^0 - C^2$ EDQ model have to be transformed to the two DOF representing the first order and second order derivatives at the related EDQ element boundary node. These two nodes can be either inside or outside the physical $C^2 - C^0 - C^2$ EDQ model. For illustration, the one of transforming the four DOF of four equivalent physical DQ nodes inside the related physical $C^2 - C^0 - C^2$ EDQ model is introduced. Figure 2.3 shows the $C^2 - C^0 - C^2$ EDQ model generated by using the equivalent DQ model with the four auxiliary nodes inside the physical EDQ element. The transformation relation between the $C^2 - C^0 - C^2$ EDQ model and the equivalent DQ model is also represented by Eq. (2.106) with

$$\{\tilde{\pi}\} = \lfloor \pi_1 \ \frac{d\pi_1}{d\xi} \ \frac{d^2\pi_1}{d\xi^2} \ \pi_4 \ \dots \ \pi_{N-3} \ \pi_N \ \frac{d\pi_N}{d\xi} \ \frac{d^2\pi_N}{d\xi^2} \rfloor^T, \tag{2.114}$$

and

$$[T] = \begin{bmatrix} 1 & 0 & 0 & 0 & \cdots & 0 & 0 & 0 & 0 \\ \bar{D}^{\xi}_{11} & \bar{D}^{\xi}_{12} & \bar{D}^{\xi}_{13} & \bar{D}^{\xi}_{14} & \cdots & \bar{D}^{\xi}_{1(N-3)} & \bar{D}^{\xi}_{1(N-2)} & \bar{D}^{\xi}_{1(N-1)} & \bar{D}^{\xi}_{1N} \\ \bar{D}^{\xi^2}_{11} & \bar{D}^{\xi^2}_{12} & \bar{D}^{\xi^2}_{13} & \bar{D}^{\xi^2}_{14} & \cdots & \bar{D}^{\xi^2}_{1(N-3)} & \bar{D}^{\xi^2}_{1(N-2)} & \bar{D}^{\xi^2}_{1(N-1)} & \bar{D}^{\xi^2}_{1N} \\ 0 & 0 & 0 & 1 & \cdots & 0 & 0 & 0 & 0 \\ \cdot & \cdot & \cdot & \cdot & \cdots & \cdot & \cdot & \cdot & \cdot \\ \cdot & \cdot & \cdot & \cdot & \cdots & \cdot & \cdot & \cdot & \cdot \\ \cdot & \cdot & \cdot & \cdot & \cdots & \cdot & \cdot & \cdot & \cdot \\ 0 & 0 & 0 & 0 & \cdots & 1 & 0 & 0 & 0 \\ 0 & 0 & 0 & 0 & \cdots & 0 & 0 & 0 & 1 \\ \bar{D}^{\xi}_{N1} & \bar{D}^{\xi}_{N2} & \bar{D}^{\xi}_{N3} & \bar{D}^{\xi}_{N4} & \cdots & \bar{D}^{\xi}_{N(N-3)} & \bar{D}^{\xi}_{N(N-2)} & \bar{D}^{\xi}_{N(N-1)} & \bar{D}^{\xi}_{NN} \\ \bar{D}^{\xi^2}_{N1} & \bar{D}^{\xi^2}_{N2} & \bar{D}^{\xi^2}_{N3} & \bar{D}^{\xi^2}_{N4} & \cdots & \bar{D}^{\xi^2}_{N(N-3)} & \bar{D}^{\xi^2}_{N(N-2)} & \bar{D}^{\xi^2}_{N(N-1)} & \bar{D}^{\xi^2}_{NN} \end{bmatrix} \tag{2.115}$$

Using $[T]$ in Eq. (2.113), the weighting coefficients for the $C^2 - C^0 - C^2$ EDQ model can be obtained with $\hat{N} = \bar{N} - 4$. For applying the $C^2 - C^0 - C^2$ EDQ model to the DQEM analysis, the physical first and second order derivatives at the two EDQ element boundary nodes are discrete EDQ parameters. Consequently, the elements of the second and $(\bar{N} - 1)$th columns of the related EDQ weighting coefficient matrix $D^{\xi^m}_{\alpha i}$ need to be multiplied by l, and the elements of the third and last columns need to be multiplied by l^2.

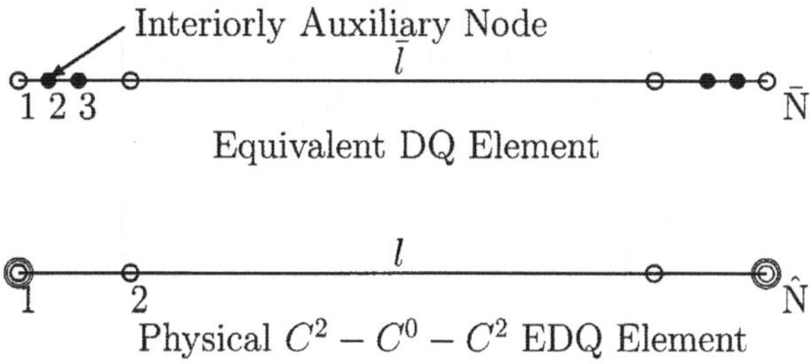

Fig. 2.3. $C^2 - C^0 - C^2$ EDQ model generated by using the equivalent DQ model with four auxiliary nodes inside the physical EDQ model.

2.7.3 $C^1 - C^0$ EDQ Model

Consider the $C^1 - C^0$ EDQ model with the first node having two DOF of the function variable and its first derivative, and each of the other nodes having only one DOF of the function variable. This EDQ model can automatically set the initial conditions of each integration step for the EDQ based step-by-step direct time integration method for solving transient structural problems.

To obtain the $C^1 - C^0$ EDQ model, one DOF of one node of the equivalent physical DQ model close to the first node of the related physical EDQ model have to be transformed to the DOF representing the first order derivative at the first node of the physical $C^1 - C^0$ EDQ model. This node can be either inside or outside the physical $C^1 - C^0$ EDQ element. For illustration, the one of transforming the DOF of one node of the equivalent physical DQ model outside the physical $C^1 - C^0$ EDQ model is introduced. Figure 2.4 shows the $C^1 - C^0$ EDQ model generated by using the equivalent DQ model with the one auxiliary node outside the physical EDQ element. The transformation relation between the $C^1 - C^0$ EDQ model and the DQ model is also represented by Eq. (2.106) with

$$\{\tilde{\pi}\} = \lfloor \pi_2 \ \tfrac{d\pi_2}{d\xi} \ \pi_3 \ \dots \ \pi_{N-1} \ \pi_N \rfloor^T, \tag{2.116}$$

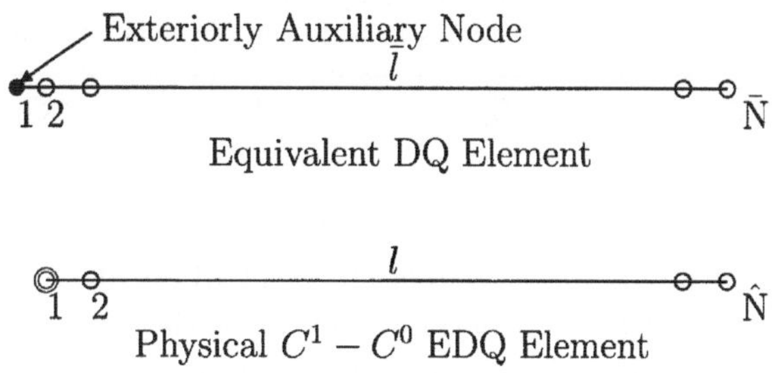

Fig. 2.4. $C^1 - C^0$ EDQ model generated by using the equivalent DQ model with one auxiliary node inside the physical EDQ model.

and

$$[T] = \begin{bmatrix} 0 & 1 & 0 & \cdots & 0 & 0 \\ \bar{D}^{\xi}_{21} & \bar{D}^{\xi}_{22} & \bar{D}^{\xi}_{23} & \cdots & \bar{D}^{\xi}_{2(N-1)} & \bar{D}^{\xi}_{2N} \\ 0 & 0 & 1 & \cdots & 0 & 0 \\ \cdot & \cdot & \cdot & \cdots & \cdot & \cdot \\ \cdot & \cdot & \cdot & \cdots & \cdot & \cdot \\ \cdot & \cdot & \cdot & \cdots & \cdot & \cdot \\ 0 & 0 & 0 & \cdots & 1 & 0 \\ 0 & 0 & 0 & \cdots & 0 & 1 \end{bmatrix} \tag{2.117}$$

Using $[T]$ in Eq. (2.113), the weighting coefficients for the $C^1 - C^0$ EDQ model can be obtained with $\hat{N} = \bar{N} - 1$. For applying the $C^1 - C^0$ EDQ model to the transient structural analysis, the physical first order derivative at the first EDQ element boundary node is a discrete EDQ parameter. Consequently, the elements of the second column of the related EDQ weighting coefficient matrix $D^{\xi^m}_{\alpha i}$ need to be multiplied by l.

Table 2.5 The first four frequency factors of a clamped square plate solved by the DQEM using the $C^1 - C^0 - C^1$ EDQ model generated by using the equivalent DQ model with two auxiliary nodes inside the physical EDQ model

Order of appr.	C_1	C_2	C_3	C_4
6	$.368344\times10^2$	$.650341\times10^2$	$.650341\times10^2$	$.901151\times10^2$
8	$.360286\times10^2$	$.796985\times10^2$	$.796985\times10^2$	
10	$.359929\times10^2$	$.730596\times10^2$	$.730596\times10^2$	$.105345\times10^3$
Leissa's sol.	$.35992\times10^2$	$.73413\times10^2$	$.73413\times10^2$	$.10827\times10^3$

Table 2.6 The first four frequency factors of a clamped square plate solved by the DQEM using the $C^1 - C^0 - C^1$ EDQ model generated by using the equivalent DQ model with two auxiliary nodes outside the physical EDQ model

Order of appr.	C_1	C_2	C_3	C_4
4	$.3475629\times10^2$			
6	$.3630833\times10^2$	$.7160144\times10^2$	$.7160144\times10^2$	$.1022950\times10^3$
8	$.3600669\times10^2$	$.7433538\times10^2$	$.7433538\times10^2$	$.1097962\times10^3$
10	$.3599020\times10^2$	$.7335022\times10^2$	$.7335022\times10^2$	$.1082930\times10^3$
Leissa's sol.	$.35992\times10^2$	$.73413\times10^2$	$.73413\times10^2$	$.10827\times10^3$

The free vibration of the clamped square plate was analyzed by using one DQEM element. The problem was solved by using the $C^1 - C^0 - C^1$ EDQ model generated by using both of the equivalent Lagrange DQ models with equally spaced nodes and with the two auxiliary nodes inside and outside the physical EDQ model. Let a, δ, D and ρ denote the edge length, thickness, flexural rigidity and mass density of a square plate. Also let the natural frequency ω_n of the nth vibration mode be expressed by $\omega_n = \frac{C_n}{a^2}\sqrt{\frac{D}{\rho\delta}}$ with C_n the frequency factor. The first four frequency factors obtained by the two DQEM models are listed in Tables 2.5 and 2.6, separately. They are compared with the results obtained by Leissa [57].

3

DQEM Analysis of One-Dimensional Elasticity Problems

This chapter introduces the DQEM analysis procedures to the solution of static deformation and vibration of nonprismatic bar structures. The mapping technique is adopted. Consequently, the computation of weighting coefficients for every element is not necessary. DQ is used to carry out the element basis discretization.

A direct assemblage procedure of assembling discrete governing equations defined at interior discrete points of an element, and discrete element boundary forces, expressed by elementary displacement parameters for establishing the natural transition condition on the inter-element boundary of two adjacent elements or natural boundary condition on the natural boundary, are directly assembled to the overall algebraic equation system based on an element by element procedure. Consequently, an element basis explicit element stiffness equation is not necessary to be formed for each element.

The solution strategies and implementation of DQEM computer program are also introduced.

3.1 Static Deformation of Bars

The governing differential equilibrium equation of one-dimensional elasticity problems is

$$\frac{d}{dx}\left(EA\frac{du}{dx}\right) = -f \tag{3.1}$$

where u is axial displacement, E is Young's modulus, A is the area of cross section and f is the distributed axial force. Let L denote the length of the bar. The boundary conditions at the natural boundary x_σ and kinetic boundary x_u of domain $0 \leq x \leq L$ are

$$EA\frac{du}{dx}\mid_{x=x_\sigma} = \bar{P}\mid_{x=x_\sigma} \tag{3.2}$$

and

$$u\mid_{x=x_u} = \bar{u}\mid_{x=x_u} \tag{3.3}$$

where $\bar{u}$ is the prescribed displacement and $\bar{P}$ is the concentrated boundary load.

3.1.1 DQEM Formulation

The governing differential equilibrium equation and boundary conditions are defined on the physical coordinate system, while the differential quadrature discretization is carried out on the natural coordinate system. Therefore, in using the differential quadrature technique to discretize the governing differential equilibrium equation and boundary conditions, the transformation operations of coordinates and derivatives of displacement between two different coordinate systems have to be carried out. Figure 3.1 shows the mapping

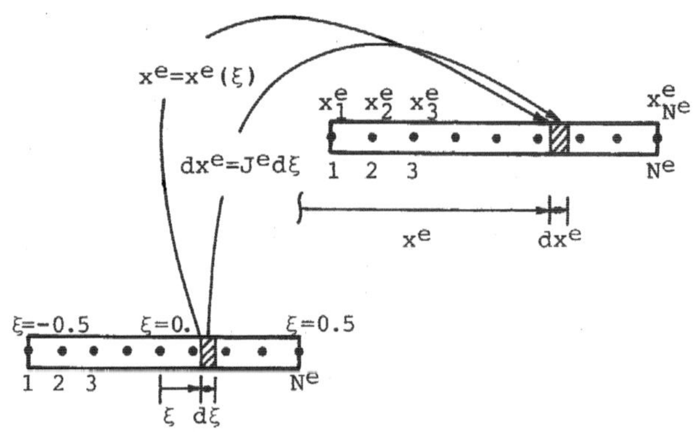

Fig. 3.1. Mapping of element configuration

of a parent element onto a physical element. The physical coordinate x^e can be written as:

$$x^e = (0.5 - \xi)x_1^e + (0.5 + \xi)x_{N^e}^e \tag{3.4}$$

where x_1^e and $x_{N^e}^e$ are the coordinates of first node and N^eth node, respectively, of a N^e-node element and ξ the natural coordinate. It should be noted that the range of the parent element is $-0.5 \leq \xi \leq 0.5$. Using Eq. (3.4), the differential of x^e can be expressed as:

$$dx^e = (x_{N^e}^e - x_1^e)d\xi = J^e d\xi \tag{3.5}$$

where $J^e = x_{N^e}^e - x_1^e = l^e$ is the Jacobian and l^e is the length of the element. From the above equation, the following relation can be obtained

$$\frac{d\xi}{dx^e} = \frac{1}{J^e} = \frac{1}{x_{N^e}^e - x_1^e} \tag{3.6}$$

Then the first and second order derivatives of element displacement u^e with respect to x^e can be written as:

$$\frac{du^e}{dx^e} = \frac{1}{J^e}\frac{du^e}{d\xi} \tag{3.7}$$

and

$$\frac{d^2u^e}{d(x^e)^2} = \frac{1}{(J^e)^2}\frac{d^2u^e}{d\xi^2} \tag{3.8}$$

The DQ model which only uses the displacements at element nodes to define the discretization of the first and second order derivatives of element displacement u^e with respect to the local coordinate x^e at node points is adopted.

The equilibrium conditions of interior node points of all elements are used in constructing a discrete algebraic system for a problem. For elements having no distributed load, the number of nodes can be 2. Since Eq. (3.1) is a second order differential equation, without using a certain technique to calculate the two equivalent nodal loads for the distributed load of a two-node linear element and include them into the natural transition conditions or natural boundary conditions, the order of approximate displacement must at least be two and the element must at least have one node point at which the discrete element equilibrium equation is defined. The DQEM linear element is equivalent to the FEM linear element. For elements having a distributed load, the number of nodes must be larger than 2 in order to get an efficient convergence. The discrete element equilibrium equation at a node point α in an element e can be expressed as

$$\frac{1}{(J^e)^2}\left[\frac{d(EA)^e_{(\alpha)}}{d\xi}\sum_{\beta=1}^{N^e} D^{e\xi}_{\alpha\beta} + (EA)^e_{(\alpha)}\sum_{\beta=1}^{N^e} D^{e\xi^2}_{\alpha\beta}\right] u^e_\beta = -f^e_\alpha,$$
$$\alpha = 2, 3, ..., N^e - 1 \tag{3.9}$$

where $D^{e\xi^m}_{\alpha\beta}$ are weighting coefficients for the mth order derivative with respect to ξ. In the above equation, the derivative of axial rigidity at the related node points can also be calculated by the DQ. It is especially useful if the distribution of EA is not continuously differentiable up to the order of its derivative. The values of EA at the N^e element nodes are used to define the DQ discretization of the derivative. Then the derivative of EA at a node α can be expressed by

$$\frac{d(EA)^e_\alpha}{d\xi} = \sum_{p=1}^{N^e} D^{e\xi}_{\alpha p}(EA)^e_p \tag{3.10}$$

The internal axial stresses at node points in the element can be expressed by using the displacements

$$\sigma^e_\alpha = \frac{E^e}{J^e}\sum_{\beta=1}^{N^e} D^{e\xi}_{\alpha\beta} u^e_\beta \tag{3.11}$$

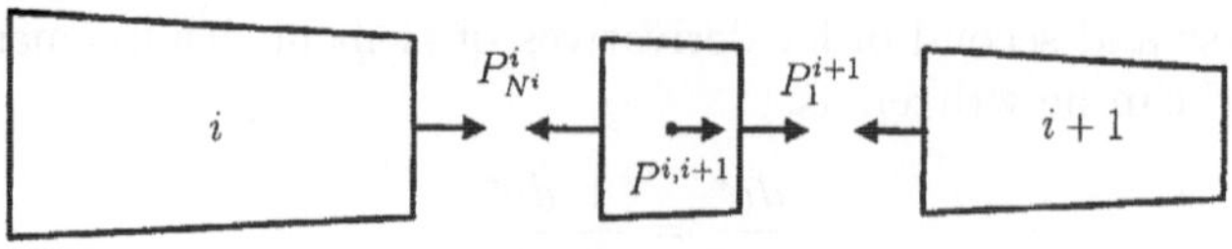

Fig. 3.2. Equilibrium of forces at the inter-element boundary

The transition conditions of two adjacent elements are the continuity of displacements and the equilibrium of axial forces at the inter-element boundary. At the inter-element boundary of two adjacent elements i and $i+1$, the condition of displacement continuity is expressed as:

$$u^i_{N^i} = u^{i+1}_1 \tag{3.12}$$

Let $P^{i,i+1}$ denote the concentrated axial force applied at the inter-element boundary $x = x^{i,i+1}$. The condition of force equilibrium can be established by referring to Fig. 3.2

$$P^i_{N^i} - P^{i+1}_1 = P^{i,i+1} \tag{3.13}$$

or

$$(EA)^i_{N^i} \frac{du^i_{N^i}}{dx^i} - (EA)^{i+1}_1 \frac{du^{i+1}_1}{dx^{i+1}} = P^{i,i+1} \tag{3.14}$$

The above equation can be discretized by using Eq. (3.7) and the DQ

$$\frac{(EA)^i_{N^i}}{J^i} \sum_{\alpha=1}^{N^i} D^{i\xi}_{N^i\alpha} u^i_\alpha - \frac{(EA)^{i+1}_1}{J^{i+1}} \sum_{\beta=1}^{N^{i+1}} D^{(i+1)\xi}_{1\beta} u^{i+1}_\beta = P^{i,i+1} \tag{3.15}$$

Letting element n be an element having one or more nodes on the natural boundary, the natural boundary condition represented by Eq. (3.2) can be discretized which shows to have the following form:

$$\frac{(EA)^n_{I^n}}{J^n} \sum_{\beta=1}^{N^n} D^{n\xi}_{I^n\beta} u^n_\beta = \bar{P} \mid_{x=x_\sigma}, \quad I^n = 1 \quad \text{or} \quad = N^n \tag{3.16}$$

Also letting element m be an element having one or more nodes on the kinematic boundary, the discrete description of Eq. (3.3) is

$$u^m_{I^m} = \bar{u} \mid_{x=x_u}, \quad I^m = 1 \quad \text{or} \quad N^m \tag{3.17}$$

3.1.2 Assemblage and Solution

Keeping the kinematic transition conditions (3.12) in mind, assembling the discrete element equilibrium equations (3.9) for elements having more than two nodes, the discrete natural transition conditions (3.15), and the discrete natural boundary conditions (3.16) for the boundary elements with natural boundary, an overall discrete equilibrium/transition/boundary equation can be obtained. It is the overall stiffness equation of the DQEM. This stiffness equation is expressed by

$$[K]\{D\} = \{R\} \tag{3.18}$$

where $[K]$ is the overall stiffness matrix, $\{D\}$ is the overall displacement vector and $\{R\}$ is the overall load vector. Considering the discrete kinematic boundary condition (3.17) and using a certain solver, the overall linear algebraic system can be solved to obtain displacements at all node points.

Like FEM, the assemblage is based on an element by element procedure. When assembling the discrete equations of element e, the discrete element equilibrium equations (3.9) and the two discrete element boundary forces, expressed by displacements, at the two element boundary nodes are directly assembled into the overall discrete equation system. This direct assembling procedure can reduce the computer costs required for the preprocessing phases. Consequently, an element basis explicit matrix equation, containing the discrete element equilibrium equations and the discrete element boundary forces placed at the first and last rows, is not necessary to be formed in the assembling process. This element basis explicit matrix equation is an element stiffness equation. Using Eqs. (3.9) and (3.11), this equation can be expressed by

$$\left[\begin{array}{ccc}
-\frac{(EA)^e}{J^e} D_{11}^{e\xi} & -\frac{(EA)^e}{J^e} D_{12}^{e\xi} & \cdots \\
\frac{1}{(J^e)^2}\left[\frac{d(EA)_2^e}{d\xi} D_{21}^{e\xi} + (EA)_2^e D_{21}^{e\xi^2}\right] & \frac{1}{(J^e)^2}\left[\frac{d(EA)_2^e}{d\xi} D_{22}^{e\xi} + (EA)_2^e D_{22}^{e\xi^2}\right] & \cdots \\
\cdot & \cdot & \cdots \\
\cdot & \cdot & \cdots \\
\cdot & \cdot & \cdots \\
\frac{1}{(J^e)^2}\left[\frac{d(EA)_{(N^e-1)}^e}{d\xi} D_{(N^e-1)1}^{e\xi} + (EA)_{(N^e-1)}^e D_{(N^e-1)1}^{e\xi^2}\right] & \frac{1}{(J^e)^2}\left[\frac{d(EA)_{(N^e-1)}^e}{d\xi} D_{(N^e-1)2}^{e\xi} + (EA)_{(N^e-1)}^e D_{(N^e-1)2}^{e\xi^2}\right] & \cdots \\
\frac{(EA)^e}{J^e} D_{N^e1}^{e\xi} & \frac{(EA)^e}{J^e} D_{N^e2}^{e\xi} & \cdots
\end{array}\right.$$

$$
\begin{matrix}
-\frac{(EA)^e}{J^e} D^{e\xi}_{1(N^e-1)} \\
\frac{1}{(J^e)^2}\left[\frac{d(EA)^e_2}{d\xi} D^{e\xi}_{2(N^e-1)} + (EA)^e_2 D^{e\xi^2}_{2(N^e-1)}\right] \\
\cdot \\
\cdot \\
\cdot \\
\frac{1}{(J^e)^2}\left[\frac{d(EA)^e_{(N^e-1)}}{d\xi} D^{e\xi}_{(N^e-1)(N^e-1)} + (EA)^e_{(N^e-1)} D^{e\xi^2}_{(N^e-1)(N^e-1)}\right] \\
\frac{(EA)^e}{J^e} D^{e\xi}_{N^e(N^e-1)}
\end{matrix}
$$

$$
\left.\begin{matrix}
-\frac{(EA)^e}{J^e} D^{e\xi}_{1N^e} \\
\frac{1}{(J^e)^2}\left[\frac{d(EA)^e_2}{d\xi} D^{e\xi}_{2N^e} + (EA)^e_2 D^{e\xi^2}_{2N^e}\right] \\
\cdot \\
\cdot \\
\cdot \\
\frac{1}{(J^e)^2}\left[\frac{d(EA)^e_{(N^e-1)}}{d\xi} D^{e\xi}_{(N^e-1)N^e} + (EA)^e_{(N^e-1)} D^{e\xi^2}_{(N^e-1)N^e}\right] \\
\frac{(EA)^e}{J^e} D^{e\xi}_{N^eN^e}
\end{matrix}\right]
\begin{Bmatrix}
u^e_1 \\ u^e_2 \\ \cdot \\ \cdot \\ \cdot \\ u^e_{N^e-1} \\ u^e_{N^e}
\end{Bmatrix}
$$

$$
= \begin{Bmatrix}
-(A\sigma)^e_1 \\ -f^e_2 \\ \cdot \\ \cdot \\ \cdot \\ -f^e_{N^e-1} \\ (A\sigma)^e_{N^e}
\end{Bmatrix} \tag{3.19}
$$

or

$$
[k^e]\{\delta^e\} = \{r^e\} \tag{3.20}
$$

where $[k^e]$ is a $N^e \times N^e$ element stiffness matrix,

$$
\{\delta^e\} = \lfloor\ u^e_1 \quad u^e_2 \quad . \quad . \quad . \quad u^e_{N^e-1} \quad u^e_{N^e}\ \rfloor^T \tag{3.21}
$$

is the element displacement vector, and

$$
\{r^e\} = \lfloor\ -(A\sigma)^e_1 \quad -f^e_2 \quad \dots \quad -f^e_{N^e-1} \quad (A\sigma)^e_{N^e}\ \rfloor^T \tag{3.22}
$$

is the element load vector. As Eq. (3.20) contains discrete resultant forces at the two element boundary nodes, equilibriums of resultant forces and external forces at the inter-element boundary of two adjacent elements and the natural boundary are exactly satisfied. Consequently, the DQEM is different from FEM which needs to form the element stiffness equation, and which neglects the exact equilibriums.

Equation (3.18) can be partitioned into the following form:

$$\begin{bmatrix} [K_{11}] & [K_{12}] \\ [K_{21}] & [K_{22}] \end{bmatrix} \begin{Bmatrix} \{D_1\} \\ \{D_2\} \end{Bmatrix} = \begin{Bmatrix} \{R_1\} \\ \{R_2\} \end{Bmatrix} \tag{3.23}$$

where $\{D_1\}$ is the column vector consisting of all unknown displacements at the nodes with $\{R_1\}$ the related load vector, $\{D_2\}$ is the column vector consisting of all prescribed nodal displacements with $\{R_2\}$ the related column vector consisting of all reaction forces. From the upper part of Eq. (3.23), the unknown displacements can be obtained by solving the following equation

$$[K_{11}]\{D_1\} = \{R_1\} - [K_{12}]\{D_2\} \tag{3.24}$$

The reaction forces can then be computed by using the lower part of Eq. (3.23)

$$\{R_2\} = [K_{21}]\{D_1\} + [K_{22}]\{D_2\} \tag{3.25}$$

3.1.3 Implementation of DQEM Computer Program

When implementing the DQEM analysis program, it is necessary to consider the effective use of computer memory units and computational efficiency. The overall stiffness matrix in the overall stiffness equation is sparse. Consequently, various techniques considering the sparseness can be incorporated into the solver to minimize the computer memory requirements, number of arithmetical operations and round off error [58–61]. Many of the developed compact storage techniques can be used to the DQEM analysis.

The linear equation systems existing in the DQEM analysis can be solved by using a certain direct or iterative solver. The most commonly used direct solvers are Gauss elimination, Cholesky decomposition and frontal method. Various techniques including various pivotal strategies, sparse implementation strategies, the domain decomposition and the parallel implementation can be considered in implementing an efficient direct solution procedure into a computer program [62–72, 82]. There are also many iterative solvers that can be used to solve a linear equation system [73–75]. Among the indirect solvers the preconditioned conjugate gradient (PCG) methods are very attractive [76–79]. In solving large linear equation systems, the PCG methods can offer promising performances due to the substantial reductions in computer memory requirements and the function of taking the advantage of vector and parallel processing strategies on computers that support these features [80]. The iterative solvers possess a relatively high degree of natural concurrency, with the predominant operations in PCG algorithms being saxpy operations, inner products and matrix-vector multiplications. Among the PCG algorithms, the stabilized and accelerated version of the biconjugate gradient method, which is an extension of the conjugate gradient method, is one of the most commonly used iterative solvers. There is also a predictor-corrector iteration procedure which cab be used to efficiently solve equation systems [84]. It is an explicit iteration procedure in the nonlinear iteration. Instead of using an assembled overall equivalent stiffness matrix, this method only uses the diagonal elements of the overall stiffness matrix to predict the incremental displacement vector in carrying out the iterative solution. Consequently, only the diagonal elements of the element equivalent stiffness matrices are needed to be calculated. Thus the computer memory requirement can be minimized.

The adoption of techniques for reducing computer memory requirements and the use of vector and parallel processing strategies on computers that support these features can raise the performance of the computer program [81–83]. When implementing the DQEM analysis program, various phases including preprocessing, calculation of elemental discrete equations, incorporation of boundary conditions, solution of system equations and postprocessing can be parallelized. However, the assembly of elemental discrete equations cannot take the advantage of parallel operation efficiently.

3.1.4 Problems

Figure 3.3 shows a bar subjected to a quartically distributed force. In the analysis, Lagrange DQ model with equally spaced nodes is used. The elements are also equally spaced. Numerical results of displacement at the free end and axial stress at the fixed end are summarized and listed in Table 3.1. It shows that the results can converge fast to exact solutions by either increasing the number of elements or the nodes per element. It also shows that the convergence performs better by increasing the nodes per element.

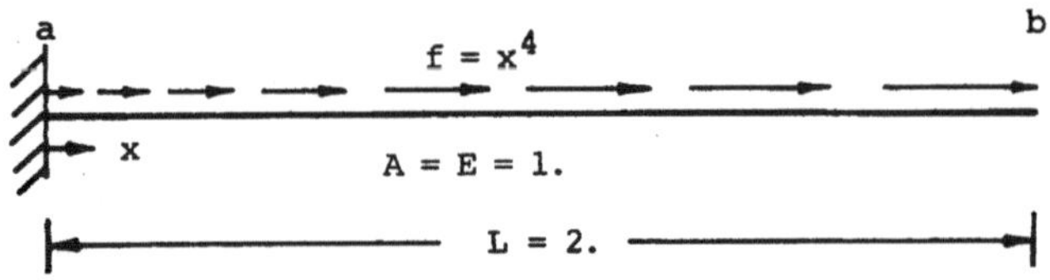

Fig. 3.3. A bar subjected to a quartically distributed force

3.2 Free Vibration of Bars

Let ω and ρ denote the natural frequency and mass density, respectively. By using U to replace u and represent the modal displacement of the vibration, and the inertia force $\rho A\omega^2 u$ to replace the distributed external load in the static equilibrium equation, the governing differential eigenvalue equation of free vibration of bars can be expressed by

$$\frac{d}{dx}\left(EA\frac{dU}{dx}\right) + \rho A\omega^2 U = 0 \tag{3.26}$$

Table 3.1. Convergence of the numerical results of a bar subjected to a quartically distributed force

Nodes per element	Number of elements	Displacement at end b	Stress at end a
3	1	$.20000\times10^1$	$.20000\times10^1$
	2	$.76250\times10^1$	$.51250\times10^1$
	4	$.98516\times10^1$	$.60703\times10^1$
	8	$.10459\times10^2$	$.63169\times10^1$
5	1	$.95000\times10^1$	$.61667\times10^1$
	2	$.10594\times10^2$	$.63854\times10^1$
	4	$.10662\times10^2$	$.63991\times10^1$
	8	$.10666\times10^2$	$.63999\times10^1$
7	1	$.10067\times10^2$	$.64000\times10^1$
	2	$.10067\times10^2$	$.64000\times10^1$
Exact solution		$.10067\times10^2$	$.64000\times10^1$

Assume that a concentrated mass $\tilde{M}^n$ is attached to the natural boundary. By using the inertia force $\tilde{M}^n\omega^2U^n$ to replace the concentrated external load applied to the natural boundary, the natural boundary conditions can be expressed as

$$EA\frac{dU}{dx} - \nu^n\tilde{M}\omega^2U = 0 \tag{3.27}$$

where ν^n equals 1 for the right boundary and equals -1 for the left boundary.

3.2.1 DQEM Formulation

For better convergence, the element nodes must be more than two with the extra nodes used to define discrete element eigenvalue equations at the related locations. The differential eigenvalue equation at a node point α in an element e can be discretized by using the DQ

$$\frac{1}{(J^e)^2}\left[\frac{d(EA)^e_{(\alpha)}}{d\xi}\sum_{\beta=1}^{N^e} D^{e\xi}_{\alpha\beta} + (EA)^e_{(\alpha)}\sum_{\beta=1}^{N^e} D^{e\xi^2}_{\alpha\beta}\right]U^e_\beta + (\rho A)^e_{(\alpha)}\omega^2U^e_\alpha = 0,$$
$$\alpha = 2,3,...,N^e-1 \tag{3.28}$$

The internal axial stresses at node points in the element can be expressed by using the modal displacement U^e_β to replace u^e_β in Eq. (3.11). Assume that a concentrated mass $\tilde{M}^{i,i+1}$ is attached to the inter-element boundary of two adjacent element i and $i+1$. By using the inertia force $\omega^2\tilde{M}^{i,i+1}U^{i,i+1}$ to replace the externally applied load $P^{i,i+1}$ in Eq. (3.13), the following dynamic equilibrium condition at the inter-element boundary can be obtained

$$\left[(EA)^i_{N^i}\frac{dU^i_{N^i}}{dx^i}-(EA)^{i+1}_1\frac{dU^{i+1}_1}{dx^{i+1}}\right]-\tilde{M}^{i,i+1}\omega^2U^{i,i+1}=0 \tag{3.29}$$

The above equation can be discretized by using DQ

$$\frac{(EA)^i_{N^i}}{J^i}\sum_{\alpha=1}^{N^i}D^{i\xi}_{N^i\alpha}U^i_\alpha-\frac{(EA)^{i+1}_1}{J^{i+1}}\sum_{\beta=1}^{N^{i+1}}D^{(i+1)\xi}_{1\beta}U^{i+1}_\beta-\tilde{M}^{i,i+1}\omega^2U^i_{N^i}=0 \tag{3.30}$$

The natural boundary condition represented by Eq. (3.27) can also be discretized which shows to have the following form:

$$\frac{(EA)^n_{I^n}}{J^n}\sum_{\beta=1}^{N^n}D^{n\xi}_{I^n\beta}U^n_\beta-\nu^n\tilde{M}\omega^2U^n_{I^n}=0,\quad I^n=1\text{or}\quad N^n \tag{3.31}$$

3.2.2 Assemblage and Solution

Keeping the kinematic transition condition (3.12) in mind, assembling the discrete element eigenvalue equations (3.28) for all elements, the discrete natural transition conditions (3.30), and the discrete natural boundary conditions (3.31), the overall discrete eigenvalue equation system can be obtained. Considering the discrete kinematic boundary condition (3.17), the overall discrete eigenvalue equation system can be expressed as

$$\left([K]-\omega^2[M]\right)\{\tilde{D}\}=\{0\} \tag{3.32}$$

where $[K]$ is the overall stiffness matrix, $[M]$ the overall mass matrix and $\{\tilde{D}\}$ the overall modal displacement vector. $[K]$ is a sparse matrix. $[M]$ is a diagonal matrix with zeros appearing on-diagonal. The overall eigenvalue system is positive semidefinite.

Like FEM, the assemblage is based on an element by element procedure. When assembling the discrete equations of element e, the discrete element eigenvalue equations (3.28), and the two discrete element boundary forces of axial forces, expressed by modal displacements, at the two element boundary nodes are directly assembled into the overall discrete eigenvalue equation system. Consequently, an element basis explicit matrix equation, containing the discrete element eigenvalue equations and the discrete element boundary forces placed at the first and last rows, is not necessary to be formed in the assembling process. This element basis explicit matrix equation is an element eigenvalue equation which can be expressed by

$$([\hat{k}^e]-\omega^2[\hat{m}^e])\{\vartheta^e\}=\{0\} \tag{3.33}$$

where $[\hat{k}^e]$ and $[\hat{m}^e]$ are $N^e \times N^e$ local element stiffness matrix and Local element mass matrix, respectively, and

$$\{\vartheta^e\} = \lfloor\ U_1^e \quad U_2^e \quad \dots \quad U_{N^e-1}^e \quad U_{N^e}^e\ \rfloor^T \tag{3.34}$$

is the local element modal displacement vector. As Eq. (3.33) contains discrete resultant forces at the two element boundary nodes, equilibriums of resultant forces and inertia forces at the inter-element boundary of two adjacent elements and the natural boundary are exactly satisfied in the assembling process. Consequently, the DQEM is different from FEM which needs to form the element eigenvalue equation, and which neglects the exact equilibriums.

3.2.3 Solution of Discrete Eigenvalue System

Equation (3.32) is a generalized eigenvalue problem with infinite frequencies existing. Premultiplication of equation (3.32) by $[K]^{-1}$ leads to

$$([A] - \lambda [I]) \{\tilde{D}\} = \{0\} \tag{3.35}$$

where $[A] = [K]^{-1}[M]$ and $\lambda = \frac{1}{\omega^2}$. Equation (3.35) can be solved by using either an exact solution technique or an approximate solution technique. If the order of the eigenvalue system is large, the approximation algorithms which calculate the eigenpairs in descending order can reduce the expense.

Some DOF can be condensed before solving Eq. (3.32). If no mass is attached to an inter-element boundary or natural boundary, the related modal displacement can also be condensed. This condensation technique is especially useful for solving the eigenvalue problem of vibration of Timoshenko beam structures by neglecting the rotary effect. The condensation treatment can reduce the size of eigenvalue system and CPU time required, drastically, with the sample solutions can be seen in Chapter 8. Considering the nonexistence of inertia forces for some component equations existing in Eq. (3.32), the equation can be rewritten as

$$\left(\begin{bmatrix} [K_{aa}] & [K_{ab}] \\ [K_{ba}] & [K_{bb}] \end{bmatrix} - \omega^2 \begin{bmatrix} [M_{aa}] & [0] \\ [0] & [0] \end{bmatrix} \right) \begin{Bmatrix} \{\tilde{D}_a\} \\ \{\tilde{D}_b\} \end{Bmatrix} = \begin{Bmatrix} \{0\} \\ \{0\} \end{Bmatrix} \tag{3.36}$$

with $[M_{aa}]$ a diagonal matrix without zeros appearing on-diagonal. From the lower part of Eq. (3.36), the following relation can be obtained

$$\{\tilde{D}_b\} = -[K_{bb}]^{-1}[K_{ba}]\{\tilde{D}_a\} \tag{3.37}$$

The substitution of Eq. (3.37) into the upper part of Eq. (3.36) yields

$$\left([\bar{K}_{aa}] - \omega^2 [M_{aa}]\right)\{\tilde{D}_a\} = \{0\} \tag{3.38}$$

where

$$\left[\bar{K}_{aa}\right] = [K_{aa}] + [K_{ab}][K_{bb}]^{-1}[K_{ba}] \tag{3.39}$$

Equation (3.38) can be treated and solved by the same procedure that transfers Eq. (3.32) into Eq. (3.35). It can also be solved by adopting the advantage of the diagonality of $[M_{aa}]$. Defining $[M_{aa}] = [L]^2$ and $\{\tilde{D}_a\} = [L]^{-1}\{Y\}$, substituting them into Eq. (3.38), and then premultiplying the resulting matrix equation by $[L]^{-1}$, the following eigenvalue problem can be obtained

$$([H] - \omega^2 [I])\{Y\} = \{0\} \tag{3.40}$$

where $[H] = [L]^{-1}[\bar{K}_{aa}][L]^{-1}$. For economically solving a large eigenvalue problem, the approximation algorithms which calculate the eigenpairs in ascending order can be used.

In addition to the above condensation procedure, the efficiency of solving an eigenvalue system resulting from a DQEM problem can be improved, further, by considering the sparseness and partial extraction, and by adopting various existing numerical techniques [85–93].

3.2.4 Problems

In the analysis, Young's modulus E, mass density ρ, area of cross section A and length of bar L are all equal to 1. The number of nodes is the same for all elements. The elements and nodes are equally spaced.

The free vibration of a fixed-free bar was solved. The natural frequencies obtained are listed in Table 3.2. They are compared with exact solutions [94]. It shows that the numerical results converge very fast by increasing either the

Table 3.2. Convergence of the first five natural frequencies of a fixed-free bar

DOF per element	Number of elements	ω_1	ω_2	ω_3	ω_4	ω_5
3	2	1.59005	4.88105			
	4	1.57579	4.83313	8.22315	10.8910	
	6	1.57303	4.77016	8.09335	11.5003	14.6432
	8	1.57205	4.74560	7.99985	11.3571	14.7698
5	2	1.57053	4.65492	7.44257	9.71602	12.0789
	4	1.57078	4.70839	7.80585	10.7509	13.6306
	6	1.57079	4.71158	7.84390	10.9437	13.9647
	8	1.57080	4.71213	7.85073	10.9785	14.0792
7	2	1.57080	4.71418	7.90301	11.3440	14.8300
	4	1.57080	4.71242	7.85505	11.0055	14.1882
	6	1.57080	4.71239	7.85399	10.9961	14.1419
	8	1.57080	4.71239	7.85399	10.9957	14.1379
Exact solution		1.57080	4.71239	7.85398	10.9956	14.1372

Table 3.3. Convergence of the first three natural frequencies of a fixed-free bar with a concentrated mass attached to the free end

DOF per element	Number of elements	ω_1	ω_2	ω_3
5	2	0.86237	3.54315	6.91639
	4	0.86059	3.44639	6.53406
7	2	0.86109	3.47654	6.60278
	4	0.86044	3.43372	6.48289
	6	0.86035	3.42814	6.45321
Exact solution		0.8602	3.4267	6.4373

number of elements or nodes per element. The problem is resolved by attaching a concentrated mass $M = 1$ to the free end. The natural frequencies obtained are listed in Table 3.3. They are compared with exact solutions [94]. It also shows that the convergence of numerical results is excellent.

4

DQEM Analysis of Euler-Bernoulli Beam Structures

Slender beam member is widely used in designing an engineering structure. Euler-Bernoulli beam theory describes the deformation behaviors of slender beam. In this chapter, the DQEM analysis models of static deflection, vibration and buckling of Euler-Bernoulli beam structures are introduced. Various EDQ models or DQ model can be used to carry out the element basis discretization.

4.1 Static Deflection of Euler-Bernoulli Beam

For reference in the sequel and for establishing notation, the equations of a nonprismatic Euler-Bernoulli beam resting on a Winkler elastic foundation are first summarized. Referring to Fig. 4.1, a x-z Cartesian coordinate system with x-axis coincident with the beam's neutral axis is used to describe the problem. The governing differential equilibrium equation is

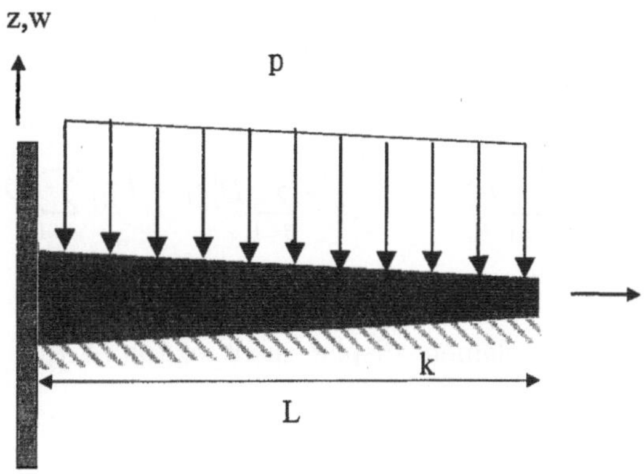

Fig. 4.1. Euler-Bernoulli beam on a Winkler foundation

$$\frac{d^2}{dx^2}\left(EI\frac{d^2w}{dx^2}\right)+kw=-p \tag{4.1}$$

where w is the transverse displacement, E is Young's modulus, I is the moment of inertia of the cross-sectional area, p is the distributed load and k is the foundation modulus. The stress resultants of bending moment and shear force are

$$M=EI\frac{d^2w}{dx^2} \tag{4.2}$$

and

$$V=\frac{d}{dx}\left(EI\frac{d^2w}{dx^2}\right), \tag{4.3}$$

respectively.

4.1.1 DQEM Formulation

For elements having no distributed load, the DOF of the element can be 4. Since Eq. (4.1) is a fourth order differential equation, without using a certain technique to calculate the four equivalent nodal loads for the distributed load of a four-DOF cubic element and include them into the natural transition conditions or natural boundary conditions, the order of approximate displacement must at least be four and the element must at least have one discrete point for defining a discrete element equilibrium equation at the discrete point. The DQEM cubic element is equivalent to the FEM cubic element. For elements having distributed load, the number of DOF must be larger than 4 in order to get an efficient convergence. For elements having five or more DOF, discrete element equilibrium equations at $\bar{N}^e$-4 discrete points are defined, with $\bar{N}^e$ the number of DOF assigned to the element. The selection of the discrete points can be arbitrary. Considering that Young's modulus is constant in an element and that the range of the parent element is $0\le\xi\le 1$, and substituting the mapping equation (3.5) and EDQ discretization equation (2.1) into Eq. (4.1), the following discrete element equilibrium equation at a discrete point α can be obtained

$$\frac{E^e}{(l^e)^4}\left[\frac{d^2I^e_{(\alpha)}}{d\xi^2}\sum_{\beta=1}^{\bar{N}^e}D^{e\xi^2}_{\alpha\beta}+2\frac{dI^e_{(\alpha)}}{d\xi}\sum_{\beta=1}^{\bar{N}^e}D^{e\xi^3}_{\alpha\beta}+I^e_{(\alpha)}\sum_{\beta=1}^{\bar{N}^e}D^{e\xi^4}_{\alpha\beta}\right]\tilde{w}^e_\beta$$
$$+k^e_{(\alpha)}w^e_\alpha=-p^e_\alpha \tag{4.4}$$

where $\tilde{w}^e_\beta$ represents the element displacement vector $\{\delta^e\}$,

$$\{\delta^e\}=\lfloor\ w^e_1\ \ .\ \ .\ \ \cdots\ \ .\ \ .\ \ w^e_{N^e}\ \ .\ \ .\ \rfloor^T \tag{4.5}$$

where w^e_1 and $w^e_{N^e}$ are lateral displacements at the first and last element nodes, respectively. It should be noted that DOF representing various orders of derivatives with respect to the coordinate variable can also be assigned to the two element boundary nodes.

The internal bending moment and shear force at a discrete point α can be expressed by

$$M_\alpha^e = \frac{E^e}{(l^e)^2}\left(I_{(\alpha)}^e \sum_{\beta=1}^{\bar{N}^e} D_{\alpha\beta}^{e\xi^2}\right)\tilde{w}_\beta^e \tag{4.6}$$

and

$$V_\alpha^e = \frac{E^e}{(l^e)^3}\left(\frac{dI_{(\alpha)}^e}{d\xi}\sum_{\beta=1}^{\bar{N}^e} D_{\alpha\beta}^{e\xi^2} + I_{(\alpha)}^e \sum_{\beta=1}^{\bar{N}^e} D_{\alpha\beta}^{e\xi^3}\right)\tilde{w}_\beta^e \tag{4.7}$$

In Eqs. (4), (6) and (7), the derivatives of the section constant I at the related discrete points can also be calculated by the DQ. It is especially useful if the distribution function of I is not continuously differentiable up to the order of its derivative for which the derivatives of I at a discrete point might not be able to be obtained by only carrying out the differential operation. The values of I at the two element boundary nodes and some interior discrete points are used to define the DQ discretizations of the derivatives. The DQ discretization for the mth order derivative of I at a discrete point α is expressed by

$$\frac{d^m I_\alpha^e}{d\xi^m} = \sum_{i=1}^{\tilde{N}^e} \bar{D}_{\alpha i}^{e\xi^m} I_i^e \tag{4.8}$$

where $\bar{D}_{\alpha i}^{e\xi^m}$ are weighting coefficients and $\tilde{N}^e$ is the number of nodes for defining the DQ discretization.

Four transition conditions at an inter-element boundary of two adjacent elements are also necessary for constructing the overall discrete equation system. These four transition conditions are the continuities of displacements and displacement gradients, and the equilibriums of moments and transverse forces at the inter-element boundary of two adjacent elements i and $i+1$. The condition of displacement continuity is expressed as:

$$w_{N^i}^i = w_1^{i+1} \tag{4.9}$$

where N^i is the right boundary node of element i. Since the DOF representing the displacement is always assigned to the element boundary nodes, the above equation is always automatically satisfied, and the DOF can be used to define a discrete natural transition condition at the inter-element boundary of two adjacent elements or a discrete natural boundary condition at the natural boundary. The continuity of displacement gradient is

$$\frac{dw_{N^i}^i}{dx^i} - \frac{dw_1^{i+1}}{dx^{i+1}} = 0 \tag{4.10}$$

If the DOF representing the deflection slope at an element boundary node is assigned to the element boundary node, Eq. (4.10) is automatically satisfied, and the DOF can be used to define a discrete natural transition condition at the inter-element boundary of two adjacent elements or a discrete natural boundary condition at the natural boundary in the DQEM analysis of flexural deflection of structures [34–35]. Otherwise, Eq. (4.10) needs to be discretized. The EDQ discretization of the above equation leads to

$$\frac{1}{l^i}\sum_{\beta=1}^{\bar{N}^i} D^{i\xi}_{N^i\alpha}\tilde{w}^i_\alpha - \frac{1}{l^{i+1}}\sum_{\beta=1}^{\bar{N}^{i+1}} D^{(i+1)\xi}_{1\beta}\tilde{w}^{i+1}_\beta = 0 \tag{4.11}$$

Then, one DOF other than the DOF representing the displacement of element boundary nodes is required for defining the above discrete conformability condition, and two more DOF are necessary for defining the remaining two discrete natural transition conditions at the inter-element boundary of two adjacent elements or two discrete natural boundary conditions at the natural boundary. A representative application has been used to the DQEM analysis of flexural deflection of structures [20,22]. Figure 4.2 shows the equilibrium of moments, in which $M^{i,i+1}$ is the concentrated moment applied on the inter-element boundary. The equilibrium of moments is

$$-M^i_{N^i} + M^{i+1}_1 = M^{i,i+1} \tag{4.12}$$

Upon the substitution of Eq. (4.6) into Eq. (4.12), the following discretized equation can be obtained

$$-\frac{E^iI^i_{N^i}}{(l^i)^2}\sum_{\alpha=1}^{\bar{N}^i} D^{i\xi^2}_{N^i\alpha}\tilde{w}^i_\alpha + \frac{E^{i+1}I^{i+1}_1}{(l^{i+1})^2}\sum_{\beta=1}^{\bar{N}^{i+1}} D^{(i+1)\xi^2}_{1\beta}\tilde{w}^{i+1}_\beta = M^{i,i+1} \tag{4.13}$$

Figure 4.2 also shows the equilibrium of transverse forces in which $P^{i,i+1}$ is the concentrated transverse load applied on the inter-element boundary, and

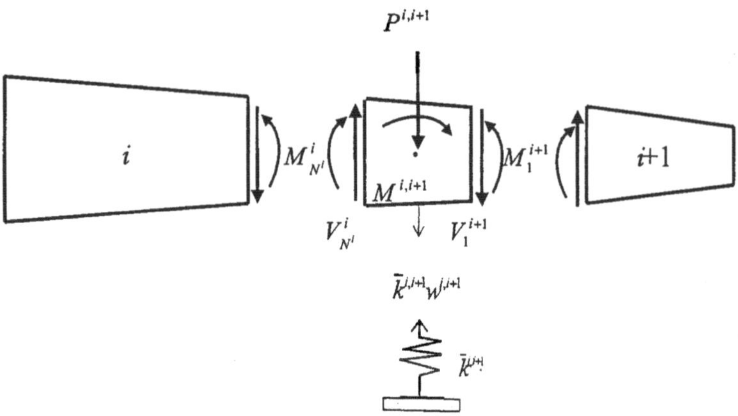

Fig. 4.2. Equilibrium of transverse forces

$\bar{k}^{i,i+1}$ is the spring constant of a concentrated spring at the inter-element boundary. Referring to the figure, the following equilibrium equation can be obtained

$$V_{N^i}^i - V_1^{i+1} - \bar{k}^{i,i+1} w^{i,i+1} = P^{i,i+1} \tag{4.14}$$

The substitution of Eq. (4.7) into Eq. (4.14) leads to the following discretized equation:

$$\begin{aligned} &\frac{E^i}{(l^i)^3}\left(\frac{dI_{N^i}^i}{d\xi}\sum_{\alpha=1}^{\bar{N}^i} D_{N^i\alpha}^{i\xi^2} + I_{N^i}^i \sum_{\alpha=1}^{\bar{N}^i} D_{N^i\alpha}^{i\xi^3}\right)\tilde{w}_\alpha^i \\ &-\frac{E^{i+1}}{(l^{i+1})^3}\left(\frac{dI_1^{i+1}}{d\xi}\sum_{\beta=1}^{\bar{N}^{i+1}} D_{1\beta}^{(i+1)\xi^2} + I_1^{i+1}\sum_{\beta=1}^{\bar{N}^{i+1}} D_{1\beta}^{(i+1)\xi^3}\right)\tilde{w}_\beta^{i+1} \\ &-\bar{k}^{i,i+1} w^{i,i+1} = P^{i,i+1} \end{aligned} \tag{4.15}$$

For pin-connected inter-element boundaries, Eqs. (4.10) and (4.12) are replaced by the following two equations

$$E^i I_{N^i}^i \frac{d^2 w_{N^i}^i}{d(x^i)^2} = 0, \quad E^{i+1} I_1^{i+1} \frac{d^2 w_1^{i+1}}{d(x^{i+1})^2} = 0 \tag{4.16}$$

The EDQ discretization of Eqs. (4.16) leads to

$$\frac{E^i I_{N^i}^i}{(l^i)^2}\sum_{\beta=1}^{\bar{N}^i} D_{N^i\beta}^{i\xi^2}\tilde{w}_\beta^i = 0, \quad \frac{E^{i+1} I_1^{i+1}}{(l^{i+1})^2}\sum_{\beta=1}^{\bar{N}^{i+1}} D_{1\beta}^{(i+1)\xi^2}\tilde{w}_\beta^{i+1} = 0 \tag{4.17}$$

The kinematic boundary conditions are $w = \bar{w}$ and $\frac{dw}{dx} = \frac{d\bar{w}}{dx}$ in which $\bar{w}$ and $\frac{d\bar{w}}{dx}$ are prescribed values, on the kinematic boundary. Letting m be an element having one or more nodes on the kinematic boundary, the kinematic boundary conditions can be rewritten as $w_{I^m}^m = \bar{w}_{I^m}^m$ and $\frac{dw_{I^m}^m}{dx^m} = \frac{d\bar{w}_{I^m}^m}{dx^m}$ in which I^m equals 1 or N^m representing the first or last node of the element. The EDQ discretization of $\frac{dw_{I^m}^m}{dx^m} = \frac{d\bar{w}_{I^m}^m}{dx^m}$ leads to

$$\frac{1}{l^m}\sum_{\beta=1}^{\bar{N}^m} D_{I^m\beta}^{m\xi}\tilde{w}_\beta^m = \frac{d\bar{w}_{I^m}^m}{dx^m}, \quad I^m = 1 \quad \text{or} \quad N^m \tag{4.18}$$

Let n be an element having one or more nodes on the natural boundary. The natural boundary condition of equilibrium of moments is

$$E^n I_{I^n}^n \frac{d^2 w_{I^n}^n}{d(x^n)^2} = \bar{M}_{I^n}^n, \quad I^n = 1 \quad \text{or} \quad N^n \tag{4.19}$$

where $\bar{M}^n$ is the moment applied on the natural boundary. The EDQ discretization of the above equation leads to

$$\frac{E^n I^n_{I^n}}{(l^n)^2} \sum_{\beta=1}^{\bar{N}^n} D^{n\xi^2}_{I^n\beta} \tilde{w}^n_\beta = \bar{M}^n_{I^n}, \quad I^n = 1 \quad \text{or} \quad N^n \tag{4.20}$$

Let $\bar{V}^n$ denote the transverse tip load applied downward on the free end. If the natural boundary is at the left end, the natural boundary condition of the equilibrium of transverse forces can be expressed as

$$-V^n_1 - \bar{k}^n w^n_1 = \bar{V}^n_1 \tag{4.21}$$

where $\bar{k}^n$ is the spring constant of a spring at the natural boundary. Upon the substitution of Eq. (4.7) into Eq. (4.21) and the use of EDQ discretization, the following discretized equation can be obtained

$$-\frac{E^n}{(l^n)^3}\left(\frac{dI^n_1}{d\xi}\sum_{\beta=1}^{\bar{N}^n} D^{n\xi^2}_{1\beta} + I^n_1 \sum_{\beta=1}^{\bar{N}^n} D^{n\xi^3}_{1\beta}\right)\tilde{w}^n_\beta - \bar{k}^n w^n_1 = \bar{V}^n_1 \tag{4.22}$$

If the natural boundary is at the right end, the natural boundary condition of the equilibrium of transverse forces can be expressed as

$$V^n_{N^n} - \bar{k}^n w^n_{N^n} = \bar{V}^n_{N^n} \tag{4.23}$$

The substitution of Eq. (4.7) into Eq. (4.23) and the use of EDQ discretization leads to the following discretized equation

$$\frac{E^n}{(l^n)^3}\left(\frac{dI^n_{N^n}}{d\xi}\sum_{\beta=1}^{\bar{N}^n} D^{n\xi^2}_{N^n\beta} + I^n_{N^n} \sum_{\beta=1}^{\bar{N}^n} D^{n\xi^3}_{N^n\beta}\right)\tilde{w}^n_\beta - \bar{k}^n w^n_{N^n} = \bar{V}^n_{N^n} \tag{4.24}$$

4.1.2 Assemblage

Keeping the compatibility condition (4.9), and conformability condition (4.11) for the inter-element boundary with the two adjacent elements having the DOF representing the deflection slopes at the inter-element boundary in mind, assembling Eq. (4.4) for elements having interior discrete points, Eq. (4.11) for the inter-element boundary with the two adjacent elements having no DOF representing the deflection slopes at the inter-element boundary, Eqs. (4.13) and (4.15) for all inter-element boundaries, Eq. (4.18) for boundary elements with kinematic boundary and having no DOF representing the deflection slope at the kinematic boundary, and Eqs. (4.20), (4.22) and (4.24) for boundary elements with natural boundary, an overall linear algebraic system can be obtained. It is the overall stiffness equation. For pin-connected inter-element boundaries, Eqs. (4.16) and (4.17) need to be used to replace the related discrete equations and assembled into the overall stiffness equation. The assemblage is based on an element by element procedure. The overall stiffness equation can be expressed as:

$$([K] + [\bar{K}])\{D\} = \{R\} \tag{4.25}$$

where $[K]$ is the overall stiffness matrix of beam, $[\bar{K}]$ the overall stiffness matrix of foundation, $\{D\}$ the overall displacement vector and $\{R\}$ the overall load vector.

Like FEM, the assemblage is based on an element by element procedure. When assembling the discrete equations of element e, the discrete element equilibrium equations (4.4), and the four discrete element boundary forces, expressed by displacement parameters, at the two element boundary nodes are directly assembled to the overall discrete equation system. Consequently, an element basis explicit matrix equation, containing the discrete element equilibrium equations and the discrete element boundary forces placed at the first and last two rows, is not necessary to be formed in the assembling process. This element basis explicit matrix equation is an element stiffness equation which can be expressed by

$$([k^e] + [\bar{k}^e])\{\delta^e\} = \{r^e\} \tag{4.26}$$

where $[k^e]$ and $[\bar{k}^e]$ are $\bar{N}^e \times \bar{N}^e$ element stiffness matrices of beam and foundation, respectively, and $\{r^e\}$ is the element load vector. The element load vector can be expressed by

$$\begin{aligned}
\{r^e\} &= \lfloor\; -V_1^e - \bar{R}^{e-1,e} \quad M_1^e \quad -p_3^e \quad \cdots \quad -p_{\bar{N}^e-2}^e \quad V_{\bar{N}^e}^e \quad -M_{\bar{N}^e}^e \;\rfloor^T, \text{ for an} \\
&\quad \text{interior element with } \bar{R}^{e-1,e} = \bar{k}^{e-1,e} w^{e-1,e} \text{ the spring force,} \\
&= \lfloor\; -V_1^1 - \bar{R}^1 \quad M_1^1 \quad -p_3^1 \quad \cdots \quad -p_{\bar{N}^1-2}^1 \quad V_{\bar{N}^1}^1 \quad -M_{\bar{N}^1}^1 \;\rfloor^T, \text{ for the} \\
&\quad \text{first or left boundary element,} \\
&= \lfloor\; -V_1^s - \bar{R}^s \quad M_1^s \quad -p_3^s \quad \cdots \quad -p_{\bar{N}^s-2}^s \quad V_{\bar{N}^s}^s - \bar{R}^s \quad -M_{\bar{N}^s}^s \;\rfloor^T, \text{ for the} \\
&\quad \text{last or right boundary element}
\end{aligned} \tag{4.27}$$

As Eq. (4.26) contains discrete resultant forces and spring forces at the two element boundary nodes, equilibriums of resultant forces, spring forces and external forces at the inter-element boundary of two adjacent elements and the natural boundary are exactly satisfied in the assembling process. Consequently, the DQEM is different from FEM which needs to form the element stiffness equation, and which neglects the exact equilibriums.

4.1.3 Problems

The static deflection of a fixed-free beam resting on a Winkler foundation and subjected to a uniformly distributed load $p = 1$ was solved. The geometrical quantities and material constants are: length of beam $= 3$, $I = 1$, $E = 1$ and $k = 1$. A is the left end and B is the right end. Lagrange DQ model is used to carry out the element basis discretization. In the analysis, the elements and node points in an element are equally spaced. Since the DOF representing the deflection slope is not assigned to the element boundary nodes, the DOF assigned to the interior node next to the related element boundary node is also used to define either a discrete transition condition or a discrete boundary condition, instead of using it to define a discrete element equilibrium equation at the node. Numerical results obtained by the DQEM are listed in Table 4.1. They are compared with exact solutions [95]. It shows that the DQEM solutions converge fast by increasing either the number of elements or nodes per element.

Table 4.1. Convergence of the numerical results of a prismatic cantilever beam subjected to a uniformly distributed load and resting on a Winkler foundation

DOF per element	Number of elements	Deflection at B	Bending moment at A	Shear force at A
5	2	$-.1382796 \times 10^1$	$-.9791807 \times 10^0$	$.1236154 \times 10^1$
	4	$-.1274248 \times 10^1$	$-.9707436 \times 10^0$	$.1302256 \times 10^1$
	6	$-.1256901 \times 10^1$	$-.9700371 \times 10^0$	$.1311689 \times 10^1$
7	2	$-.1243156 \times 10^1$	$-.9683331 \times 10^0$	$.1323559 \times 10^1$
	4	$-.1243531 \times 10^1$	$-.9697217 \times 10^0$	$.1319060 \times 10^1$
	6	$-.1243556 \times 10^1$	$-.9698504 \times 10^0$	$.1318822 \times 10^1$
9	2	$-.1243483 \times 10^1$	$-.9698806 \times 10^0$	$.1318828 \times 10^1$
	4	$-.1243563 \times 10^1$	$-.9698853 \times 10^0$	$.1318765 \times 10^1$
	6	$-.1243563 \times 10^1$	$-.9698852 \times 10^0$	$.1318764 \times 10^1$
Exact solution		$-.1243562 \times 10^1$	$-.9698853 \times 10^0$	$.1318764 \times 10^1$

In solving the next two problems, two degrees of freedom representing the lateral displacement and deflection slope are assigned to each of the element boundary nodes, while only one DOF representing the lateral displacement is assigned to an interior node. Explicit Lagrange DQ weighting coefficients with the grid points equally spaced are used to generate the $C^1 - C^0 - C^1$ EDQ model. Consider a $\bar{N}^e$-DOF element and define $\Delta\xi = 1./(\bar{N}^e - 1)$. The interior discrete points for defining the element-based discrete equilibrium equations are located at $\xi = (p-1)\Delta\xi$, $p = 3, ..., \bar{N}^e - 2$.

A nonprismatic cantilever beam having a rectangular cross section subjected to a transverse tip load which can be seen in Fig. 4.3. was solved. In solving the problem, the elements are equally spaced. Let A_0 and I_0 denote the area and moment of inertia of the cross section at the fixed end $x = 0$, respectively. The distributions of A and I are expressed as $A = A_0(1-0.1x/l)$ and $I = I_0(1 - 0.1x/l)^3$, respectively. The beam model for carrying out the numerical test has the values of E, A_0, I_0 and L all equal to 1. The value of transverse tip load P is also equal to 1. Numerical results obtained by using the DQEM are listed in Table 4.2. It shows that the results converge fast to

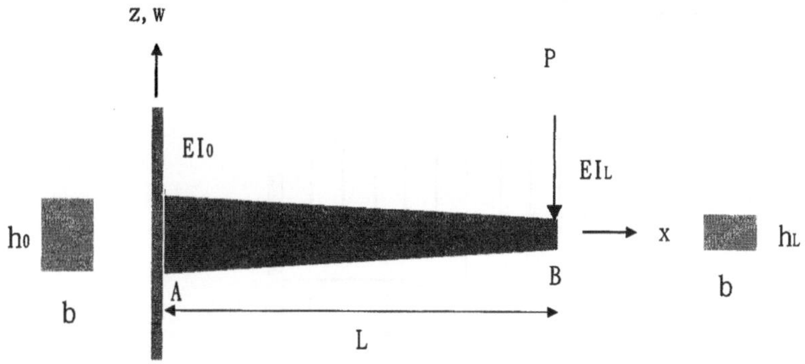

Fig. 4.3. A nonprismatic cantilever beam subjected to a transverse tip load

Table 4.2. Convergence of the numerical results of a nonprismatic cantilever beam subjected to a transverse tip load

DOF per element	Number of elements	Deflection at B	Bending moment at A	Shear force at A
5	1	–.6012931	–1.2801724	.4655172
	2	–.4773527	–1.0302632	.8958810
	3	–.4597839	–1.0118832	.9552016
7	1	–.4653140	–1.0351111	.9416616
	2	–.4489403	–1.0015277	.9957239
	3	–.4478343	–1.0002687	.9991200
9	1	–.4490091	–1.0029813	.9953998
	2	–.4475927	–1.0000359	.9999012
	3	–.4475611	–1.0000028	.9999907
11	1	–.4476460	–1.0001849	.9997271
	2	–.4475586	–1.0000006	.9999983
	3	–.4475579	–1.0000000	.9999999
Exact solution		–.4475579	–1.0000000	1.000000

close to the analytical solutions by either increasing the number of elements or the number of DOF per element [96].

The static deflection of a fixed beam composed of two prismatic segments and subjected to distributed loads and resting on a Winkler foundation which is shown in Fig. 4.4 was also solved. The values of parameters used for the analysis are: $E_1 = L_1 = 1$, $I_1 = P_1 = 1$; $E_2 = I_2 = L_2 = P_2 = 1$, $k = 1$. In the analysis, two DQEM elements are used to model the composed beam. The convergence can be assured by increasing the order of approximation. Numerical results obtained are listed in Table 4.3.

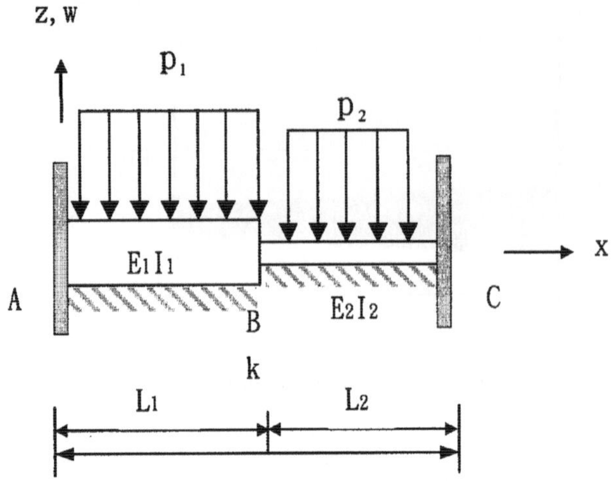

Fig. 4.4. A fixed beam composed of two prismatic elements and subjected to distributed loads

Table 4.3. Convergence of the numerical results of a composed beam subjected to a distributed load

DOF per element	Deflection at B	Bending moment at A	Bending moment at B	Bending moment at C
5	−.0434515	−.6271896	.2385751	−.3713788
7	−.0431278	−.6245045	.2364095	−.3697859
9	−.0431745	−.6249026	.2367386	−.3700341
11	−.0431745	−.6249029	.2367389	−.3700342

4.2 Free Vibration of Euler-Bernoulli Beam

In this analysis model, the effect of rotary inertia is neglected. By using W to replace w and represent the modal displacement, and using the inertia force $\rho A\omega^2 W$ to replace the distributed load -p in the static equilibrium equation, the governing differential eigenvalue equation of free vibration is

$$\frac{d^2}{dx^2}\left[EI\frac{d^2W(x)}{dx^2}\right]+kW(x)-\rho A\omega^2W=0 \tag{4.28}$$

4.2.1 DQEM Formulation

In the DQEM formulation, the DOF of an element must be five or more. The discrete element eigenvalue equations are defined at $\bar{N}^e$-4 discrete points.

The selection of the discrete points can be arbitrary. The following discrete element eigenvalue equation at a discrete point α is expressed by

$$\frac{E^e}{(l^e)^4}\left[\frac{d^2 I^e_{(\alpha)}}{d\xi^2}\sum_{\beta=1}^{\bar{N}^e} D^{e\xi^2}_{\alpha\beta} + 2\frac{dI^e_{(\alpha)}}{d\xi}\sum_{\beta=1}^{\bar{N}^e} D^{e\xi^3}_{\alpha\beta} + I^e_{(\alpha)}\sum_{\beta=1}^{\bar{N}^e} D^{e\xi^4}_{\alpha\beta}\right]\tilde{W}^e_\beta$$
$$+k^e_{(\alpha)}W^e_\alpha - \rho A^e_{(\alpha)}\omega^2 W^e_\alpha = 0 \qquad (4.29)$$

The two kinematic transition conditions have been discretized and expressed by Eqs. (4.9) and (4.11). Assume that a concentrated mass $\tilde{M}^{i,i+1}$ is attached to the inter-element boundary of two adjacent element i and $i+1$. Neglecting the effect of the rotary inertia of the concentrated mass, the discrete natural transition condition of dynamic equilibrium of moments can be obtained from Eq. (4.13)

$$-\frac{E^i I^i_{N^i}}{(l^i)^2}\sum_{\beta=1}^{\bar{N}^i} D^{i\xi^2}_{N^i\beta}\tilde{W}^i_\beta + \frac{E^{i+1}I^{i+1}_1}{(l^{i+1})^2}\sum_{\beta=1}^{\bar{N}^{i+1}} D^{(i+1)\xi^2}_{1\beta}\tilde{W}^{i+1}_\beta = 0 \qquad (4.30)$$

By using the inertia force $\tilde{M}^{i,i+1}\omega^2 W^{i,i+1}$ to replace the externally applied load -$P^{i,i+1}$ in Eq. (4.15), the discrete natural transition condition of dynamic equilibrium of lateral forces can be obtained

$$\frac{E^i}{(l^i)^3}\left(\frac{dI^i_{N^i}}{d\xi}\sum_{\beta=1}^{\bar{N}^i} D^{i\xi^2}_{N^i\beta} + I^i_{N^i}\sum_{\beta=1}^{\bar{N}^i} D^{i\xi^3}_{N^i\beta}\right)\tilde{W}^i_\beta$$
$$-\frac{E^{i+1}}{(l^{i+1})^3}\left(\frac{dI^{i+1}_1}{d\xi}\sum_{\beta=1}^{\bar{N}^{i+1}} D^{(i+1)\xi^2}_{1\beta} + I^{i+1}_1\sum_{\beta=1}^{\bar{N}^{i+1}} D^{(i+1)\xi^3}_{1\beta}\right)\tilde{W}^{i+1}_\beta$$
$$-\bar{k}^{i,i+1}W^{i,i+1} + \tilde{M}^{i,i+1}\omega^2 W^{i,i+1} = 0 \qquad (4.31)$$

The discrete kinematic boundary conditions are the same as those for the static deflection analysis. Let $\tilde{M}^n$ denote the concentrated mass attached to the free end. The discrete natural boundary condition of dynamic equilibrium of moments can be expressed by

$$\frac{E^n I^n_I}{(l^n)^2}\sum_{\beta=1}^{\bar{N}^n} D^{n\xi^2}_{I\beta}\tilde{W}^n_\beta = 0, \quad I = 1 \quad \text{or} \quad N^n \qquad (4.32)$$

By using the inertia force $\tilde{M}^n\omega^2 W^n$ to replace the tip load $\bar{V}^n$ in Eqs. (4.22) and (4.24), the discrete natural boundary conditions of dynamic equilibrium of transverse forces can be obtained. If the natural boundary is at the left end, the discrete equation is expressed by

$$-\frac{E^n}{(l^n)^3}\left(\frac{dI^n_1}{d\xi}\sum_{\beta=1}^{\bar{N}^n} D^{n\xi^2}_{1\beta} + I^n_1\sum_{\beta=1}^{\bar{N}^n} D^{n\xi^3}_{1\beta}\right)\tilde{W}^n_\beta - \bar{k}^n W^n_1 + \tilde{M}^n\omega^2 W^n_1 = 0 \quad (4.33)$$

If the natural boundary is at the right end, the discrete equation is expressed by

$$\frac{E^n}{(l^n)^3}\left(\frac{dI^n_{N^n}}{d\xi}\sum_{\beta=1}^{\bar{N}^n} D^{n\xi^2}_{N^n\beta} + I^n_{N^n}\sum_{\beta=1}^{\bar{N}^n} D^{n\xi^3}_{N^n\beta}\right)\tilde{W}^n_\beta - \bar{k}^n W^n_{N^n}$$
$$+\tilde{M}^n\omega^2 W^n_{N^n} = 0 \tag{4.34}$$

4.2.2 Assemblage

Keeping Eq. (4.9), and Eq. (4.10) for the inter-element boundary with the two adjacent elements having the DOF representing the modal deflection slopes at the inter-element boundary in mind, assembling Eq. (4.29) for all elements, Eq. (4.11) for the inter-element boundary with the two adjacent elements having no DOF representing the modal deflection slopes at the inter-element boundary, Eqs. (4.30) and (4.31) for all inter-element boundaries, Eq. (4.18) for boundary elements with kinematic boundary and having no DOF representing the modal deflection slope at the kinematic boundary, and Eqs. (4.32), (4.33) and (4.34) for boundary elements with natural boundary, an overall linear eigenvalue equation system can be obtained. The overall eigenvalue equation system considering the discrete kinematic boundary conditions is represented by Eq. (3.32).

4.2.3 Problems

A problem involves the transverse vibration of a nonprismatic cantilever beam having a circular cross section and a concentrated mass at the free end and resting on a Winkler foundation was solved [97]. In carrying out the DQEM vibration analysis, elements and nodes in an element are equally spaced. The structure is shown in Fig. 4.5. Let A_0 and I_0 denote the area and the moment of inertia of the cross section at the fixed end $z = 0$, respectively. Also let L denote the length of the beam. The distributions of A and I are expressed as $A = A_0(1 - x/L)^2$ and $I = I_0(1 - x/L)^4$. In this analysis, the rotary inertia of the concentrated mass is neglected. The beam model for carrying out the numerical test has the values of E, ρ, A_0 and I_0 all equal to 1. The length of the beam is 0.5. The value of the concentrated mass is also 0.5. In the analysis, the DOF of modal displacement and deflection slope are assigned to an element boundary node, while only one DOF of modal displacement is assigned to an interior node. At the interior nodes, discrete element eigenvalue equations are defined. The $C^1 - C^0 - C^1$ EDQ model and positions of interior nodes are the same as those used to the static deflection analysis of nonprismatic beams. Numerical results for the beam with and without a Winkler foundation with foundation modulus $k = 1$ are summarized and listed in Tables 4.4 and 4.5. It shows that the results converge fast to close to the analytical solutions

Table 4.4. The first three natural frequencies of a nonprismatic cantilever beam having a circular cross section and a concentrated mass at the free end

DOF per element	Number of elements	ω_1	ω_2	ω_3
5	1	$.31118363\times10^1$	$.48164473\times10^2$	
	2	$.50548209\times10^1$	$.32830290\times10^2$	$.18038164\times10^3$
	3	$.49184053\times10^1$	$.43682655\times10^2$	$.13032745\times10^3$
7	1	$.41386874\times10^1$	$.51784248\times10^2$	$.14659378\times10^3$
	2	$.47396084\times10^1$	$.51435204\times10^2$	$.14782394\times10^3$
	3	$.47533172\times10^1$	$.52045419\times10^2$	$.15735178\times10^3$
9	1	$.45799844\times10^1$	$.52373517\times10^2$	$.14498368\times10^3$
	2	$.47458288\times10^1$	$.52187430\times10^2$	$.15353384\times10^3$
	3	$.47478046\times10^1$	$.52295521\times10^2$	$.15344593\times10^3$
11	1	$.47122791\times10^1$	$.52329384\times10^2$	$.15423109\times10^3$
	2	$.47474227\times10^1$	$.52302511\times10^2$	$.15367730\times10^3$
	3	$.47475890\times10^1$	$.52318144\times10^2$	$.15359321\times10^3$
Analytical solution		$.47476052\times10^1$	$.52319471\times10^2$	$.15359835\times10^3$

Table 4.5. The first three natural frequencies of a nonprismatic cantilever beam having a circular cross section and a concentrated mass at the free end and resting on a Winkler foundation

DOF per element	Number of elements	ω_1	ω_2	ω_3
5	1	$.31137898\times10^1$	$.48182799\times10^2$	
	2	$.50683376\times10^1$	$.32866029\times10^2$	$.18038547\times10^3$
	3	$.49357297\times10^1$	$.43707340\times10^2$	$.13033635\times10^3$
7	1	$.41479616\times10^1$	$.51804248\times10^2$	$.14659459\times10^3$
	2	$.47578132\times10^1$	$.51454238\times10^2$	$.14783099\times10^3$
	3	$.47721754\times10^1$	$.52064281\times10^2$	$.15735808\times10^3$
9	1	$.45954499\times10^1$	$.52392705\times10^2$	$.14499096\times10^3$
	2	$.47646426\times10^1$	$.52206256\times10^2$	$.15354071\times10^3$
	3	$.47666911\times10^1$	$.52314553\times10^2$	$.15345261\times10^3$
11	1	$.47303334\times10^1$	$.52348451\times10^2$	$.15423802\times10^3$
	2	$.47663027\times10^1$	$.52321560\times10^2$	$.15368398\times10^3$
	3	$.47664746\times10^1$	$.52337202\times10^2$	$.15359990\times10^3$

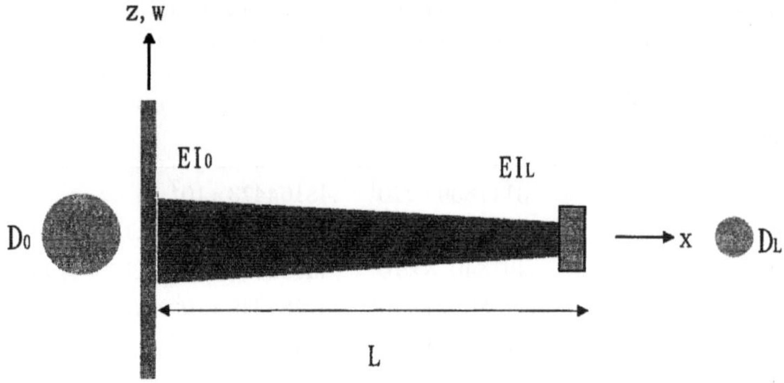

Fig. 4.5. A nonprismatic cantilever beam having a concentrated mass at the free end

obtained by Lau by either increasing the number of elements or the number of DOF per element [98].

The other problem solved involves the transverse vibration of a fixed-roller supported ASTM Standard Strong steel pipe having a concentrated mass at the mid-span and resting on a Winkler elastic foundation. The nominal diameter of the pipe is 10 *in* with the outside diameter 10.75 *in* and inside diameter 10.02 *in*. The area of cross section is 11.9 in^2, while the moment of inertia of the cross section is 161 in^4. The length of the pipe is 40 *ft*. The weight of the concentrated mass and the weight of the pipe per feet are 1000. *lb* and 40.48 *lb*, respectively. The Young's modulus of the pipe is $E = 2.9 \times 10^7$ *psi*. In carrying out the DQEM analysis, two elements were used to model the pipe with the mid-span being the inter-element boundary. The rotary inertia of the concentrated mass is neglected. The convergence can be assured by the increase of DOF per element. The vibration without Winkler foundation was first solved. In this analysis, Hermite EDQ model was used to carry out the element basis discretization. Consequently, each node has two DOF representing the modal displacement and modal deflection slope. The nodes for defining the Hermite EDQ model are equally spaced, while the arrangement of interior discrete points is the same as that of the first problem. Numerical results of the first four natural frequencies are summarized and listed in Table 4.6. It shows fast convergence. In solving the problem with Winkler foundation, a Chebyshev EDQ model was used to carry out the element basis discretization. The $N^e - 2$ interior nodes are used to define the discrete element eigenvalue equations. The value of foundation modulus is 7.5 lb/in. Numerical results of the first four natural frequencies are summarized and listed in Table 4.7. It also shows excellent convergence.

Table 4.6. The first four natural frequencies of a fixed-roller supported beam having a concentrated mass at the midspan (cycles/sec)

DOF per element	ω_1	ω_2	ω_3	ω_4
6	$.1297143\times10^2$	$.4137066\times10^2$	$.8333984\times10^2$	$.1295109\times10^3$
8	$.1349680\times10^2$	$.4722486\times10^2$	$.9010459\times10^2$	$.1739265\times10^3$
10	$.1347260\times10^2$	$.4560830\times10^2$	$.9261325\times10^2$	$.1628337\times10^3$
12	$.1347292\times10^2$	$.4562152\times10^2$	$.9229664\times10^2$	$.1629238\times10^3$
14	$.1347287\times10^2$	$.4560231\times10^2$	$.9235354\times10^2$	$.1629862\times10^3$

Table 4.7. The first four natural frequencies of a fixed-roller supported beam having a concentrated mass at the mid-span and resting on a Winkler foundation (cycles/sec)

DOF per element	ω_1	ω_2	ω_3	ω_4
5	$.1647340\times10^2$	$.4363615\times10^2$	$.1697997\times10^3$	
7	$.1560780\times10^2$	$.4611839\times10^2$	$.8375120\times10^2$	$.1328202\times10^3$
9	$.1570467\times10^2$	$.4641352\times10^2$	$.9433294\times10^2$	$.1723055\times10^3$
11	$.1570331\times10^2$	$.4639374\times10^2$	$.9260515\times10^2$	$.1626785\times10^3$
13	$.1570333\times10^2$	$.4639476\times10^2$	$.9268929\times10^2$	$.1632215\times10^3$
15	$.1570333\times10^2$	$.4639474\times10^2$	$.9268660\times10^2$	$.1631978\times10^3$

4.3 Buckling of Euler-Bernoulli Beam

Considering that the beam is subjected to a compressive force P and resting on a Winkler foundation, the differential eigenvalue equation of buckling equilibrium is expressed by

$$\frac{d^2}{dx^2}\left[EI\frac{d^2w(x)}{dx^2}\right]+kw(x)+P\frac{d^2w}{dx^2}=0 \tag{4.35}$$

In the above equation, w represents the displacement of the buckling mode.

4.3.1 DQEM Formulation

In the DQEM formulation, the DOF of an element must be five or more. The discrete element eigenvalue equations are defined at $\bar{N}^e$-4 discrete points. The selection of the discrete points can be arbitrary. The following discrete element eigenvalue equation at a discrete point α is expressed by

$$\frac{E^e}{(l^e)^4}\left[\frac{d^2 I^e_{(\alpha)}}{d\xi^2}\sum_{\beta=1}^{\bar{N}^e} D^{e\xi^2}_{\alpha\beta} + 2\frac{dI^e_{(\alpha)}}{d\xi}\sum_{\beta=1}^{\bar{N}^e} D^{e\xi^3}_{\alpha\beta} + I^e_{(\alpha)}\sum_{\beta=1}^{\bar{N}^e} D^{e\xi^4}_{\alpha\beta}\right]\tilde{w}^e_\beta$$

$$+\frac{P}{(l^e)^2}\sum_{\beta=1}^{\bar{N}^e} D^{e\xi^2}_{\alpha\beta}\tilde{w}^e_\beta + k^e_{(\alpha)}w^e_\alpha = 0 \tag{4.36}$$

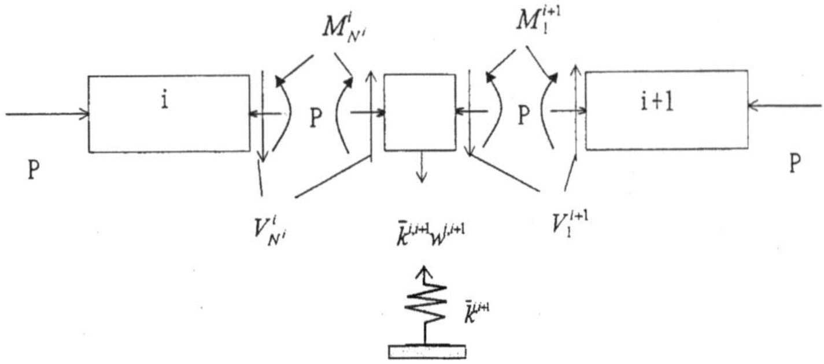

Fig. 4.6. Equilibriums at the inter-element boundary

The two kinematic transition conditions have been discretized and expressed by Eqs. (4.9) and (4.11). The discrete natural boundary condition of moment equilibrium is expressed by Eq. (4.30).Referring to Fig. 4.6, the discrete natural transition condition of equilibrium of lateral forces can also be obtained

$$-\frac{E^i}{(l^i)^3}\left(\frac{dI^i_{N^i}}{d\xi}\sum_{\beta=1}^{\bar{N}^i} D^{i\xi^2}_{N^i\beta} + I^i_{N^i}\sum_{\beta=1}^{\bar{N}^i} D^{i\xi^3}_{N^i\beta}\right)\tilde{w}^i_\beta - \frac{P}{l^i}\sum_{\beta=1}^{\bar{N}^i} D^{i\xi^2}_{N^i\beta}\tilde{w}^i_\beta$$

$$+\frac{E^{i+1}}{(l^{i+1})^3}\left(\frac{dI^{i+1}_{N^{i+1}}}{d\xi}\sum_{\beta=1}^{\bar{N}^{i+1}} D^{(i+1)\xi^2}_{N^{i+1}\beta} + I^{i+1}_1\sum_{\beta=1}^{\bar{N}^{i+1}} D^{(i+1)\xi^3}_{1\beta}\right)\tilde{w}^{i+1}_\beta$$

$$+\bar{k}^{i,i+1}w^{i,i+1} + \frac{P}{l^{i+1}}\sum_{\beta=1}^{\bar{N}^{i+1}} D^{(i+1)\xi^2}_{N^{i+1}\beta}\tilde{w}^{i+1}_\beta = 0 \tag{4.37}$$

The discrete kinematic boundary conditions are the same as those for the static deflection analysis. The discrete natural boundary condition of moment equilibrium is expressed by Eq. (4.29). The discrete natural boundary condition of equilibrium of transverse forces is expressed as

$$\frac{E^n}{(l^n)^3}\left(\frac{dI^n_1}{d\xi}\sum_{\beta=1}^{\bar{N}^n} D^{n\xi^2}_{1\beta} + I^n_1\sum_{\beta=1}^{\bar{N}^n} D^{n\xi^3}_{1\beta}\right)\tilde{w}^n_\beta + \bar{k}^n w^n_1 + \frac{P}{l^n}\sum_{\beta=1}^{\bar{N}^n} D^{n\xi^2}_{N^n\beta}\tilde{w}^n_\beta = 0 \tag{4.38}$$

if the natural boundary is at the left end, and

$$-\frac{E^n}{(l^n)^3}\left(\frac{dI^n_{N^n}}{d\xi}\sum_{\beta=1}^{\bar{N}^n} D^{n\xi^2}_{N^n\beta} + I^n_{N^n}\sum_{\beta=1}^{\bar{N}^n} D^{n\xi^3}_{N^n\beta}\right)\tilde{w}^n_\beta + \bar{k}^n w^n_{N^n}$$

$$-\frac{P}{l^n}\sum_{\beta=1}^{\bar{N}^n} D^{n\xi^2}_{N^n\beta}\tilde{w}^n_\beta = 0 \tag{4.39}$$

if the natural boundary is at the right end.

4.3.2 Assemblage

The assemblage of discrete equations for the DQEM buckling analysis of Euler-Bernoulli beam structures can be carried out following the procedures introduced in Subsection 4.2.2 regarding the assemblage of discrete equations for the DQEM vibration analysis of Euler-Bernoulli beam structures. Considering the discrete kinematic boundary conditions, the overall discrete eigenvalue equation system can be expressed as

$$\left(\left[\tilde{K}\right] - P\left[K_g\right]\right)\{\tilde{D}\} = \{0\} \tag{4.40}$$

where $[\tilde{K}] = [K] + [\bar{K}]$, $-P\left[K_g\right]$ is the overall initial stress matrix, and $\{\tilde{D}\}$ is the overall displacement vector. $[K_g]$ is a sparse matrix.

Equation (4.40) is a generalized eigenvalue problem. Pre-multiplication of equation (4.40) by $\left[\tilde{K}\right]^{-1}$ leads to

$$\left([\bar{A}] - \bar{\lambda}\,[I]\right)\{\tilde{D}\} = \{0\} \tag{4.41}$$

where $[\bar{A}] = [\tilde{K}]^{-1}\left[K_g\right]$ and $\bar{\lambda} = \frac{1}{P}$. Some DOF can be eliminated before solving Eq. (4.40). The related solution procedures are introduced in Subsection 3.2.2.

4.3.3 Problems

The buckling analysis of a nonprismatic overhanging bar with length L subjected to a compressive force P at the free end was also carried out [99]. The bar structure is shown in Fig. 4.7. At the middle point there is a roller support. The bar is a solid truncated cone with I expressed as $I = \frac{I_0}{2}\left[2^{\frac{1}{4}} - \left(2^{\frac{1}{4}} - 1\right)\frac{x}{L}\right]^4$, where I_0 is the value of I at the fixed end $x = 0$. The $C^1 - C^0 - C^1$ EDQ model used for the static deflection analyses is adopted for carrying out the element basis discretization. Defining $\bar{C} = \frac{PL^2}{EI_0}$ as the load factor, numerical results of the first three critical load factors are summarized

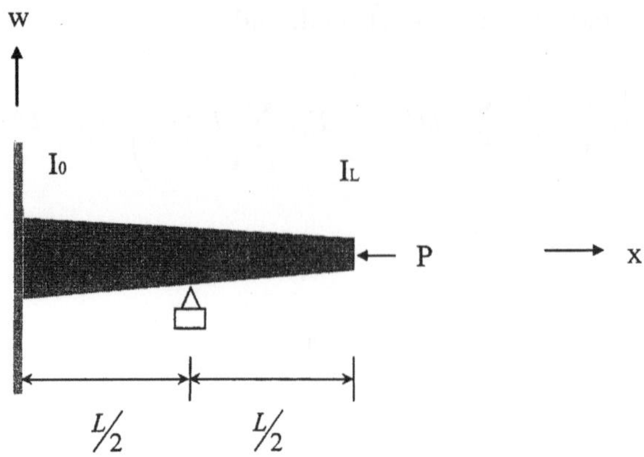

Fig. 4.7. Buckling of a nonprismatic overhanging bar

Table 4.8. The lowest critical load factor $\bar{C}_1$ of a compressed nonprismatic overhanging bar

DOF per element	Number of elements	$\bar{C}_1$	$\bar{C}_2$	$\bar{C}_3$
5	2	$.4651052\times10^1$	$.2699958\times10^2$	
	4	$.4440624\times10^1$	$.3410679\times10^2$	$.7358817\times10^2$
	6	$.4393859\times10^1$	$.3501234\times10^2$	$.8847569\times10^2$
7	2	$.4397124\times10^1$	$.3221310\times10^2$	
	4	$.4358650\times10^1$	$.3452757\times10^2$	$.8546602\times10^2$
	6	$.4356912\times10^1$	$.3452336\times10^2$	$.8351979\times10^2$
9	2	$.4355639\times10^1$	$.3494950\times10^2$	
	4	$.4356483\times10^1$	$.3453623\times10^2$	$.8418209\times10^2$
	6	$.4356497\times10^1$	$.3454465\times10^2$	$.8431407\times10^2$

and listed in Table 4.8. It shows that the convergence performance is also excellent for solving the nonprismatic overhanging bar problem.

A compressed cantilever beam composed of two prismatic segments having different cross sections, shown in Fig. 4.8, was also solved. The length of the bar is L with portion 1 and portion 2 being $0.6L$ and $0.4L$ long, respectively. Portion 2 has a fixed end, while portion 1 has the free end. The section constant I of portion 2 is I_2, while I of portion 1 is $I_1 = 0.4I_2$. The Hermite EDQ model that used to the vibration analysis is adopted for carrying out the element basis discretization. Defining $\bar{C} = \frac{PL^2}{EI_2}$ as the load factor, numerical results of the first three critical load factors are summarized and listed

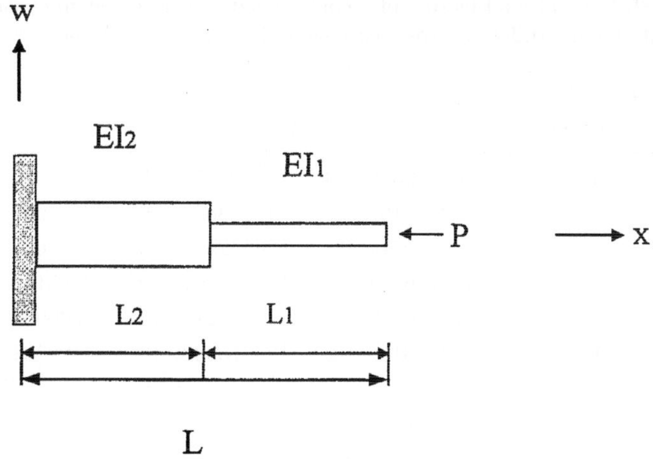

Fig. 4.8. Buckling of a nonprismatic cantilever bar composed of two prismatic segments

in Table 4.9. It also shows that the DQEM buckling analysis can efficiently converge to the analytical solutions. Numerical results of the analytical solution are calculated by using the Newton iteration procedure to the related characteristic equation resulting from solving the eigenvalue problem of the governing differential equation [100–101].

The problem by resting the composed cantilever beam on a Winkler foundation was also solved. In solving the problem, only two elements were used to model the structure. A Chebyshev EDQ model was used for the element

Table 4.9. Critical load factors of a compressed cantilever beam composed of two segments having different cross sections

DOF per element	Number of elements	$\bar{C}_1$	$\bar{C}_2$	$\bar{C}_3$
6	6	$.1672846\times10$	$.1177810\times10^2$	$.3009922\times10^2$
	10	$.1670577\times10$	$.1231134\times10^2$	$.3118579\times10^2$
8	2	$.1668739\times10$	$.1426503\times10^2$	$.2380288\times10^2$
	6	$.1669341\times10$	$.1265800\times10^2$	$.3203857\times10^2$
	10	$.1669346\times10$	$.1264754\times10^2$	$.3198959\times10^2$
10	2	$.1669351\times10$	$.1259495\times10^2$	$.3268258\times10^2$
	6	$.1669347\times10$	$.1264593\times10^2$	$.3197285\times10^2$
Analytical solution		$.1669347\times10$	$.1264599\times10^2$	$.3198116\times10^2$

Table 4.10. Critical load factors of a compressed cantilever beam consisting of two segments having different cross sections and resting on a Winkler foundation

Number of elements	DOF per element	$\bar{C}_1$	$\bar{C}_2$	$\bar{C}_3$
2	5	$.204080589\times10$	$.770401757\times10$	
	7	$.173135162\times10$	$.147423536\times10^2$	$.368890794\times10^2$
	9	$.173750776\times10$	$.126942928\times10^2$	$.359061315\times10^2$
	11	$.173746133\times10$	$.127427997\times10^2$	$.318516725\times10^2$
	13	$.173746149\times10$	$.127418218\times10^2$	$.320090482\times10^2$
	15	$.173746148\times10$	$.127418333\times10^2$	$.319978039\times10^2$

basis discretization. Defining $\bar{C} = \frac{PL^2}{EI_2}$ as the load factor, numerical results of the first three critical load factors for $\frac{kL^4}{EI} = .5$ are summarized and listed in Table 4.10. It also shows that the convergence performance of using Chebyshev EDQ model is excellent.

5

DQEM Analysis of Static Deflection of Three-Dimensional Trusses

The DQEM truss analysis model uses various DQ models to the element basis local discretization. The discrete local element equilibrium equations and element boundary forces at the two element boundary nodes can be transformed using transformation matrices defined by the local and global coordinates. Only displacements at the two element boundary nodes are transformed into the global coordinate system. The equilibriums of forces at joints are considered in constructing the overall stiffness equation. The discrete local element equilibrium equations are defined on the spaces of local coordinates.

5.1 Discrete Element Equations

5.1.1 Discrete Element Equilibrium Equations

The differential equilibrium equation of a nonprismatic bar in the axial direction is

$$E^e \frac{d}{d\bar{z}^e}\left(A^e \frac{d\bar{u}^e}{d\bar{z}^e}\right) = -\bar{f}^e \tag{5.1}$$

where $\bar{w}^e$ is the axial displacement, $\bar{z}^e$ the local physical coordinate, E^e Young's modulus, A^e the area of cross section and $\bar{f}^e$ the distributed axial force.

The DQ model which only uses the axial displacements at element nodes to represent the DQ discretization is adopted. In carrying out the element-based DQ discretization, $\bar{z}^e_{N^e}$ which is equal to the element length can be calculated by using the global coordinates of node 1 and node N^e. Let (x^e_1, y^e_1, z^e_1) and $(x^e_{N^e}, y^e_{N^e}, z^e_{N^e})$ denote the global coordinates of node 1 and node N^e, respectively. The element length l^e is expressed as:

$$l^e = \left[(x^e_{N^e} - x^e_1)^2 + (y^e_{N^e} - y^e_1)^2 + (z^e_{N^e} - z^e_1)^2\right]^{1/2} = \bar{z}^e_{N^e} \tag{5.2}$$

For the developed three-dimensional DQEM truss analysis model, the two-node prismatic element can be used to model prismatic truss members having no distributed load. The two-node prismatic element is the same as the two-node finite element truss element. However, generic truss problems might have axially distributed external causes. Without using a certain technique to obtain the equivalent nodal loads and include them into the natural transition conditions or boundary conditions, the element must have three or more nodes. This method can better describe local responses caused by locally, highly nonlinear distributed external causes either through the use of adaptive discretization or by using more elements to discretize certain specific structural members. This DQEM truss analysis model is also efficient for solving vibrations of truss structures. There are $N^e - 2$ discrete equilibrium equations in an element for N^e being larger than two. The selection of the $N^e - 2$ nodes, at which discrete equilibrium equations are defined, is flexible. Here, nodes $2, 3, ..., N^e - 2$ and $N^e - 1$ are used. Consider that the range of the natural coordinate ζ is $0 \leq \zeta \leq 1$. Using DQ, Eq. (5.1) can be discretized

$$\frac{E^e}{(l^e)^2}\left(\frac{dA^e_{(\alpha)}}{d\zeta}\sum_{\beta=1}^{N^e} D^{e\zeta}_{\alpha\beta} + A^e_{(\alpha)}\sum_{\beta=1}^{N^e} D^{e\zeta^2}_{\alpha\beta}\right)\bar{w}^e_\beta = -\bar{f}^e_\alpha, \quad \alpha = 2, 3, ..., N^e - 1 \tag{5.3}$$

In the above equation, the first derivatives of A^e with respect to ζ at element nodes can also be calculated by DQ. It is especially useful if the distribution function of A^e is not continuously differentiable up to the order of its derivative. The above equation can be rewritten to obtain the following discrete local element equilibrium equation

$$[\bar{\kappa}^e]\{\bar{\delta}^e\} = \{\bar{r}^e\} \tag{5.4}$$

where

$$[\bar{\kappa}^e] = \frac{E^e}{(l^e)^2}\begin{bmatrix} \frac{dA^e_1}{d\zeta}D^{e\zeta}_{11} + A^e_1 D^{e\zeta^2}_{11} & \frac{dA^e_1}{d\zeta}D^{e\zeta}_{12} + A^e_1 D^{e\zeta^2}_{12} & \cdots & \frac{dA^e_1}{d\zeta}D^{e\zeta}_{1N^e} + A^e_1 D^{e\zeta^2}_{1N^e} \\ \frac{dA^e_2}{d\zeta}D^{e\zeta}_{21} + A^e_2 D^{e\zeta^2}_{21} & \frac{dA^e_2}{d\zeta}D^{e\zeta}_{22} + A^e_2 D^{e\zeta^2}_{22} & \cdots & \frac{dA^e_2}{d\zeta}D^{e\zeta}_{2N^e} + A^e_2 D^{e\zeta^2}_{2N^e} \\ \vdots & \vdots & \vdots & \vdots \\ \frac{dA^e_{N^e}}{d\zeta}D^{e\zeta}_{N^e1} + A^e_{N^e} D^{e\zeta^2}_{N^e1} & \frac{dA^e_{N^e}}{d\zeta}D^{e\zeta}_{N^e2} + A^e_{N^e} D^{e\zeta^2}_{N^e2} & \cdots & \frac{dA^e_{N^e}}{d\zeta}D^{e\zeta}_{N^eN^e} + A^e_{N^e} D^{e\zeta^2}_{N^eN^e} \end{bmatrix} \tag{5.5}$$

is a $(N^e - 2) \times N^e$ local element stiffness coefficient matrix,

$$\{\bar{\delta}^e\} = \lfloor\ \bar{w}_1^e \quad \bar{w}_2^e \quad \bar{w}_3^e \quad ... \quad \bar{w}_{N^e-1}^e \quad \bar{w}_{N^e}^e\ \rfloor^T \tag{5.6}$$

is the $N^e \times 1$ local element displacement vector, and

$$\{\bar{f}^e\} = \lfloor\ -\bar{f}_2^e \quad -\bar{f}_3^e \quad ... \quad -\bar{f}_{N^e-2}^e \quad -\bar{f}_{N^e-1}^e\ \rfloor^T \tag{5.7}$$

is the $(N^e - 2) \times 1$ local element distributed force vector.

5.1.2 Coordinate Transformations

In order to solve truss problems, the discrete local element equilibrium equations of all discrete elements have to be assembled into the overall algebraic equation system. In the overall algebraic equation system, displacement components of nodes 1 and N^e are in directions of global coordinates, therefore the six displacement components at the two element boundary nodes have to be transformed into the global coordinate system. The transformation of a vector can be accomplished through the use of a transformation matrix which is defined by the direction cosines of the nine direction angles between the three local coordinate axes $\bar{x}$, $\bar{y}$, $\bar{z}$ and the three global coordinate axes x, y, z. Using the direction cosines, a transformation relation of the displacement vectors in the two coordinate systems can be constructed

$$\begin{bmatrix} \cos(\bar{x},x) & \cos(\bar{x},y) & \cos(\bar{x},z) \\ \cos(\bar{y},x) & \cos(\bar{y},y) & \cos(\bar{y},z) \\ \cos(\bar{z},x) & \cos(\bar{z},y) & \cos(\bar{z},z) \end{bmatrix} \begin{Bmatrix} u \\ v \\ w \end{Bmatrix} = \begin{Bmatrix} \bar{u} \\ \bar{v} \\ \bar{w} \end{Bmatrix} \tag{5.8}$$

The 3×3 rotational transformation matrix in the above equation can be expressed by using the notation $[t_r^e]$

$$[t_r^e] = \begin{bmatrix} \cos(\bar{x},x) & \cos(\bar{x},y) & \cos(\bar{x},z) \\ \cos(\bar{y},x) & \cos(\bar{y},y) & \cos(\bar{y},z) \\ \cos(\bar{z},x) & \cos(\bar{z},y) & \cos(\bar{z},z) \end{bmatrix} \tag{5.9}$$

A transformation relation between the element displacement vector in local-global coordinate system and the element displacement vector in local coordinate system can be obtained through the use of the third row, $\lfloor t_{r3}^e \rfloor$, of $[t_r^e]$

$$[T^e]\{\delta^e\} = \{\bar{\delta}^e\} \tag{5.10}$$

where $\{\bar{\delta}^e\}$ is the local element displacement vector expressed by Eq. (5.6),

$$\{\delta^e\} = \lfloor\ u_1^e \quad v_1^e \quad w_1^e \quad \bar{w}_2^e \quad \bar{w}_3^e \quad ... \quad \bar{w}_{N^e-1}^e \quad u_{N^e}^e \quad v_{N^e}^e \quad w_{N^e}^e\ \rfloor^T \tag{5.11}$$

is the local-global element displacement vector, and

$$[T^e] = \begin{bmatrix} \lfloor t_{r3}^e \rfloor & & & & & \\ & 1 & & & & \\ & & \cdot & & & \\ & & & \cdot & & \\ & & & & 1 & \\ & & & & & \lfloor t_{r3}^e \rfloor \end{bmatrix} \tag{5.12}$$

is the element transformation matrix. Employing Eq. (5.10) in Eq. (5.4), the following matrix equation can be obtained

$$[\bar{\kappa}^e][T^e]\{\delta^e\} = \{\bar{f}^e\} \tag{5.13}$$

The above equation can be rewritten as

$$[\kappa^e]\{\delta^e\} = \{\bar{f}^e\} \tag{5.14}$$

where

$$[\kappa^e] = [\bar{\kappa}^e][T^e] \tag{5.15}$$

is the local-global element stiffness coefficient matrix. Equation (5.14) is the discrete local-global element equilibrium equation in which the element displacement vector is defined on both of local and global coordinate systems and the local element distributed force vector is defined on the local coordinate system.

5.1.3 Discrete Element Internal Forces

The internal forces of a discrete element at nodes can be obtained using the technique of DQ discretization. Denote $\bar{F}^e_{\bar{z}}$ the axial force at an arbitrary point $\bar{z}^e$ in the element. $\bar{F}^e_{\bar{z}}$ is expressed as

$$\bar{F}^e_{\bar{z}} = E^e A^e \frac{d\bar{w}^e}{d\bar{z}^e} \tag{5.16}$$

Using DQ in Eq. (5.16), axial forces at a node point α can be obtained:

$$\bar{F}^e_{\bar{z}\alpha} = \frac{E^e A^e_{(\alpha)}}{l^e} \sum_{\beta=1}^{N^e} D^{e\varsigma}_{\alpha\beta} \bar{w}^e_\beta = \frac{E^e A^e_{(\alpha)}}{l^e} \lfloor D^{e\varsigma}_\alpha \rfloor \{\bar{\delta}^e\} \tag{5.17}$$

where

$$\lfloor D^{e\varsigma}_\alpha \rfloor = \lfloor\ D^{e\varsigma}_{\alpha 1} \quad D^{e\varsigma}_{\alpha 2} \quad \cdots \quad D^{e\varsigma}_{\alpha(N^e-1)} \quad D^{e\varsigma}_{\alpha N^e}\ \rfloor \tag{5.18}$$

Employing Eq. (5.10) in Eq. (5.17), $\bar{F}^e_{\bar{z}\alpha}$ can be related to the local-global element displacement vector, which is written as:

$$\bar{F}^e_{\bar{z}\alpha} = \frac{E^e A^e_{(\alpha)}}{l^e} \lfloor D^{e\varsigma}_\alpha \rfloor [T^e]\{\delta^e\} \tag{5.19}$$

5.2 Discrete Condition Equations of Joints

DQEM requires that all condition equations at joints are satisfied. The condition equations include compatibility conditions and equilibrium of external and internal forces at joints. The condition equations have to be expressed as discrete forms using DQ.

5.2.1 Discrete Joint Compatibility Conditions

Let M^j denote the number of elements connected to joint j. Also let I^{m^j} denote the element node number of the m^jth element connected to the joint. Then I^{m^j} is equal to 1 or N^{m^j}, with N^{m^j} being the largest node number of the m^jth element. Let $\{d^{m^j}_{I^{m^j}}\} = \lfloor u^{m^j}_{I^{m^j}} \; v^{m^j}_{I^{m^j}} \; w^{m^j}_{I^{m^j}} \rfloor^T$ and $\{d^j\} = \lfloor u^j \; v^j \; w^j \rfloor^T$ represent the globally nodal displacement vector of node I^{m^j} of the m^jth element and the global joint displacement vector of joint j, respectively. Then the compatibility conditions, which are kinematic conditions, of joint j can be expressed as follows:

$$\left\{d^{1^j}_{I^{1^j}}\right\} = \left\{d^{2^j}_{I^{2^j}}\right\} = \cdots = \left\{d^{m^j}_{I^{m^j}}\right\} = \cdots = \left\{d^{M^j}_{I^{M^j}}\right\} = \left\{d^j\right\} \tag{5.20}$$

5.2.2 Discrete Joint Equilibrium Conditions

For a three-dimensional DQEM truss analysis model, the equilibrium conditions of external and internal forces at joints also have to be satisfied. Each equilibrium condition is either a natural transition condition or a natural boundary condition. Let ν^{m^j} denote an indicator defined by the local element node number of an element at the joint. ν^{m^j} is defined as:

$$\nu^{m^j} = \begin{cases} +1, & \text{if } I^{m^j} = N^{m^j} \\ -1, & \text{if } I^{m^j} = 1 \end{cases} \tag{5.21}$$

Let $\{P^j\}$ denote the vector of concentrated forces applied at joint j. And consider the inverse transformation of a vector by using the transformation matrix defined by Eq. (5.8), the axially nodal force of the m^jth element at joint j can be transformed into the global coordinates. Let $\{V^{m^j}\} = \lfloor V^{m^j}_x \; V^{m^j}_y \; V^{m^j}_z \rfloor^T$ denote the globally nodal force vector of the m^jth element at joint j. The three translational equilibrium conditions of joint j can be expressed as follows:

$$\sum_{m^j=1}^{M^j} \nu^{m^j} \{V^{m^j}\} = \{P^j\} \tag{5.22}$$

The globally nodal force vector can be related to the locally internal force $\bar{F}^{m^j}_{\bar{z}I^{m^j}}$

$$\{V^{m^j}\} = \lfloor t^{m^j}_{r1} \rfloor^T \bar{F}^{m^j}_{\bar{z}I^{m^j}} \tag{5.23}$$

where $\lfloor t^{m^j}_{r1} \rfloor^T$ is the transpose of the first row of $[t^{m^j}_r]$. Using Eqs. (5.19) and (5.23), Eq. (5.22) can be rewritten as

$$\sum_{m^j=1}^{M^j} \nu^{m^j} \lfloor t^{m^j}_{r1} \rfloor^T \frac{\left(E^{m^j} A^{m^j}\right)_{(I^{m^j})}}{l^{m^j}} \left\lfloor D^{m^j \varsigma}_{I^{m^j}} \right\rfloor \left[T^{m^j}\right] \left\{\delta^{m^j}\right\} = \{P^j\} \tag{5.24}$$

5.2.3 Prescribed Joint Displacements

A joint might have one or more prescribed displacement components. Let $\{\hat{d}_p^j\}$ and $\{d_p^j\}$ denote the prescribed joint displacement vector and the corresponding joint displacement vector of joint j, respectively. The condition equations of prescribed displacements can be obtained from the following vector equation:

$$\{d_p^j\} = \left\{\hat{d}_p^j\right\} \tag{5.25}$$

5.2.4 Inclined Roller

Let $(\tilde{x}, \tilde{y}, \tilde{z})$ denote the three coordinate axes of a local rectangular Cartesian coordinate system with $\tilde{x}$ and $\tilde{y}$ axes located on the inclined surface and $\tilde{z}$ axis outward normal to the inclined surface, respectively. Also let $\{\tilde{d^j}\} = \lfloor \tilde{u}^j \; \tilde{v}^j \; \tilde{w}^j \rfloor^T$ denote the related prescribed displacement vector with the three components in $\tilde{x}$, $\tilde{y}$ and $\tilde{x}$ coordinate directions. The transformation matrix $\{\tilde{t}_r\}$ can be defined by using $\tilde{x}$, $\tilde{y}$ and $\tilde{z}$ to replace $\bar{x}$, $\bar{y}$ and $\bar{z}$ in Eq. (5.8). Then the kinematic condition equations can be expressed by the following matrix equation

$$\left[\tilde{t}_r\right]\{d^j\} = \{\tilde{d^j}\} \tag{5.26}$$

Let $\tilde{P}_{\tilde{x}}^j$ and $\tilde{P}_{\tilde{y}}^j$ denote the external forces in $\tilde{x}$ and $\tilde{y}$ directions, respectively. Also let $\lfloor\tilde{t}_{r1}\rfloor$ and $\lfloor\tilde{t}_{r2}\rfloor$ denote the first and second rows of $[\tilde{t}_r]$, respectively. The two natural condition equations which are translational equilibrium equations can be expressed by the following matrix equation

$$[\tilde{t}_{r12}] \sum_{m^j=1}^{M^j} \nu^{m^j}\{V^{m^j}\} = [\tilde{t}_{r12}]\{P^j\} = \{\tilde{P}^j\} \tag{5.27}$$

where $\{\tilde{P}^j\} = \lfloor \tilde{P}_{\tilde{x}}^j \; \tilde{P}_{\tilde{y}}^j \rfloor^T$, and $[\tilde{t}_{r12}] = \begin{bmatrix} \lfloor\tilde{t}_{r1}\rfloor \\ \lfloor\tilde{t}_{r2}\rfloor \end{bmatrix}$. By using Eqs. (5.22), (5.23) and (5.24), Eq. (5.27) can be rewritten as

$$\begin{aligned} &[\tilde{t}_{r12}] \sum_{m^j=1}^{M^j} \nu^{m^j}[\tilde{t}_{r12}]\lfloor t_{r1}^{m^j}\rfloor^T \frac{(E^{m^j}A^{m^j})_{(I^{m^j})}}{l^{m^j}} \lfloor D_{I^{m^j}}^{m^j\zeta}\rfloor \left[T^{m^j}\right]\{\delta^{m^j}\} \\ &\quad = \{\tilde{P}^j\} \end{aligned} \tag{5.28}$$

5.3 Assemblage

With the discrete joint compatibility conditions in mind, then by assembling all discrete local-global element equilibrium equations (5.14) for elements having more than two nodes, joint equilibrium conditions (5.24) and (5.28), for the general joints and inclined rollers, respectively, and the prescribed joint displacement conditions (5.25) and (5.26) for the general joints and inclined rollers, respectively, the overall algebraic system represented by Eq. (3.18) can be obtained. Like FEM, the assemblage is based on an element by element procedure. The discrete local-global element equilibrium equations Eq. (5.14) for elements having more than two nodes, and discrete element boundary forces existing in Eqs. (5.24) and (5.28) defined at the two element boundary nodes and expressed by displacements, for all elements, are directly assembled to the overall discrete equation system. Consequently, an element basis explicit matrix equation is not necessary to be formed in the assembling process.

This element basis explicit matrix equation is an element stiffness equation. It contains the relations between the two axial forces and the local element displacement vector expressed by Eq. (5.6) and placed at the first and last rows, and component equations of the discrete local element equilibrium equation (5.14), for an element having more that two nodes, placed at the remaining $N^e - 2$ rows. This element stiffness equation can be expressed by

$$\left[\hat{k}^e\right]\{\bar{\delta}^e\} = \{\hat{r}^e\} \tag{5.29}$$

where $[\hat{k}^e]$ is a $N^e \times N^e$ local element stiffness matrix with the first and last rows filled by coefficients for calculating the two discrete element boundary forces, and

$$\{\hat{r}^e\} = \lfloor -\bar{F}^e_{\bar{z}1} \quad -\bar{f}^e_2 \quad -\bar{f}^e_3 \quad \ldots \quad -\bar{f}^e_{N^e-2} \quad -\bar{f}^e_{N^e-1} \quad \bar{F}^e_{\bar{z}N^e} \rfloor^T \tag{5.30}$$

is the local element load vector. Using the element transformation matrix $[T^e]$ in Eq. (5.29), the following local-global element stiffness equation can be obtained

$$[\tilde{k}^e]\{\delta^e\} = \{\tilde{r}^e\} \tag{5.31}$$

where

$$[\tilde{k}^e] = [T^e]^T[\hat{k}^e][T^e] \tag{5.32}$$

is the transformed local-global element stiffness matrix and

$$\{\tilde{r}^e\} = [T^e]^T\{\hat{r}^e\} \tag{5.33}$$

is the transformed local-global element load vector. By assembling Eq. (5.31) for all elements having more than two nodes and considering Eqs. (5.22) and (5.27), for the general joints and inclined rollers, respectively, the overall stiffness equation represented by Eq. (3.18) can also be obtained.

5.4 Problems

In the analyses, the Lagrange DQ model with equally spaced grid nodes is used for the element-based discretization. Only the degree of freedom of the axial displacement is assigned to each node. The interior nodes are used to define discrete equilibrium equations while the two element boundary nodes are used to define the joint condition equations.

Figure 5.1 shows a fixed-free I-bar subjected to a uniformly distributed axial force. z axis is coincident with the centroid line. The cross sections at A and B are shown in Fig. 5.2. The web has the same thickness as the flange. The variation of width is expressed by $b(z) = b_o(1 - z/L + z^2/2L^2)$ with $b_o = 40.\ mm$ and $L = 1000.\ mm$. The variation of depth is expressed by $d(z) = d_o(1 - z/L + z^2/2L^2)$ with $d_o = 80.\ mm$. The values of Young's modulus is $E = 206000\ N/mm^2$. In the analysis, the elements are equally spaced. Numerical results obtained are summarized and listed in Table 5.1. They are compared with exact solutions. By increasing either the number of elements or degrees of freedom per element, the results converge very fast to the exact solutions.

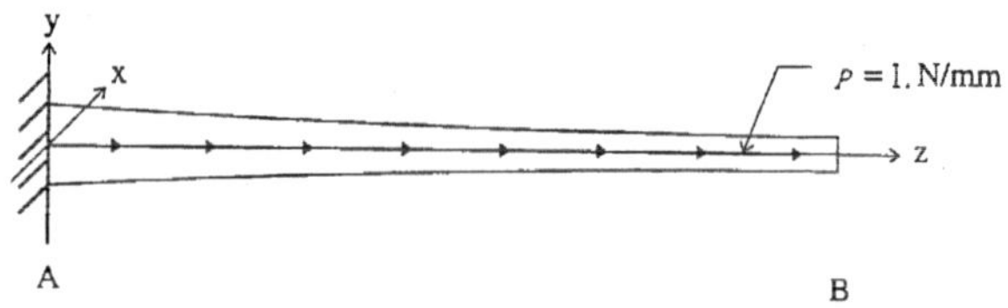

Fig. 5.1. A nonprismatic bar subjected to a distributed load

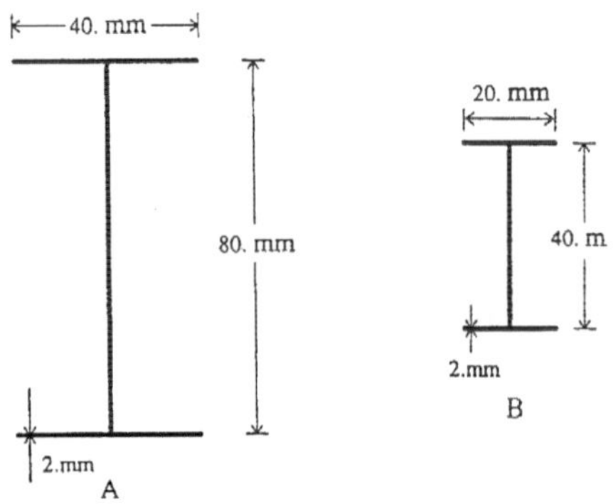

Fig. 5.2. The cross sections at A and B

The two-dimensional truss structure shown in Fig. 5.3 has 29 prismatic members and a support settlement and is subjected to concentrated and distributed loads. In the DQEM analysis, 29 elements were used to model the structure. Elements 3 and 4 which are subjected to linearly distributed loads are four-node elements. Three-node element is used to represent elements 17,

Table 5.1. The results of a nonprismatic I-bar subjected to a uniformly distributed axial load

Element type	Number of elements	w (mm) (at B)	P (N) (at A)	$\log\frac{\lvert w-w_{exact}\rvert}{\lvert w_{exact}\rvert}$ (at B)
5-node	2	$.1052838\times10^{-1}$	$.9998271\times10^{3}$	–2.894704
	4	$.1051554\times10^{-1}$	$.9999883\times10^{3}$	–4.273620
	6	$.1051508\times10^{-1}$	$.9999977\times10^{3}$	–5.021809
9-node	2	$.1051500\times10^{-1}$	$.1000000\times10^{4}$	–5.720779
	4	$.1051498\times10^{-1}$	$.1000000\times10^{4}$	<–7.000000
Exact solution		$.1051498\times10^{-1}$	$.1000000\times10^{4}$	

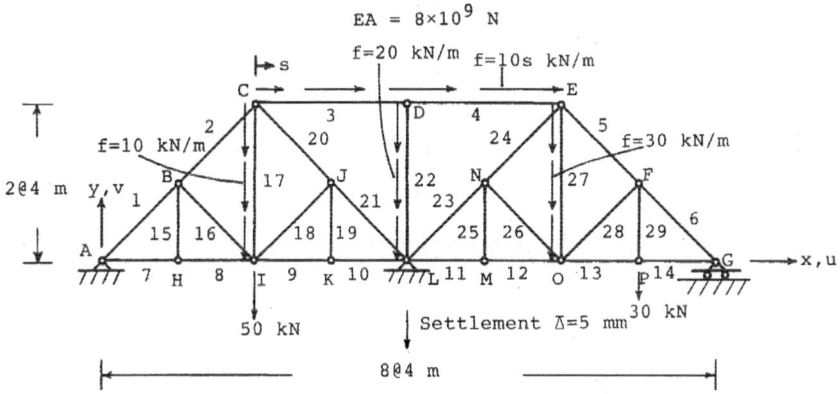

Fig. 5.3. A truss structure subjected to concentrated and axially distributed loads, and a support settlement

22 and 27 which are subjected to uniformly distributed loads. All other elements are two-node elements. The results of displacement and axial force are plotted, and shown in Figs. 5.4 and 5.5, respectively. Constant, linear and nonlinear distributions of axial forces were obtained. Since the satisfied mechanics relations are enough, the results are exact.

The three-dimensional truss structure shown in Fig. 5.6 is composed of eighteen prismatic members. Only concentrated loads are applied at four joints. Thus the two-node element is used to represent each of the eighteen members. Numerical results of member internal forces are summarized and listed in Table 5.2. They are exact values.

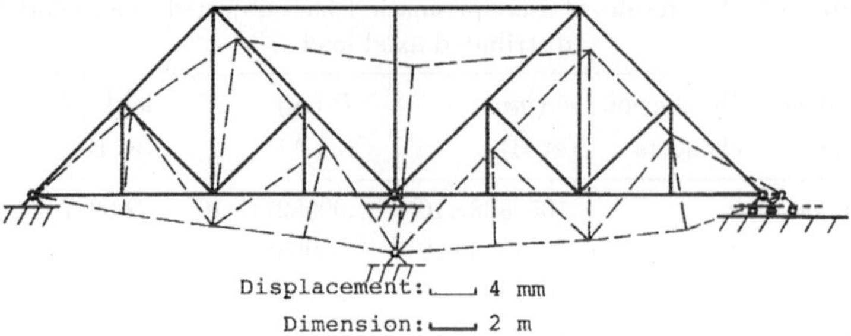

Fig. 5.4. Displacement diagram

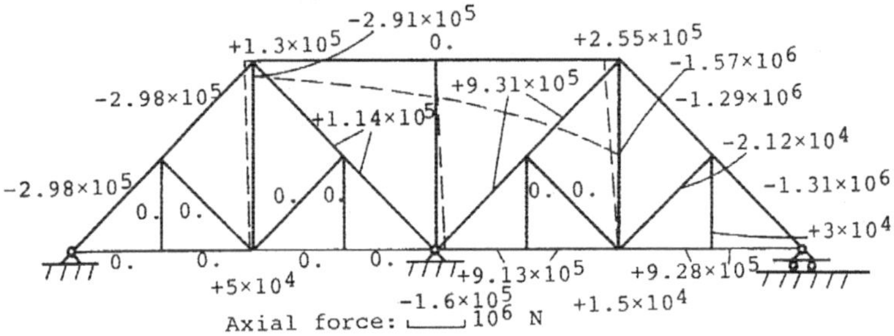

Fig. 5.5. Axial force diagram

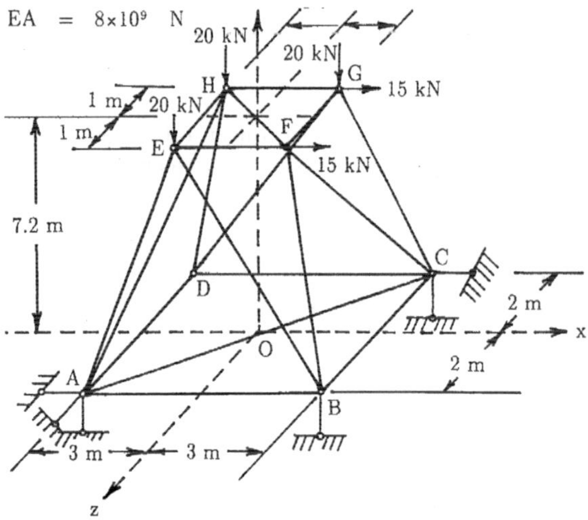

Fig. 5.6. A three-dimensional truss structure

Table 5.2. The results of a three-dimensional truss problem

Member	Member force (kN)	Exact solution
AB	14.59	14.59
CD	–7.92	–7.92
BC	3.61	3.61
AD	0.00	0.00
AC	0.25	0.25
EF	14.17	14.17
GH	–5.00	–5.00
FG	–2.78	–2.78
EH	–2.78	–2.78
FH	1.00	1.00
AE	2.06	2.06
BF	–4.13	–4.13
CG	–40.21	–40.21
DH	–19.59	–19.59
BE	–26.13	–26.13
CF	4.42	4.42
DG	22.56	22.56
AH	–1.11	–1.11

6

DQEM Analysis of Static Deflection of Three-Dimensional Frames

The DQEM frame analysis model uses various EDQ models which can automatically set compatibility and conformability conditions at the joints to the element basis local discretization. The discrete local element equilibrium equations and element boundary forces at the two element boundary nodes can be transformed using various transformation matrices. Only displacements at the two element boundary nodes are transformed into the global coordinate system. The equilibriums of forces at joints are considered in constructing the overall stiffness equation. The discrete local element equilibrium equations are defined on the local coordinates. All discrete equations are directly assembled into the overall stiffness equation.

6.1 Fundamental Relations of Nonprismatic Beam

For a nonprismatic beam of length l, assume that the material is isotropic and homogeneous with Young's modulus E and shear modulus G. Figure 6.1

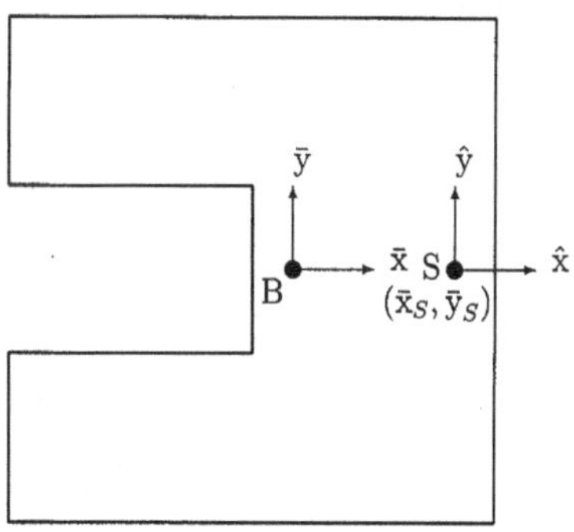

Fig. 6.1. The centroid and shear center on a cross section

shows a cross section of a beam with B the centroid and S the shear center. By locating the origin of the local coordinate system $(\bar{x}, \bar{y}, \bar{z})$ at the centroid of the cross section and orienting the two axes on the cross section to the principal directions, various section constants can be defined. Then, the coordinates $(\bar{x}_S, \bar{y}_S)$ defining the shear center S can be calculated by using certain cross section constants. Let $\bar{u}_S$ and $\bar{v}_S$ denote the two lateral displacement components at the shear center in $\hat{x}_S$ and $\hat{y}_S$ directions, respectively. Also let $\bar{w}$ and $\bar{\theta}_{\bar{z}}$ denote the average axial displacement and angle of twist. Neglecting the effect of warping torsion, the equilibrium equations of the beam considering the flexural, axial and torsional deformations are expressed by:

$$E\frac{d^2}{d\bar{z}^2}\left(I_{\bar{x}\bar{x}}\frac{d^2\bar{u}_S}{d\bar{z}^2}\right) = q_{\bar{x}} - \frac{dm_{\bar{y}}}{d\bar{z}}, \quad E\frac{d^2}{d\bar{z}^2}\left(I_{\bar{y}\bar{y}}\frac{d^2\bar{v}_S}{d\bar{z}^2}\right) = q_{\bar{y}} + \frac{dm_{\bar{x}}}{d\bar{z}},$$

$$E\frac{d}{d\bar{z}}\left(A\frac{d\bar{w}}{d\bar{z}}\right) = -p, \quad G\frac{d}{d\bar{z}}\left(J\frac{d\bar{\theta}_{\bar{z}}}{d\bar{z}}\right) = -m_{\bar{z}}^{(s)} \tag{6.1}$$

The boundary conditions are:

$$EI_{\bar{x}\bar{x}}\frac{d^2\bar{u}_S}{d\bar{z}^2} = \bar{M}_{\bar{y}} \quad \text{or} \quad \delta\left(\frac{d\bar{u}_S}{d\bar{z}}\right) = 0,$$

$$-E\frac{d}{d\bar{z}}\left(I_{\bar{x}\bar{x}}\frac{d^2\bar{u}_S}{d\bar{z}^2}\right) = \bar{V}_{\bar{x}} + m_{\bar{y}} \quad \text{or} \quad \delta\bar{u}_S = 0,$$

$$EI_{\bar{y}\bar{y}}\frac{d^2\bar{v}_S}{d\bar{z}^2} = -\bar{M}_{\bar{x}} \quad \text{or} \quad \delta\left(\frac{d\bar{v}_S}{d\bar{z}}\right) = 0,$$

$$-E\frac{d}{d\bar{z}}\left(I_{\bar{y}\bar{y}}\frac{d^2\bar{v}_S}{d\bar{z}^2}\right) = \bar{V}_{\bar{y}} - m_{\bar{x}} \quad \text{or} \quad \delta\bar{v}_S = 0,$$

$$EA\frac{d\bar{w}}{d\bar{z}} = \bar{P} \quad \text{or} \quad \delta\bar{w} = 0,$$

$$GJ\frac{d\bar{\theta}_{\bar{z}}}{d\bar{z}} = \bar{M}_{\bar{z}}^{(s)} \quad \text{or} \quad \delta\bar{\theta}_{\bar{z}} = 0 \tag{6.2}$$

where J is the torsional constant. In Eqs. (6.1) and (6.2), the following relations are used

$$I_S = I_p + A(\bar{x}_S^2 + \bar{y}_S^2), \quad m_{\bar{z}}^{(s)} = m_{\bar{z}} + (\bar{y}_S q_{\bar{x}} - \bar{x}_S q_{\bar{y}}),$$

$$\bar{M}_{\bar{z}}^{(s)} = \bar{M}_{\bar{z}} + (\bar{y}_S \bar{V}_{\bar{x}} - \bar{x}_S \bar{V}_{\bar{y}}) \tag{6.3}$$

where $I_{\bar{x}\bar{x}}$ and $I_{\bar{y}\bar{y}}$ are moments of inertia of the cross section with respect to $\bar{y}$ and $\bar{x}$ axes, respectively, $I_p = I_{\bar{x}\bar{x}} + I_{\bar{y}\bar{y}}$ is the polar moment of inertia, $q_{\bar{x}}$ and $q_{\bar{y}}$ are distributed external forces in $\bar{x}$ and $\bar{y}$ axes, respectively, $m_{\bar{x}}$ and $m_{\bar{y}}$ are intensities of distributed external moments with respect to $\bar{x}$ and $\bar{y}$ axes, respectively, $m_{\bar{z}}$ is the intensity of distributed torque, $\bar{V}_{\bar{x}}$ and $\bar{V}_{\bar{y}}$ are lateral forces in $\bar{x}$ and $\bar{y}$ directions, respectively, $\bar{M}_{\bar{x}}$ and $\bar{M}_{\bar{y}}$ are the external moments with respect to $\bar{x}$ and $\bar{y}$ axes, respectively, and $\bar{M}_{\bar{z}}$ is the external torque, applied on the natural boundary.

6.2 Discrete Element Equations

6.2.1 Discrete Element Equilibrium Equations

The fundamental relations are referred to the physical coordinate system while the EDQ discretization is carried out on the natural coordinate system. Therefore, in using the EDQ technique to discretize the fundamental relations, the transformation operations of coordinates and derivatives of displacements between two different coordinate systems, have to be carried out. Let (x_1^e, y_1^e, z_1^e) and $(x_{N^e}^e, y_{N^e}^e, z_{N^e}^e)$ denote the global coordinates of node 1 and node N^e which are two end nodes, respectively. The element length l^e can thus be calculated by using Eq. (5.2).

The element degrees of freedom for defining the axial, flexural and torsional discretizations must equal the corresponding element-basis discrete fundamental relations for constructing the overall discrete equation system. In addition to the degrees of freedom for representing displacement components, the degrees of freedom for representing derivatives of a local displacement component with respect to $\bar{z}^e$ at an element boundary node can also be assigned to that element boundary node. The selection of derivatives can be flexible [35,102]. In order to automatically set the kinematic transition conditions by only using the degrees of freedom assigned to the element boundary nodes, the degrees of freedom representing the two first order derivatives of flexural deflections must be assigned to the element boundary nodes. In the present DQEM frame analysis model, only the degrees of freedom which are necessary for automatically setting the kinematic transition conditions are assigned to the element boundary nodes.

Since the highest order of derivatives of displacement parameters existing in the third and fourth of Eqs. (6.1) is two, without using a specific technique to calculate the two sets of equivalent nodal forces, separately, and include them into the related natural transition conditions or natural boundary conditions, the order of approximate axial displacement and/or angle of twist must at least be two and the element must at least have one discrete point for defining the discrete axial and/or torsional equilibrium equations, separately, for the element having distributed axial and/or torsional loads. However, the order of approximate axial displacement and/or angle of twist can be one and no interior discrete point is necessary for the element having no distributed axial and/or torsional loads. The DQEM linear element is equivalent to the FEM linear element. Since the highest order of derivatives of the two lateral displacements is four, without using a specific technique to calculate the two sets of four equivalent nodal forces for the two distributed loads, separately, and include them into the related natural transition conditions or natural boundary conditions, the order of the two approximate lateral displacements must at least be four, and each of the related two equilibrium equations needs at least one discrete point for defining its discrete equation. The DQEM cubic element is equivalent to the FEM Hermite cubic element. The discrete points for defining the discrete equilibrium equations can be either in the interior of the element or on the element boundary.

In the present frame analysis model, only interior discrete points are used to define discrete element equilibrium equations.

Figure 6.2 shows the element with the two element boundary nodes and four representative interior nodes with the assigned deformation parameters, $\bar{u}_S$, $\bar{v}_S$, $\bar{w}$ and $\bar{\theta}_{\bar{z}}$, used to define various EDQ discretizations at the discrete points. In the numerical simulation, the nodes used to define the axial discretization, the nodes used to define the flexural discretizations and the nodes used to define the torsional discretization can be different.

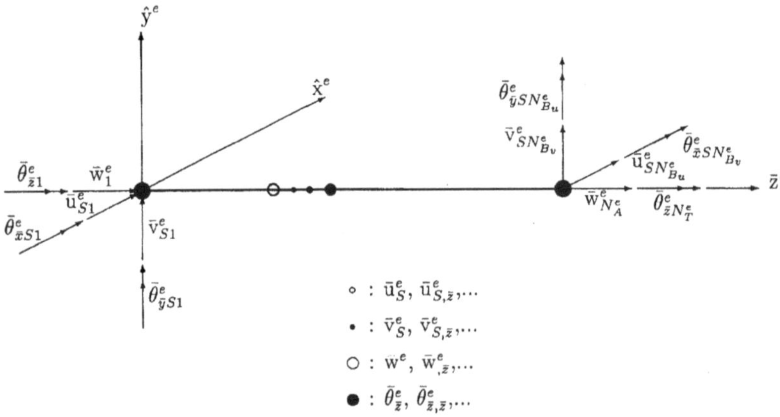

Fig. 6.2. The two element boundary nodes and four representative interior nodes with the assigned deformation parameters used to define various EDQ discretizations at the discrete points

Figure 6.3 shows the element with each of the four interior nodes a representative node for defining the element basis EDQ discretization of a derivative of a displacement parameter among $\bar{u}_S$, $\bar{v}_S$, $\bar{w}$ and $\bar{\theta}_{\bar{z}}$. The deformation parameters at an interior node used to define the EDQ discretization can be

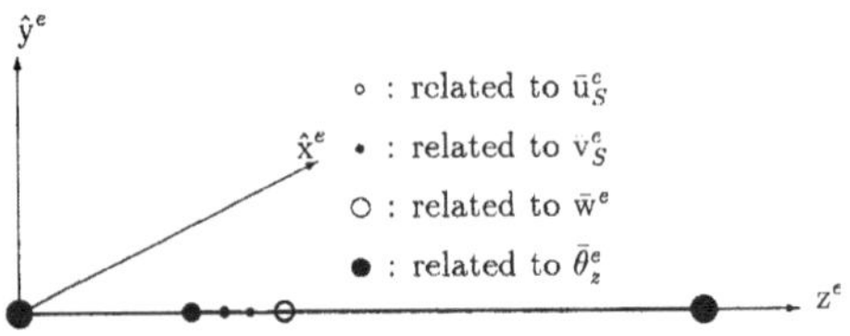

Fig. 6.3. Representative interior discrete points at which various discrete equilibrium equations are defined

the displacement parameter and/or it derivatives at that interior nodes. Two or more nodes among these four interior nodes can be located at the same place. Let N_A^e, $N_{B_u}^e$, $N_{B_v}^e$ and N_T^e denote the numbers of nodes for defining the axial discretization, the flexural discretization in $\bar{x}^e$ direction, the flexural discretization in $\bar{y}^e$ direction and the torsional discretization, respectively, $\bar{N}_A^e$, $\bar{N}_{B_u}^e$, $\bar{N}_{B_v}^e$ and $\bar{N}_T^e$ denote the corresponding element degrees of freedom, $\hat{N}_A^e$, $\hat{N}_{B_u}^e$, $\hat{N}_{B_v}^e$ and $\hat{N}_T^e$ denote the numbers of the corresponding interior discrete points for defining the related discrete equilibrium equations plus the two element boundary nodes, and $D_{\gamma k}^{e\zeta^m}$, $D_1^{e\zeta^m}$, $D_{\beta j}^{e\zeta^m}$ and $D_{\delta l}^{e\zeta^m}$ denote the corresponding weighting coefficients. Also let the origin of the local coordinate system be located at node 1 and the range of the natural coordinate ζ be $0. \leq \zeta \leq 1$. Then by using the EDQ, the discrete equation of the first of Eqs. (6.1) at a discrete point α in element e can be expressed as

$$\frac{E^e}{(l^e)^4}\left[\frac{d^2 I_{\bar{x}\bar{x}(\alpha)}^e}{d\zeta^2}\sum_{i=1}^{\bar{N}_{B_u}^e} D_{\alpha i}^{e\zeta^2} + 2\frac{dI_{\bar{x}\bar{x}(\alpha)}^e}{d\zeta}\sum_{i=1}^{\bar{N}_{B_u}^e} D_{\alpha i}^{e\zeta^3} + I_{\bar{x}\bar{x}(\alpha)}^e\sum_{i=1}^{\bar{N}_{B_u}^e} D_{\alpha i}^{e\zeta^4}\right]\tilde{u}_{Si}^e$$

$$= q_{\bar{x}\alpha}^e - \frac{1}{l^e}\frac{dm_{\bar{y}\alpha}^e}{d\zeta}, \quad \alpha = 2, ..., \hat{N}_{B_u}^e - 1 \tag{6.4}$$

The discrete equation of the second of Eqs. (6.1) at a discrete point β can be similarly expressed as

$$\frac{E^e}{(l^e)^4}\left[\frac{d^2 I_{\bar{y}\bar{y}(\beta)}^e}{d\zeta^2}\sum_{j=1}^{\bar{N}_{B_v}^e} D_{\beta j}^{e\zeta^2} + 2\frac{dI_{\bar{y}\bar{y}(\beta)}^e}{d\zeta}\sum_{j=1}^{\bar{N}_{B_v}^e} D_{\beta j}^{e\zeta^3} + I_{\bar{y}\bar{y}(\beta)}^e\sum_{j=1}^{\bar{N}_{B_v}^e} D_{\beta j}^{e\zeta^4}\right]\tilde{v}_{Sj}^e$$

$$= q_{\bar{y}\beta}^e + \frac{1}{l^e}\frac{dm_{\bar{x}\beta}^e}{d\zeta}, \quad \beta = 2, ..., \hat{N}_{B_v}^e - 1 \tag{6.5}$$

The discrete equation of the third of Eqs. (6.1) at a discrete point γ can be expressed as

$$\frac{E^e}{(l^e)^2}\left[\frac{dA_{(\gamma)}^e}{d\zeta}\sum_{k=1}^{\bar{N}_A^e} D_{\gamma k}^{e\zeta} + A_{(\gamma)}^e\sum_{k=1}^{\bar{N}_A^e} D_{(\gamma)k}^{e\zeta^2}\right]\tilde{w}_k^e = -p_\gamma^e, \quad \gamma = 2, ..., \hat{N}_A^e - 1 \tag{6.6}$$

And the discrete equation of the last of Eqs. (6.1) at a discrete point δ can be expressed as

$$\frac{G^e}{(l^e)^2}\left[\frac{dJ_{(\delta)}^e}{d\zeta}\sum_{l=1}^{\bar{N}_T^e} D_{\delta l}^{e\zeta} + J_{(\delta)}^e\sum_{l=1}^{\bar{N}_T^e} D_{\delta l}^{e\zeta^2}\right]\tilde{\theta}_{\bar{z}l}^e = -m_{\bar{z}\delta}^{(s)e}, \quad \delta = 2, ..., \hat{N}_T^e - 1 \tag{6.7}$$

In Eqs. (6.4) to (6.7), the derivatives of section constants, $m_{\bar{y}}^e$ and $m_{\bar{x}}^e$ at the related discrete points can also be calculated by the DQ. It is especially useful if the distribution function of a section constant, $m_{\bar{x}}^e$ or $m_{\bar{y}}^e$ is not continuously differentiable up to the order of its derivative. The values of section constants, $m^e_{\bar{x}}$ and $m^e_{\bar{y}}$ at the two element boundary nodes and certain interior discrete points are used to define the DQ discretizations of the derivatives of these quantities. Let $\bar{\phi}(\zeta)$ denote the distribution of these quantities. Then the DQ discretization for the mth order derivative of $\bar{\phi}$ at a discrete point α is expressed by

$$\frac{d^m \bar{\phi}^e_\alpha}{d\zeta^m} = \sum_{\bar{i}=1}^{\tilde{N}^e_D} \bar{D}^{e\zeta^m}_{\alpha\bar{i}} \bar{\Phi}^e_{\bar{i}} \tag{6.8}$$

where $\bar{D}^{e\zeta^m}_{\alpha\bar{i}}$ are weighting coefficients and $\tilde{N}^e_D$ is the number of points for defining the DQ discretization. Let $\tilde{N}^e_A$, $\tilde{N}^e_{I1}$, $\tilde{N}^e_{I2}$, $\tilde{N}^e_J$ and $\tilde{N}^e_{b1}$ denote the numbers of nodes for defining the discretizations of the derivatives of A^e, $I^e_{\bar{x}\bar{x}}$, $I^e_{\bar{y}\bar{y}}$, J^e, $m^e_{\bar{x}}$ and $m^e_{\bar{y}}$, respectively. Then the derivatives at the related discrete points can be expressed by the following DQ discretization equations

$$\begin{aligned}
&\frac{dA^e_\gamma}{d\zeta} = \sum_{\bar{k}=1}^{\tilde{N}^e_A} \bar{D}^\zeta_{\gamma\bar{k}} A^e_{\bar{k}}, \qquad \frac{dI^e_{\bar{x}\bar{x}\alpha}}{d\zeta} = \sum_{\bar{i}=1}^{\tilde{N}^e_{I1}} \bar{D}^\zeta_{\alpha\bar{i}} I^e_{\bar{x}\bar{x}\bar{i}}, \\
&\frac{d^2 I^e_{\bar{x}\bar{x}\alpha}}{d\zeta^2} = \sum_{\bar{i}=1}^{\tilde{N}^e_{I1}} \bar{D}^{\zeta^2}_{\alpha\bar{i}} I^e_{\bar{x}\bar{x}\bar{i}}, \qquad \frac{dI^e_{\bar{y}\bar{y}\beta}}{d\zeta} = \sum_{\bar{j}=1}^{\tilde{N}^e_{I2}} \bar{D}^\zeta_{\beta\bar{j}} I^e_{\bar{y}\bar{y}\bar{j}}, \\
&\frac{d^2 I^e_{\bar{y}\bar{y}\beta}}{d\zeta^2} = \sum_{\bar{j}=1}^{\tilde{N}^e_{I2}} \bar{D}^{\zeta^2}_{\beta\bar{j}} I^e_{\bar{y}\bar{y}\bar{j}}, \qquad \frac{dJ^e_\delta}{d\zeta} = \sum_{\bar{l}=1}^{\tilde{N}^e_{I3}} \bar{D}^\zeta_{\delta\bar{l}} J^e_{\bar{l}}, \\
&\frac{dm^e_{\bar{x}\beta}}{d\zeta} = \sum_{t=1}^{\tilde{N}^e_{b1}} \bar{D}^\zeta_{\beta\bar{t}} m^e_{\bar{x}\bar{t}}, \qquad \frac{dm^e_{\bar{y}\alpha}}{d\zeta} = \sum_{t=1}^{\tilde{N}^e_{b1}} \bar{D}^\zeta_{\alpha\bar{t}} m^e_{\bar{y}\bar{t}}
\end{aligned} \tag{6.9}$$

Using Eqs. (6.4) to (6.7), the discrete local element equilibrium equation represented by Eq. (5.4) can be constructed. In the discrete local element equilibrium equation, the local element stiffness coefficient matrix $[\bar{\kappa}^e]$ is a $(\hat{N}^e_{B_u} + \hat{N}^e_{B_v} + \hat{N}^e_A + \hat{N}^e_T - 8) \times (\bar{N}^e_{B_u} + \bar{N}^e_{B_v} + \bar{N}^e_A + \bar{N}^e_T)$ matrix with the values of elements in the matrix depending on geometrical constants, material constants and weighting coefficients. The local element displacement vector $\{\bar{\delta}^e\}$ is a $(\bar{N}^e_{B_u} + \bar{N}^e_{B_v} + \bar{N}^e_A + \bar{N}^e_T) \times 1$ vector expressed as

$$\begin{aligned}
\{\bar{\delta}^e\} = \lfloor\ & \bar{u}^e_{S1} \quad \bar{v}^e_{S1} \quad \bar{w}^e_1 \quad \bar{\theta}^e_{\bar{x}S2} \quad \bar{\theta}^e_{\bar{y}S2} \quad \bar{\theta}^e_{\bar{z}1} \quad \cdots \quad \bar{u}^e_{S(\bar{N}^e_{B_u}-1)} \\
& \bar{v}^e_{S(\bar{N}^e_{B_v}-1)} \quad \bar{w}^e_{\bar{N}^e_A} \quad \bar{\theta}^e_{\bar{x}S\bar{N}^e_{B_v}} \quad \bar{\theta}^e_{\bar{y}S\bar{N}^e_{B_u}} \quad \bar{\theta}^e_{\bar{z}\bar{N}^e_T}\ \rfloor^T
\end{aligned} \tag{6.10}$$

The local element distributed force vector $\{\bar{r}^e\}$ is a $(\hat{N}^e_{B_u} + \hat{N}^e_{B_v} + \hat{N}^e_A + \hat{N}^e_T - 8) \times 1$ vector expressed as

$$\begin{aligned}
\{\bar{r}^e\} = \lfloor\ & -p^e_2 \quad \bar{q}^e_{\bar{x}2} \quad \bar{q}^e_{\bar{y}2} \quad -m^{(s)e}_{\bar{z}2} \quad \cdots \quad -p^e_{\hat{N}^e_A-1} \\
& \bar{q}^e_{\bar{x}(\hat{N}^e_{B_u}-1)} \quad \bar{q}^e_{\bar{y}(\hat{N}^e_{B_v}-1)} \quad -m^{(s)e}_{\bar{z}(\hat{N}^e_T-1)}\ \rfloor^T
\end{aligned} \tag{6.11}$$

In Eq. (6.10), we see that each of the two element boundary nodes has six degrees of freedom. In Eq. (6.11), we see that the maximum number of discrete element equilibrium equations defined at an interior discrete point is four. It should also be noted that $\{\bar{\delta}^e\}$ twelve more elements than $\{\bar{r}\}$. In Eq. (6.10), $\bar{\theta}^e_{\bar{x}Si}$ and $\bar{\theta}^e_{\bar{y}Si}$ Represent the rotation angles $-\frac{d\bar{v}^e_S}{d\bar{z}^e} = -\frac{1}{l^e}\frac{d\bar{v}^e_S}{d\zeta}$ and $\frac{d\bar{u}^e_S}{d\bar{z}^e} = \frac{1}{l^e}\frac{d\bar{u}^e_S}{d\zeta}$, respectively, at node i. In Eq. (6.11), $\bar{q}^e_{\bar{x}i}$, $\bar{q}^e_{\bar{y}i}$ and $m^{(s)e}_{\bar{z}i}$ represent $q^e_{\bar{x}} - \frac{1}{l^e}\frac{dm^e_{\bar{y}}}{d\zeta}$, $q^e_{\bar{y}} + \frac{1}{l^e}\frac{dm^e_{\bar{x}}}{d\zeta}$ and $m^{(s)e}_{\bar{z}}$, respectively, at node i. If no axially distributed force is applied, interior discrete points can be neglected and only two discrete points of the two element boundary nodes are necessary which are used to define natural transition conditions or natural boundary conditions involving axial forces. Then no discrete axial equilibrium equation needs to be included in the discrete local element equilibrium equation (5.4). If the element is axially rigid, $\bar{w}^e_i$ in $\{\bar{\delta}^e\}$ and the related columns in $[\bar{\kappa}^e]$ can also be eliminated. Similarly, if no laterally distributed force in a certain direction is applied the discrete flexural equilibrium equations in that direction can be neglected. If the element is flexurally rigid in a certain direction the lateral displacements and their gradients in $\{\bar{\delta}^e\}$ and the related columns in $[\bar{\kappa}^e]$ can also be eliminated. Furthermore, if no distributed torque is applied the discrete torsional equilibrium equations can be neglected. If the element is torsional rigid, the angles of twist and the angles of twist per unit length in $\{\bar{\delta}^e\}$ and the related columns in $[\bar{\kappa}^e]$ can be eliminated.

6.2.2 Coordinate Transformations

All elements connected to a joint must have the same position, which is the joint node, to define the global displacements at that position and express the displacements of the element boundary nodes as the global displacements of that position. Thus the coordinate transformations have to be carried out. The local displacements $\bar{U}(\bar{x},\bar{y},\bar{z})$, $\bar{V}(\bar{x},\bar{y},\bar{z})$ and $\bar{W}(\bar{x},\bar{y},\bar{z})$ of a point $(\bar{x},\bar{y},\bar{z})$ on the cross section can be related to the local displacements of the centroid $\bar{u}(\bar{z})$, $\bar{v}(\bar{z})$, the average axial displacement $\bar{w}(\bar{z})$ and the local angle of twist $\bar{\theta}_{\bar{z}}(\bar{z})$ through the following equations

$$\bar{U}(\bar{x},\bar{y},\bar{z}) = \bar{u}(\bar{z}) - \bar{y}\bar{\theta}_{\bar{z}}(\bar{z}), \quad \bar{V}(\bar{x},\bar{y},\bar{z}) = \bar{v}(\bar{z}) + \bar{x}\bar{\theta}_{\bar{z}}(\bar{z}),$$
$$\bar{W}(\bar{x},\bar{y},\bar{z}) = \bar{w}(\bar{z}) - \bar{x}\bar{u}_{,\bar{z}}(\bar{z}) - \bar{y}\bar{v}_{,\bar{z}}(\bar{z}) \tag{6.12}$$

If the joint is rigid and has a finite size, the relations between the local displacements of the shear center and the local displacements of a point R which is the joint node used to define the transition conditions are necessary to be constructed. If the rigid joint does not have a finite size, the transition conditions can also be defined at a point which is not the shear center. Let O be a certain point on the cross section. Assuming that the deformation of torsion is negligible as compared to the deformation of bending, the relations between the local displacements $\bar{u}_S(\bar{z})$, $\bar{v}_S(\bar{z})$, $\bar{w}(\bar{z})$, $\bar{\theta}_{\bar{x}S}$, $\bar{\theta}_{\bar{y}S}$ and $\theta_{\bar{z}}$, and the displacements $\bar{u}_O$, $\bar{v}_O$, $\bar{w}_O$, $\bar{\theta}_{\bar{x}O}$, $\bar{\theta}_{\bar{y}O}$ and $\bar{\theta}_{\bar{z}O}$ of point O can be obtained by using Eqn. (6.12)

$$\{\bar{d}_O\} = [t_{SO}]\{\bar{d}_S\} \tag{6.13}$$

where

$$\{\bar{d}_O\} = \lfloor\ \bar{u}_O \quad \bar{v}_O \quad \bar{w}_O \quad \bar{\theta}_{\bar{x}O} \quad \bar{\theta}_{\bar{y}O} \quad \bar{\theta}_{\bar{z}O}\ \rfloor^T \tag{6.14}$$

is the local displacement vector of point O,

$$\{\bar{d}_S\} = \lfloor\ \bar{u}_S \quad \bar{v}_S \quad \bar{w} \quad \bar{\theta}_{\bar{x}S} \quad \bar{\theta}_{\bar{y}S} \quad \bar{\theta}_{\bar{z}}\ \rfloor^T \tag{6.15}$$

and

$$[t_{SO}] = \begin{bmatrix} 1 & 0 & 0 & 0 & 0 & -\bar{y}_{SO} \\ 0 & 1 & 0 & 0 & 0 & \bar{x}_{SO} \\ 0 & 0 & 1 & \bar{y}_O & -\bar{x}_O & 0 \\ 0 & 0 & 0 & 1 & 0 & 0 \\ 0 & 0 & 0 & 0 & 1 & 0 \\ 0 & 0 & 0 & 0 & 0 & 1 \end{bmatrix} \tag{6.16}$$

The relations between the local displacements of point O and the local displacements $\bar{u}_R$, $\bar{v}_R$, $\bar{w}_R$, $\bar{\theta}_{\bar{x}R}$, $\bar{\theta}_{\bar{y}R}$ and $\bar{\theta}_{\bar{z}R}$ of point R can also be defined by using the concept of kinematics of rigid body motion. Then the local displacements of O can be expressed by the local displacements of R through the following matrix equation

$$\{\bar{d}_O\} = [t_t]\,\{\bar{d}_R\} \tag{6.17}$$

where

$$\{\bar{d}_R\} = \lfloor\ \bar{u}_R \quad \bar{v}_R \quad \bar{w}_R \quad \bar{\theta}_{\bar{x}R} \quad \bar{\theta}_{\bar{y}R} \quad \bar{\theta}_{\bar{z}R}\ \rfloor^T \tag{6.18}$$

$$[t_{RO}] = \begin{bmatrix} 1 & 0 & 0 & 0 & -\bar{z}_{OR} & \bar{y}_{OR} \\ 0 & 1 & 0 & \bar{z}_{OR} & 0 & -\bar{x}_{OR} \\ 0 & 0 & 1 & -\bar{y}_{OR} & \bar{x}_{OR} & 0 \\ 0 & 0 & 0 & 1 & 0 & 0 \\ 0 & 0 & 0 & 0 & 1 & 0 \\ 0 & 0 & 0 & 0 & 0 & 1 \end{bmatrix} \tag{6.19}$$

Using Eqs. (6.13) and (6.17), the local displacements of S can be related to the local displacements of R through the following matrix equation

$$\{\bar{d}_S\} = [t_t]\{\bar{d}_R\} \tag{6.20}$$

where

$$[t_t] = [t_{SO}]^{-1}[t_{RO}] \tag{6.21}$$

is the translational transformation matrix. In $[t_{SO}]$ and $[t_{RO}]$, the relations $\bar{x}_{SO} = \bar{x}_O - \bar{x}_S$, $\bar{y}_{SO} = \bar{y}_O - \bar{y}_S$, $\bar{x}_{OR} = \bar{x}_R - \bar{x}_O$, $\bar{y}_{OR} = \bar{y}_R - \bar{y}_O$, and $\bar{z}_{OR} = \bar{z}_R - \bar{z}_O$ are used.

In the overall algebraic system, displacement components at joints must be in directions of global coordinates while the displacement components at the interior nodes can be in the directions of global or local coordinates. In the present numerical simulation, only the displacement components at joints are transformed to express them as the global joint displacements. Therefore the local element displacement vector $\{\bar{\delta}^e\}$ have to be transformed to obtain a local-global element displacement vector. The transformation of a vector can be accomplished through the use of the rotational transformation matrix represented by Eq. (5.9). The relations between the global displacements and the local displacements of the joint node R can be expressed by the following equation

$$[t_R]\{d_R\} = \{\bar{d}_R\} \tag{6.22}$$

where

$$[t_R] = \begin{bmatrix} [t_r^e] & \\ & [t_r^e] \end{bmatrix} \tag{6.23}$$

is the joint transformation matrix with $[t_r^e]$ the rotational transformation matrix defined by Eq. (5.9), and

$$\{d_R\} = \lfloor \; u_R \quad v_R \quad w_R \quad \theta_{xR} \quad \theta_{yR} \quad \theta_{zR} \; \rfloor^T \tag{6.24}$$

the global displacement vector of the joint node R. Using Eqs. (6.20) and (6.22), the transformation relation between the local displacement vector of the shear center and the global displacement vector of the joint node R can be expressed by the following equation

$$\{\bar{d}_S\} = [t^e]\{d_R\} \tag{6.25}$$

where

$$[t^e] = [t_t][t_R] \tag{6.26}$$

A transformation relation between the element displacement vector in global coordinate system and the element displacement vector in local coordinate system can be obtained through the use of $[t^e]$

$$[T^e]\{\delta^e\} = \{\bar{\delta}^e\} \tag{6.27}$$

where $\{\bar{\delta}^e\}$ is the local element displacement vector expressed by Eq. (6.10),

$$\begin{aligned}\{\delta^e\} = \lfloor \; & u^e_{R1} \quad v^e_{R1} \quad w^e_{R1} \quad \theta^e_{xR1} \quad \theta^e_{yR1} \quad \theta^e_{zR1} \quad \cdots \\ & u^e_{RN^e} \quad v^e_{RN^e} \quad w^e_{RN^e} \quad \theta^e_{xRN^e} \quad \theta^e_{yRN^e} \quad \theta^e_{zRN^e} \; \rfloor^T\end{aligned} \tag{6.28}$$

the local-global element displacement vector, and

$$[T^e] = \begin{bmatrix} [t^e] & & & & & & & \\ & 1 & & & & & & \\ & & 1 & & & & & \\ & & & \cdot & & & & \\ & & & & \cdot & & & \\ & & & & & \cdot & & \\ & & & & & & 1 & & \\ & & & & & & & 1 & \\ & & & & & & & & [t^e] \end{bmatrix} \tag{6.29}$$

the element transformation matrix. Employing Eq. (6.27) in Eq. (5.4), the discrete local-global element equilibrium equation represented by Eq. 5.14 can be obtained.

6.2.3 Discrete Element Internal Forces

The internal forces of a discrete element at an arbitrary point can be calculated. Figure 6.4 shows the internal forces at the two element boundary nodes and an interior point. Denote $\bar{F}^e_{\bar{z}}$ the axial force in the element. $\bar{F}^e_{\bar{z}}$ is expressed as $\bar{F}^e_{\bar{z}} = (EA)^e \frac{d\bar{w}^e}{d\bar{z}^e}$. Using the EDQ discretization, the axial force at an arbitrary point γ can be obtained:

$$\bar{F}^e_{\bar{z}\gamma} = \frac{E^e A^e_{(\gamma)}}{l^e} \sum_{k=1}^{\bar{N}^e_A} D^{e\zeta}_{\gamma k} \tilde{w}^e_k = \frac{E^e A^e_{(\gamma)}}{l^e} \lfloor D^{e\zeta}_{\bar{w}\gamma} \rfloor \{\bar{\delta}^e\} \tag{6.30}$$

where

$$\lfloor D^{e\zeta}_{\bar{w}_\gamma} \rfloor = \lfloor\, 0 \quad 0 \quad D^{e\zeta}_{\gamma 1} \quad \dots \quad D^{e\zeta}_{\gamma \bar{N}^e_A} \quad 0 \quad 0 \quad 0 \,\rfloor \tag{6.31}$$

Employing Eq. (6.28) in Eq. (6.31), $\bar{F}^e_{\bar{z}\gamma}$ can be related to the local-global element displacement vector, which is written as:

$$\bar{F}^e_{\bar{z}\gamma} = \frac{E^e A^e_{(\gamma)}}{l^e} \lfloor D^{e\zeta}_{\bar{w}\gamma} \rfloor [T^e] \{\delta^e\} \tag{6.32}$$

The distributions of bending moments in the element are

$$\bar{M}^e_{\bar{x}} = -E^e I^e_{\bar{y}\bar{y}} \frac{d^2 \bar{v}^e_S}{d(\bar{z}^e)^2}, \quad \bar{M}^e_{\bar{y}} = E^e I^e_{\bar{x}\bar{x}} \frac{d^2 \bar{u}^e_S}{d(\bar{z}^e)^2}$$

Using the EDQ discretization, bending moments $\bar{M}^e_{\bar{x}}$ at point β and $\bar{M}^e_{\bar{y}}$ at point α can be obtained:

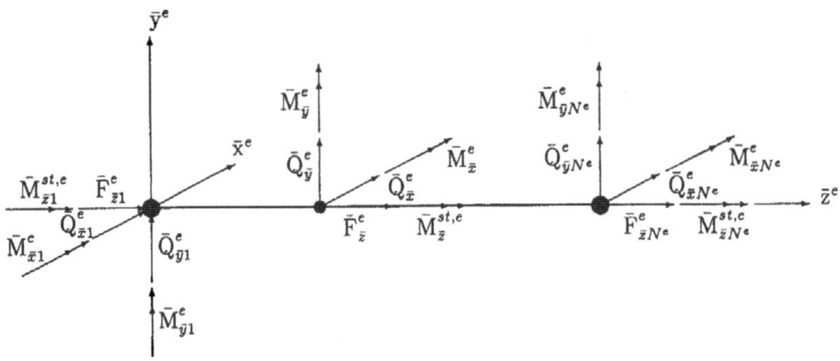

Fig. 6.4. Internal forces at the two element boundary sections and an interior cross section

$$\bar{M}^e_{\bar{x}\beta} = -\frac{E^e I^e_{\bar{y}\bar{y}(\beta)}}{(l^e)^2}\sum_{j=1}^{\bar{N}^e_{B_v}} D^{e\zeta^2}_{\beta j}\tilde{v}^e_{Sj} = -\frac{E^e I^e_{\bar{y}\bar{y}(\beta)}}{(l^e)^2}\lfloor D^{e\zeta^2}_{\bar{v}\beta}\rfloor\{\bar{\delta}^e\},$$

$$\bar{M}^e_{\bar{y}(\alpha)} = \frac{E^e I^e_{\bar{x}\bar{x}(\alpha)}}{(l^e)^2}\sum_{i=1}^{\bar{N}^e_{B_u}} D^{e\zeta^2}_{\alpha i}\tilde{u}^e_{Si} = \frac{E^e I^e_{\bar{x}\bar{x}(\alpha)}}{(l^e)^2}\lfloor D^{e\zeta^2}_{\bar{u}\alpha}\rfloor\{\bar{\delta}^e\} \tag{6.33}$$

where

$$\lfloor D^{e\zeta^2}_{\bar{u}\alpha}\rfloor = \lfloor\ D^{e\zeta^2}_{\alpha 1}\quad 0\quad 0\quad \ldots\quad 0\quad l^e D^{e\zeta^2}_{\alpha\bar{N}^e_{B_u}}\quad 0\ \rfloor,$$

$$\lfloor D^{e\zeta^2}_{\bar{v}\beta}\rfloor = \lfloor\ 0\quad D^{e\zeta^2}_{\beta 1}\quad 0\quad \ldots\quad -l^e D^{e\zeta^2}_{\beta\bar{N}^e_{B_v}}\quad 0\quad 0\ \rfloor \tag{6.34}$$

Employing Eq. (6.27) in Eqs. (6.33), $\bar{M}^e_{\bar{x}\beta}$ and $\bar{M}^e_{\bar{y}\alpha}$ can be related to the local-global element displacement vector, which are written as:

$$\bar{M}^e_{\bar{x}\beta} = -\frac{E^e I^e_{\bar{y}\bar{y}(\beta)}}{(l^e)^2}\lfloor D^{e\zeta^2}_{\bar{v}\beta}\rfloor[T^e]\{\delta^e\},\quad \bar{M}^e_{\bar{y}\alpha} = \frac{E^e I^e_{\bar{x}\bar{x}(\alpha)}}{(l^e)^2}\lfloor D^{e\zeta^2}_{\bar{u}\alpha}\rfloor[T^e]\{\delta^e\} \tag{6.35}$$

The distributions of internal shear forces in the element are

$$\bar{Q}^e_{\bar{x}} = -E^e\frac{d}{d\bar{z}^e}\left[I^e_{\bar{x}\bar{x}}\frac{d^2\bar{u}^e_S}{d(\bar{z}^e)^2}\right],\quad \bar{Q}^e_{\bar{y}} = -E^e\frac{d}{d\bar{z}^e}\left[I^e_{\bar{y}\bar{y}}\frac{d^2\bar{v}^e_S}{d(\bar{z}^e)^2}\right]$$

Using the EDQ discretization, shear forces $\bar{Q}^e_{\bar{x}}$ at point α and $\bar{Q}^e_{\bar{y}}$ at point β can be obtained:

$$\begin{aligned}\bar{Q}^e_{\bar{x}\alpha} &= -\frac{E^e}{(l^e)^3}\left[\frac{dI^e_{\bar{x}\bar{x}(\alpha)}}{d\zeta}\sum_{i=1}^{\bar{N}^e_{B_u}} D^{e\zeta^2}_{\alpha i} + I^e_{\bar{x}\bar{x}(\alpha)}\sum_{i=1}^{\bar{N}^e_{B_u}} D^{e\zeta^3}_{\alpha i}\right]\tilde{u}^e_{Si}\\ &= -\frac{E^e}{(l^e)^3}\left[\frac{dI^e_{\bar{x}\bar{x}(\alpha)}}{d\zeta}\lfloor D^{e\zeta^2}_{\bar{u}\alpha}\rfloor + I^e_{\bar{x}\bar{x}(\alpha)}\lfloor D^{e\zeta^3}_{\bar{u}\alpha}\rfloor\right]\{\bar{\delta}^e\},\end{aligned}$$

$$\begin{aligned}\bar{Q}^e_{\bar{y}\beta} &= -\frac{E^e}{(l^e)^3}\left[\frac{dI^e_{\bar{y}\bar{y}(\beta)}}{d\zeta}\sum_{j=1}^{\bar{N}^e_{B_v}} D^{e\zeta^2}_{\beta i} + I^e_{\bar{y}\bar{y}(\beta)}\sum_{j=1}^{\bar{N}^e_{B_v}} D^{e\zeta^3}_{\beta j}\right]\tilde{v}^e_{Sj}\\ &= -\frac{E^e}{(l^e)^3}\left[\frac{dI^e_{\bar{y}\bar{y}(\beta)}}{d\zeta}\lfloor D^{e\zeta^2}_{\bar{v}\beta}\rfloor + I^e_{\bar{y}\bar{y}(\beta)}\lfloor D^{e\zeta^3}_{\bar{v}\beta}\rfloor\right]\{\bar{\delta}^e\}\end{aligned} \tag{6.36}$$

where

$$\lfloor D^{e\zeta^3}_{\bar{u}\alpha}\rfloor = \lfloor\ D^{\zeta^3}_{\alpha 1}\quad 0\quad 0\quad \ldots\quad 0\quad l^e D^{e\zeta^3}_{\alpha\bar{N}^e_{B_u}}\quad 0\ \rfloor,$$

$$\lfloor D^{e\zeta^3}_{\bar{v}\beta}\rfloor = \lfloor\ 0\quad D^{\zeta^3}_{\beta 1}\quad 0\quad \ldots\quad -l^e D^{e\zeta^3}_{\beta\bar{N}^e_{B_v}}\quad 0\quad 0\ \rfloor \tag{6.37}$$

Employing Eq. (6.27) in Eqs. (6.37), $\bar{Q}^e_{\bar{x}\alpha}$ and $\bar{Q}^e_{\bar{y}\beta}$ can be related to the local-global element displacement vector, which are written as

$$\bar{Q}^e_{\bar{x}\alpha} = -\frac{E^e}{(l^e)^3}\left[\frac{dI^e_{\bar{x}\bar{x}(\alpha)}}{d\zeta}\lfloor D^{e\zeta^2}_{\bar{u}\alpha}\rfloor + I^e_{\bar{x}\bar{x}(\alpha)}\lfloor D^{e\zeta^3}_{\bar{u}\alpha}\rfloor\right][T^e]\{\delta^e\},$$

$$\bar{Q}^e_{\bar{y}\beta} = -\frac{E^e}{(l^e)^3}\left[\frac{dI^e_{\bar{y}\bar{y}(\beta)}}{d\zeta}\lfloor D^{e\zeta^2}_{\bar{v}\beta}\rfloor + I^e_{\bar{y}\bar{y}(\beta)}\lfloor D^{e\zeta^3}_{\bar{v}\beta}\rfloor\right][T^e]\{\delta^e\} \tag{6.38}$$

The distribution of Saint Venant torsion moment $\bar{M}^{st,e}_{\bar{z}}$ in the element is $\bar{M}^{st,e}_{\bar{z}} = (GJ)^e\frac{d\bar{\theta}^e_{\bar{z}}}{d\bar{z}^e}$. Using the EDQ discretization, $\bar{M}^{st,e}_{\bar{z}}$ at point δ can be obtained

$$\bar{M}^{st,e}_{\bar{z}\delta} = \frac{G^e J^e_{(\delta)}}{l^e}\sum_{l=1}^{\bar{N}^e_T} D^{e\zeta}_{\delta l}\tilde{\theta}^e_{\bar{z}l} = \frac{G^e J^e_{(\delta)}}{l^e}\lfloor D^{e\zeta}_{\bar{\theta}_{\bar{z}}\delta}\rfloor\{\bar{\delta}^e\} \tag{6.39}$$

where

$$\lfloor D^{e\zeta}_{\bar{\theta}_{\bar{z}}\delta}\rfloor = \lfloor\, 0 \quad 0 \quad 0 \quad 0 \quad 0 \quad D^{e\zeta}_{\alpha 1} \quad \ldots \quad 0 \quad 0 \quad D^{e\zeta}_{\alpha\bar{N}^e_T}\,\rfloor \tag{6.40}$$

6.3 Discrete Condition Equations of Joints

The DQEM requires that all condition equations at joints are satisfied. The condition equations include compatibility conditions, conformability conditions, and equilibriums of external and internal forces at joints. The condition equations have to be expressed as discrete forms using EDQ.

6.3.1 Discrete Joint Compatibility Conditions

Let M^j denote the number of elements connected to joint j. Also let I^{m^j} denote the element node number of the m^jth element connected to the joint. Then I^{m^j} is equal to 1 or the other element boundary node N^{m^j} of the m^jth element. Then the compatibility conditions, which are kinematic transition conditions, of joint j can be expressed as follows:

$$\left\{d^{1^j}_{I^{1^j}}\right\} = \left\{d^{2^j}_{I^{2^j}}\right\} =\cdots \left\{d^{m^j}_{I^{m^j}}\right\} =\cdots \left\{d^{M^j}_{I^{M^j}}\right\} = \{d^j\} \tag{6.41}$$

where $\{d^{m^j}_{I^{m^j}}\}$ and $\{d^j\}$ represent the globally nodal displacement vector of node I^{m^j} of the m^jth element and the global joint displacement vector of joint j, respectively.

6.3.2 Discrete Joint Equilibrium Conditions

For the DQEM frame analysis model, the equilibrium conditions of external and internal forces at joints also have to be satisfied. Each equilibrium condition is either a natural transition condition or a natural boundary condition. Let $\bar{V}^{m^j}_{\bar{x}} = \bar{Q}^{m^j}_{\bar{x}} - m^{m^j}_{\bar{y}}$ and $\bar{V}^{m^j}_{\bar{y}} = \bar{Q}^{m^j}_{\bar{y}} + m^{m^j}_{\bar{x}}$ denote the two lateral forces,

$\bar{F}_{\bar{z}}^{m^j}$ denote the axial force, $\bar{M}_{\bar{x}}^{m^j}$ and $\bar{M}_{\bar{y}}^{m^j}$ denote the two bending moments, and $\tilde{M}_{\bar{z}}^{m^j}$ denote the twisting moment, of the m^jth element at joint j. Also let P_x^j, P_y^j and P_z^j denote the three concentrated forces, and M_x^j, M_y^j and M_z^j denote the three concentrated moments, applied at the joint. Using the inverse transformation of Eq. (5.8) the force vectors formed by the nodal forces at joint j for all elements connected to the joint can be transformed into the global coordinate system. Let $\{V^{m^j}\} = \lfloor V_x^{m^j}\ V_y^{m^j}\ F_z^{m^j}\ M_x^{m^j}\ M_y^{m^j}\ M_z^{m^j} \rfloor^T$ denote the globally nodal force vector. Then the equilibrium conditions of joint j can be expressed as the following matrix equation:

$$\sum_{m^j=1}^{M^j} \nu^{m^j}\{V^{m^j}\} = \{P^j\} \tag{6.42}$$

where $\{P^j\} = \lfloor P_x^j\ P_y^j\ P_z^j\ M_x^j\ M_y^j\ M_z^j \rfloor^T$ and the value of ν^{m^j} is expressed by Eq. (5.21). The globally nodal force vector can be related to locally internal force vector by the following equation

$$\{V^{m^j}\} = [t^{m^j}]^T\left(\{\bar{Q}^{m^j}\} - \{\bar{P}^{m^j}\}\right) \tag{6.43}$$

where $\{\bar{Q}^{m^j}\} = \lfloor \bar{Q}_{\bar{x}}^{m^j}\ \bar{Q}_{\bar{y}}^{m^j}\ \bar{F}_{\bar{z}}^{m^j}\ \bar{M}_{\bar{x}}^{m^j}\ \bar{M}_{\bar{y}}^{m^j}\ \bar{M}_{\bar{z}}^{m^j} \rfloor^T$ is the locally element internal force vector at the joint, and $\{\bar{P}^{m^j}\} = \lfloor m_{\bar{y}}^{m^j}\ -m_{\bar{x}}^{m^j}\ 0\ 0\ 0\ 0 \rfloor^T$. Using Eqs. (6.32), (6.35), (6.38) and (6.43), Eq. (42) can be written as

$$\sum_{m^j=1}^{M^j} \nu^{m^j}[t^{m^j}]^T\left([S^{m^j}][T^{m^j}]\{\delta^{m^j}\} - \{\bar{P}^{m^j}\}\right) = \{P^j\} \tag{6.44}$$

where $[S^{m^j}]$ is a matrix having six rows expressed by

$$\lfloor S_1^{m^j} \rfloor = -\frac{E^{m^j}}{(l^{m^j})^3}\left(\frac{dI_{\bar{x}\bar{x}I^{m^j}}^{m^j}}{d\zeta}\lfloor D_{\bar{u}I^{m^j}}^{m^j\zeta^2} \rfloor + I_{\bar{x}\bar{x}I^{m^j}}^{m^j}\lfloor D_{\bar{u}I^{m^j}}^{m^j\zeta^3} \rfloor\right),$$

$$\lfloor S_2^{m^j} \rfloor = -\frac{E^{m^j}}{(l^{m^j})^3}\left(\frac{dI_{\bar{y}\bar{y}I^{m^j}}^{m^j}}{d\zeta}\lfloor D_{\bar{v}I^{m^j}}^{m^j\zeta^2} \rfloor + I_{\bar{y}\bar{y}I^{m^j}}^{m^j}\lfloor D_{\bar{v}I^{m^j}}^{m^j\zeta^3} \rfloor\right),$$

$$\lfloor S_3^{m^j} \rfloor = \frac{E^{m^j}A_{I^{m^j}}^{m^j}}{l^{m^j}}\lfloor D_{\bar{w}I^{m^j}}^{m^j\zeta} \rfloor,\quad \lfloor S_4^{m^j} \rfloor = -\frac{E^{m^j}I_{\bar{y}\bar{y}I^{m^j}}^{m^j}}{(l^{m^j})^2}\lfloor D_{\bar{v}I^{m^j}}^{m^j\zeta^2} \rfloor,$$

$$\lfloor S_5^{m^j} \rfloor = \frac{E^{m^j}I_{\bar{x}\bar{x}I^{m^j}}^{m^j}}{(l^{m^j})^2}\lfloor D_{\bar{u}I^{m^j}}^{m^j\zeta^2} \rfloor,\quad \lfloor S_6^{m^j} \rfloor = \frac{G^{m^j}J_{I^{m^j}}^{m^j}}{l^{m^j}}\lfloor D_{\bar{\theta}_{\bar{z}}I^{m^j}}^{m^j\zeta} \rfloor \tag{6.45}$$

and

$$\{\bar{P}^j\} = \lfloor\ m_{\bar{y}I^{m^j}}\quad -m_{\bar{x}I^{m^j}}\quad 0\quad 0\quad 0\quad 0\ \rfloor^T$$

6.3.3 Prescribed Joint Displacement or Rotation

A joint might have one or more prescribed displacement components. Let $\{\hat{d}_p^j\}$ and $\{d_p^j\}$ denote the vector formed by the prescribed displacement components and the vector formed by the corresponding displacement components of the joint, respectively. The condition equations of prescribed displacements can be obtained from the following vector equation: $\{d_p^j\} = \{\hat{d}_p^j\}$.

6.3.4 Inclined Roller

There are six kinematic condition equations. Among the six condition equations of the inclined roller, the three translational condition equations need to be transformed while the three rotational conditions are the same as those of a joint. The transformation procedure has been stated in Chapter 4 regarding the analysis of truss structures. For this frame analysis model, the kinematic condition equations can be expressed by the following matrix equation

$$\left[\dot{\tilde{t}}_r\right]\{d^j\} = \{\tilde{d}^j\} \tag{6.46}$$

where

$$\left[\dot{\tilde{t}}_r\right] = \begin{bmatrix} [\tilde{t}_r] & [0] \\ [0] & [I] \end{bmatrix} \tag{6.47}$$

is the transformation matrix, $[0]$ is a 3×3 zero matrix, $[I]$ is a 3×3 unit matrix, and $\{\tilde{d}^j\} = \lfloor \tilde{u}\ \tilde{v}\ \tilde{w}\ \bar{\theta}_x\ \bar{\theta}_y\ \bar{\theta}_z \rfloor^T$ is the prescribed displacement vector with the three translational components in $\tilde{x}$, $\tilde{y}$ and $\tilde{z}$ coordinate directions, and the three rotational components in x, y and z coordinate directions.

There are five natural conditions which can be represented by the following equation

$$\left[\ddot{\tilde{t}}_r\right] \sum_{m^j=1}^{M^j} \nu^{m^j} \{V^{m^j}\} = \{\tilde{P}^j\} \tag{6.48}$$

where

$$\left[\ddot{\tilde{t}}_r\right] = \begin{bmatrix} [\tilde{t}_{r12}] & [0] \\ [0]^T & [I] \end{bmatrix} \tag{6.49}$$

is a 5×5 transformation matrix with $[\tilde{t}_{r12}]^T$ the 2×3 matrix used in Eqs. (5.26) and (5.27), $[0]$ a 2×3 zero matrix and $[I]$ a 3×3 unit matrix, and $\{\tilde{P}^j\} = \lfloor \tilde{P}^j_{\tilde{x}}\ \tilde{P}^j_{\tilde{y}}\ \bar{M}_x\ \bar{M}_y\ \bar{M}_z \rfloor^T$. The five natural conditions can be expressed by the element nodal displacement vectors at the joint on the inclined surface

$$\left[\ddot{\tilde{t}}_r\right] \sum_{m^j=1}^{M^j} \nu^{m^j} [t^{m^j}]^T \left([S^{m^j}][T^{m^j}]\{\delta^{m^j}\} - \{\bar{P}^{m^j}\}\right) = \{\tilde{P}^j\} \tag{6.50}$$

6.4 Assemblage

With the discrete joint compatibility conditions in mind, then by assembling all discrete local-global element equilibrium equations represented by Eq. (5.14) for elements having interior discrete points, joint equilibrium condition equations (6.44) and (6.50), for the general joints and inclined rollers, respectively, and the prescribed joint displacement conditions, the overall algebraic system represented by Eq. (3.18) can be obtained. Like FEM, the assemblage is based on an element by element procedure. The discrete local-global element equilibrium equations (5.14) for element having interior discrete points, and discrete element boundary forces existing in Eqs. (6.44) and (6.50) defined at the two element boundary nodes and expressed by displacements, for all elements, are directly assembled to the overall discrete equation system.

Another approach can be used to assemble all discrete fundamental relations. This approach includes the twelve discrete equations for defining the internal element boundary forces in the local element stiffness equation to form another matrix equation. In this new matrix equation, each of the first six component equations represents one equation for defining an internal force corresponding to each individual degree of freedom assigned to node 1 while each of the last six component equations represents one equation for defining an internal force corresponding to each individual degree of freedom assigned to the other element boundary node. The component equations of the discrete local element equilibrium equation (5.14), for an element having interior discrete points, are placed at the remaining rows. This matrix equation is represented by Eq. (5.29). In Eq. (5.29), $[\hat{\kappa}^e]$ is a $(\bar{N}^e_{B_u} + \bar{N}^e_{B_v} + \bar{N}^e_T + \bar{N}^e_A) \times (\bar{N}^e_{B_u} + \bar{N}^e_{B_v} + \bar{N}^e_T + \bar{N}^e_A)$ matrix with the first and last six rows filled by coefficients for calculating the twelve element boundary forces,

$$\begin{aligned}\{\hat{r}^e\} = \lfloor & -(\bar{V}^e_{\bar{x}1} + m_{\bar{y}1}) \quad -(\bar{V}^e_{\bar{y}1} - m_{\bar{x}1}) \quad -\bar{F}^e_{\bar{z}1} \quad -\bar{M}^e_{\bar{x}1} \quad -\bar{M}^e_{\bar{y}1} \\ & -\bar{M}^e_{\bar{z}1} \quad \dots \quad \bar{V}^e_{\bar{x}N^e} + m_{\bar{y}N^e} \quad \bar{V}^e_{\bar{y}N^e} - m_{\bar{x}N^e} \\ & \bar{F}^e_{\bar{z}N^e} \quad \bar{M}^e_{\bar{x}N^e} \quad \bar{M}^e_{\bar{y}N^e} \quad \bar{M}^e_{\bar{z}N^e} \rfloor^T \end{aligned} \tag{6.51}$$

is the local element load vector. Using the element transformation matrix $[T^e]$ represented by Eq. (6.29) in Eq. (5.29), the local-global element stiffness equation can be obtained. The transformed local-global element load vector, $\{\tilde{r}^e\}$, represented by Eq. (5.33) are defined on the local-global coordinate system. Then, the overall algebraic equation system represented by Eq. (3.18) can be obtained by assembling the transformed local-global element stiffness equation, represented by Eq. (5.29), for all elements and by considering the joint equilibrium equations, (6.42) and (6.48) for the general joints and inclined rollers, respectively.

6.5 Problems

For the analyses, the axial discretization adopts a two-noded element represented by a Lagrange DQ model. The first problem solved involves the analysis of a two-dimensional frame structure subjected to uniformly distributed and concentrated loads, and a support settlement. The structure is shown in Fig. 6.5 in which joint A is a roller support. In the DQEM analysis, 10 elements are used to represent the 10 beam members which form the frame structure. Elements 3 and 4 which are subjected to uniformly distributed lateral loads are three-noded elements with the interior middle node having one degree of freedom representing the locally lateral displacement. These two elements use the $C^1 - C^0 - C^1$ EDQ model generated by using the equivalent Lagrange DQ models with equally spaced nodes and with the two auxiliary nodes inside the physical EDQ model. A discrete lateral equilibrium equation is defined at this node. The other eight elements are two-noded elements adopting the Hermite EDQ model. In carrying out the element basis discretization, the order of EDQ discretization adopted is based on the external cause applied on the element. Consequently, different elements might adopt different orders of EDQ approximation. By using this adaptive DQEM discretization, exact results can be obtained. Certain numerical results are listed in Table 6.1.

The second problem solved involves a 3-D frame structure subjected to distributed loads which is shown in Fig. 6.6. All three members are circular bars. Three elements are used to model the three members with each element boundary node having three translational degrees of freedom and three rotational degrees of freedom. Element AB has five equally spaced nodes

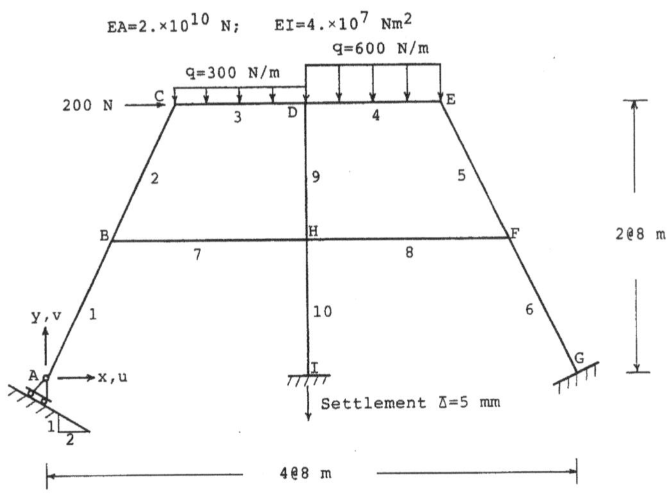

Fig. 6.5. A two-dimensional frame structure subjected to uniformly distributed and concentrated loads, and a support settlement

Table 6.1. The results of a two-dimensional frame structure subjected to uniformly distributed and concentrated loads, and a support settlement

Location		C	E
Displ. (mm)	u	$-.464238\times10^{-3}$	$-.467829\times10^{-3}$
	(Exact sol.)	$(-.464238\times10^{-3})$	$(-.467829\times10^{-3})$
	v	$.226480\times10^{-3}$	$-.240607\times10^{-3}$
	(Exact sol.)	$(.226480\times10^{-3})$	$(-.240607\times10^{-3})$
Location		I	G
Moment (N·m)	$\bar{M}_{\bar{y}}$	$.109176\times10^{4}$	$.453136\times10^{3}$
	(Exact sol.)	$(.109176\times10^{4})$	$(.453136\times10^{3})$
Shear (N)	$\bar{Q}_{\bar{x}}$	$.234755\times10^{3}$	$.120953\times10^{3}$
	(Exact sol.)	$(.234755\times10^{3})$	$(.120953\times10^{3})$

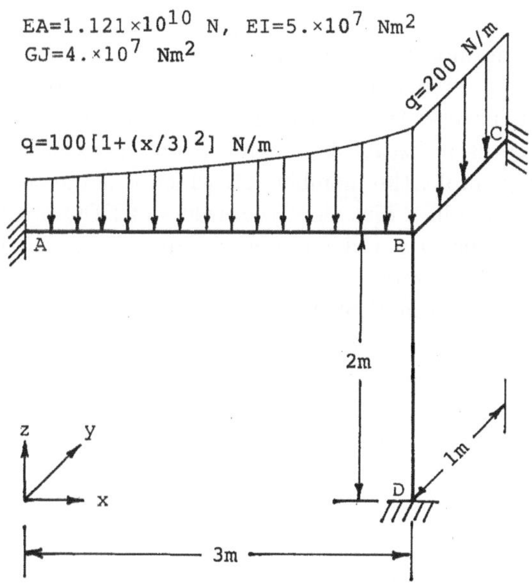

Fig. 6.6. A $3-D$ frame structure subjected to distributed loads

while in addition to the two element boundary nodes element BC has one center node. Each interior node has one degree of freedom used to represent the displacement in z direction and define a discrete flexural equilibrium equation in that direction. These two elements uses the $C^1-C^0-C^1$ EDQ model generated by using the equivalent Lagrange DQ models with equally spaced nodes and with the two auxiliary nodes inside the physical EDQ model.

Table 6.2. The results of a 3-D frame structure subjected to distributed loads

Location		B	Exact sol.
Displ. (mm)	u	$-.923014\times10^{-5}$	$-.923014\times10^{-5}$
	v	$-.630199\times10^{-6}$	$-.630199\times10^{-6}$
	w	$-.459265\times10^{-4}$	$-.459265\times10^{-4}$
Rot. angle (rad)	θ_x	$-.906850\times10^{-7}$	$-.906850\times10^{-7}$
	θ_y	$-.504007\times10^{-7}$	$-.504007\times10^{-7}$
	θ_z	$-.973273\times10^{-8}$	$-.973273\times10^{-8}$
Location		**D**	**Exact sol.**
Moment (N·m)	$\bar{M}_{\bar{x}}$	$.911577\times10^{1}$	$.911577\times10^{1}$
	$\bar{M}_{\bar{y}}$	$.434781\times10^{0}$	$.434781\times10^{0}$
	$\bar{M}_{\bar{z}}$	$-.194655\times10^{0}$	$-.194655\times10^{0}$

Element BD has no interior node and adopts the Hermite EDQ model with two grid nodes. No discrete equilibrium equation needs to be defined in this element. The EDQ discretization also adopts the optimum approach with which the order of EDQ discretization is dependent on the external cause applied. Considering, for example, the member AB which is subjected to a quadratically distributed force in z direction, the orders of EDQ approximations for the flexural deflections in z and y directions are six and three, respectively. By using this DQEM discretization, exact results can be obtained. Certain numerical data are listed in Table 6.2.

7

DQEM Analysis of Vibration of Frames Considering Warping Torsion

This DQEM vibration analysis model considers the warping torsion of frame members. The element basis discretization uses various EDQ models which can automatically set compatibility and conformability conditions at the joints. The discrete local element eigenvalue equations and element boundary forces at the two element boundary nodes can be transformed using various transformation matrices. Only modal displacements at the two element boundary nodes are transformed into the global coordinate system. The equilibriums of dynamic forces at joints are considered in constructing the overall eigenvalue equation. The discrete local element eigenvalue equations are defined on the local coordinates. All discrete equations are directly assembled into the overall eigenvalue equation.

7.1 Fundamental Relations

Assume that the material is isotropic and homogeneous. A representative cross section and the local coordinates are shown in Fig. 6.1. For the static deflection analysis model introduced in Chapter 6, the effect of warping torsion is neglected. In the vibration analysis model, the effect of warping torsion is considered [103]. Let $\hat{\omega}(\bar{x}, \bar{y})$ denote the warping function defined on the cross section. The warping function can be defined by using Saint Venant's torsion theory. If the beam is thin-walled, the Leibnitz sectorial formula can also be used to define the warping function. Let $I_{\hat{\omega}}$ denote the first moment of sectorial area. Also let $\bar{\omega}$ denote the normalized warping function. Then $\bar{\omega}$ can be calculated by the following equation using the cross section area A and $\hat{\omega}$: $\bar{\omega} = \hat{\omega} - I_{\hat{\omega}}/A$.

Let $I_{\bar{\omega}\bar{x}}$, $I_{\bar{\omega}\bar{x}}$, J and $I_{\bar{\omega}\bar{\omega}}$ denote the two warping cross products, Saint Venant's torsional constant and warping torsional constant with respect to the centroid of the cross section, respectively. These section constants can be defined by the following relations:

$$I_{\bar{\omega}\bar{x}} = \iint_A \bar{\omega}\bar{x}d\bar{x}d\bar{y}, \quad I_{\bar{\omega}\bar{y}} = \iint_A \bar{\omega}\bar{y}d\bar{x}d\bar{y},$$

$$J = \iint_A \left[\left(\frac{\partial \bar{\omega}}{\partial \bar{x}} - \bar{y} \right)^2 + \left(\frac{\partial \bar{\omega}}{\partial \bar{y}} + \bar{x} \right)^2 \right] d\bar{x}d\bar{y}, \quad I_{\bar{\omega}\bar{\omega}} = \iint_A \bar{\omega}^2 d\bar{x}d\bar{y} \tag{7.1}$$

The coordinates, $(\bar{x}_S, \bar{y}_S)$ of the shear center can be defined by the following relations

$$\bar{x}_S = -\frac{I_{\bar{\omega}\bar{y}}}{I_{yy}}, \quad \bar{y}_S = \frac{I_{\bar{\omega}\bar{x}}}{I_{xx}} \tag{7.2}$$

The polar moment of inertia I_s and warping torsional constant $I_{\bar{\omega}\bar{\omega}}^{(s)}$ with respect to the shear center can be defined by the following relations

$$I_s = I_{xx} + I_{yy} + A(x_S^2 + y_S^2), \quad I_{\bar{\omega}\bar{\omega}}^{(s)} = I_{\bar{\omega}\bar{\omega}} - \bar{y}_S^2 I_{\bar{x}\bar{x}} - \bar{x}_S^2 I_{\bar{y}\bar{y}} \tag{7.3}$$

Let $\bar{U}_S$ and $\bar{V}_S$ denote the two lateral modal displacements of the shear center, $\bar{W}$ denote the average axial modal displacement and $\bar{\Theta}_{\bar{z}}$ denote the modal angle of twist of the cross section, with respect to the local coordinate system. Also let ω denote the natural frequency. For a nonprismatic beam of length l, the dynamic equilibrium equations considering elastic restoring forces and inertia forces can be expressed by the following differential eigenvalue equations:

$$E\frac{d^2}{d\bar{z}^2}\left(I_{\bar{x}\bar{x}}\frac{d^2\bar{U}_S}{d\bar{z}^2}\right) - \rho A\omega^2\bar{U}_S + \rho\omega^2\frac{d}{d\bar{z}}\left(I_{\bar{x}\bar{x}}\frac{d\bar{U}_S}{d\bar{z}}\right) - \bar{y}_S\rho A\omega^2\bar{\Theta}_{\bar{z}} = 0,$$

$$E\frac{d^2}{d\bar{z}^2}\left(I_{\bar{y}\bar{y}}\frac{d^2\bar{V}_S}{d\bar{z}^2}\right) - \rho A\omega^2\bar{V}_S + \rho\omega^2\frac{d}{d\bar{z}}\left(I_{\bar{y}\bar{y}}\frac{d\bar{V}_S}{d\bar{z}}\right) + \bar{x}_S\rho A\omega^2\bar{\Theta}_{\bar{z}} = 0,$$

$$-E\frac{d}{d\bar{z}}\left(A\frac{d\bar{W}}{d\bar{z}}\right) - \rho\omega^2 A\bar{W} = 0,$$

$$-\bar{y}_S\rho A\omega^2\bar{U}_S + E\frac{d^2}{d\bar{z}^2}\left(I_{\bar{\omega}\bar{\omega}}^{(s)}\frac{d^2\bar{\Theta}_{\bar{z}}}{d\bar{z}^2}\right) + \bar{x}_S\rho A\omega^2\bar{V}_S - G\frac{d}{d\bar{z}}\left(J\frac{d\bar{\Theta}_{\bar{z}}}{d\bar{z}}\right)$$

$$-\rho\omega^2 I_s\bar{\Theta}_{\bar{z}} + \rho\omega^2\frac{d}{d\bar{z}}\left(I_{\bar{\omega}\bar{\omega}}^{(s)}\frac{d^3\bar{\Theta}_{\bar{z}}}{d\bar{z}}\right) = 0 \tag{7.4}$$

Assume that no rigid body is attached to the ends, the natural boundary conditions are:

$$EI_{\bar{x}\bar{x}}\frac{d^2\bar{U}_S}{d\bar{z}^2} = 0, \quad E\frac{d}{d\bar{z}}\left(I_{\bar{x}\bar{x}}\frac{d^2\bar{U}_S}{d\bar{z}^2}\right) + \rho I_{\bar{x}\bar{x}}\omega^2\frac{\bar{U}_S}{d\bar{z}} = 0, \quad EI_{\bar{y}\bar{y}}\frac{d^2\bar{V}_S}{d\bar{z}^2} = 0,$$

$$E\frac{d}{d\bar{z}}\left(I_{\bar{y}\bar{y}}\frac{d^2\bar{V}_S}{d\bar{z}^2}\right) + \rho I_{\bar{y}\bar{y}}\omega^2\frac{d\bar{V}_S}{d\bar{z}} = 0, \quad EA\frac{d\bar{W}}{d\bar{z}} = 0, \quad EI_{\bar{\omega}\bar{\omega}}^{(s)}\frac{d^2\bar{\Theta}_{\bar{z}}}{d\bar{z}^2} = 0,$$

$$-E\frac{d}{d\bar{z}}\left(I^{(s)}_{\bar{\omega}\bar{\omega}}\frac{d^2\bar{\Theta}_{\bar{z}}}{d\bar{z}^2}\right)+GJ\frac{d\bar{\Theta}_{\bar{z}}}{d\bar{z}}-\rho I^{(s)}_{\bar{\omega}\bar{\omega}}\omega^2\frac{d\bar{\Theta}_{\bar{z}}}{d\bar{z}}=0 \tag{7.5}$$

The kinematic boundary conditions are:

$$\frac{d\bar{U}_S}{d\bar{z}}=\hat{\Theta}_{\bar{y}S},\quad \bar{U}_S=\hat{U}_S,\quad -\frac{d\bar{V}_S}{d\bar{z}}=\hat{\Theta}_{\bar{x}S},\quad \bar{V}_S=\hat{V}_S,\quad \bar{W}=\hat{W},$$

$$\frac{d\bar{\Theta}_{\bar{z}}}{d\bar{z}}=\hat{\beta}_{\bar{z}},\quad \bar{\Theta}_{\bar{z}}=\hat{\Theta}_{\bar{z}} \tag{7.6}$$

on the kinematic boundary, where $\hat{\Theta}_{\bar{y}S}$, $\hat{U}_S$, $\hat{\Theta}_{\bar{x}S}$, $\hat{V}_S$, $\hat{W}$, $\hat{\beta}_{\bar{z}}$ and $\hat{\Theta}_{\bar{z}}$ are prescribed values.

7.2 Discrete Element Equations

7.2.1 Discrete Element Eigenvalue Equations

The element degrees of freedom for defining the axial, flexural and torsional discretizations must equal the corresponding element basis discrete fundamental relations for constructing the overall discrete equation system. In addition to the degrees of freedom for representing modal displacement parameters, the degrees of freedom for representing derivatives of a local component of modal displacement with respect to $\bar{z}^e$ at an element boundary node can also be assigned to that element boundary node. The selection of derivatives can be flexible. In order to automatically set the kinematic transition conditions and kinematic boundary conditions by only using the degrees of freedom assigned to the element boundary nodes, the degrees of freedom representing the two first order derivatives of flexural deflections must be assigned to the element boundary nodes. In the present DQEM frame vibration analysis model, only the degrees of freedom which are necessary for automatically setting the kinematic transition conditions and kinematic boundary conditions are assigned to the element boundary nodes.

Since the highest order of the spacial derivatives of axial modal displacement existing in the third one of Eqs. (7.4) is two, without using a specific technique to calculate the respective two equivalent nodal inertia forces at the two element boundary nodes, and include them into the related natural transition conditions or natural boundary conditions, the order of approximate axial modal displacement must at least be two and the element must at least have one discrete point for defining the discrete element eigenvalue equations of axial vibration. The DQEM linear element is equivalent to the FEM linear element. Since the highest order of the spacial derivatives of the two lateral modal displacements and the modal angle of twist is four, without using a specific technique to calculate the respective four equivalent nodal inertia forces at the two element boundary nodes, and include them into the related natural transition conditions or natural boundary conditions, the orders of approximate lateral modal displacements and modal angle of twist must at least be four and each of the related three differential eigenvalue equations needs at least one discrete point for defining one of its discrete element eigenvalue equation. The DQEM cubic element is equivalent to the FEM Hermite cubic element. The discrete points for defining the discrete element eigenvalue equations can be either in the interior of the element or on the element boundary.

For the present DQEM vibration analysis model of frame structures, only interior discrete points are used to define the discrete element eigenvalue equations. Figure 7.1 shows the element with the two element boundary nodes and four representative interior nodes with the assigned modal displacement parameters, $\bar{U}_S$, $\bar{V}_S$, $\bar{W}$ and $\bar{\Theta}_{\bar{z}}$, used to define various EDQ discretizations at the discrete points.

Figure 7.2 shows the element with each of the four interior nodes a representative node for defining the element basis EDQ discretization of a derivative of a modal displacement among $\bar{U}_S$, $\bar{V}_S$, $\bar{W}$ and $\bar{\Theta}_{\bar{z}}$. Let the origin of the local coordinate system be located at node 1 and the range of the natural coordinate ζ be $0. \leq \zeta \leq 1$. Using EDQ, the discrete equation of the first of

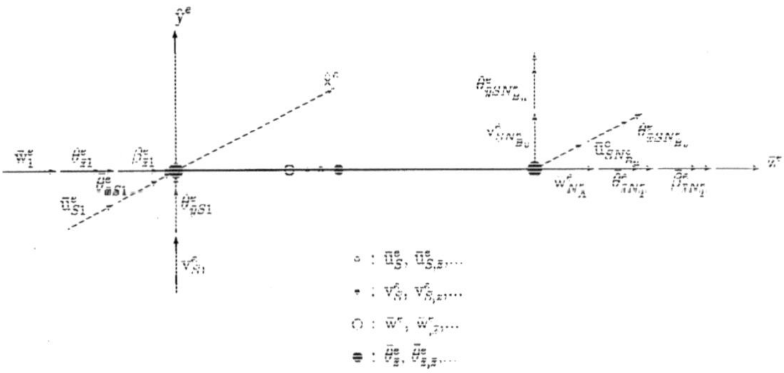

Fig. 7.1. The two element boundary nodes and four representative interior nodes with the assigned modal deformation parameters used to define various EDQ discretizations at the discrete points

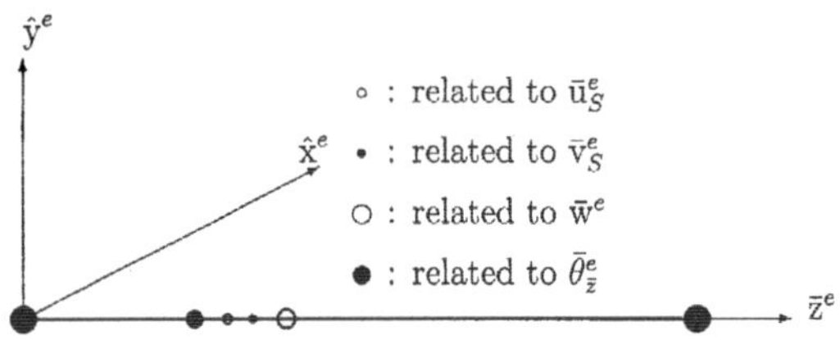

Fig. 7.2. Representative interior discrete points at which various discrete element eigenvalue equations are defined

Eqs. (7.4) at a discrete point α in element e can be expressed as

$$\frac{E^e}{(l^e)^4}\left[\frac{d^2I^e_{\bar{x}\bar{x}(\alpha)}}{d\zeta^2}\sum_{i=1}^{\bar{N}^e_{B_u}} D^{e\zeta^2}_{\alpha i} + 2\frac{dI^e_{\bar{x}\bar{x}(\alpha)}}{d\zeta}\sum_{i=1}^{\bar{N}^e_{B_u}} D^{e\zeta^3}_{\alpha i} + I^e_{\bar{x}\bar{x}(\alpha)}\sum_{i=1}^{\bar{N}^e_{B_u}} D^{e\zeta^4}_{\alpha i}\right]\tilde{U}^e_{Si}$$
$$-\rho^e\omega^2\left\{\left[A^e_{(\alpha)}\delta_{\alpha i} - \frac{1}{(l^e)^2}\left(\frac{dI^e_{\bar{x}\bar{x}(\alpha)}}{d\zeta}\sum_{i=1}^{\bar{N}^e_{B_u}} D^{e\zeta}_{\alpha i} + I^e_{\bar{x}\bar{x}(\alpha)}\sum_{i=1}^{\bar{N}^e_{B_u}} D^{e\zeta^2}_{\alpha i}\right)\right]\tilde{U}^e_{Si}\right.$$
$$\left.+\bar{y}^e_{S(\alpha)}A^e_{(\alpha)}\delta_{\alpha l}\tilde{\Theta}^e_{\bar{z}l}\right\} = 0, \quad \alpha = 2, ..., \hat{N}^e_{B_u} - 1 \tag{7.7}$$

The discrete equation of the second of Eqs. (7.4) at a discrete point β can be similarly expressed as

$$\frac{E^e}{(l^e)^4}\left[\frac{d^2I^e_{\bar{y}\bar{y}(\beta)}}{d\zeta^2}\sum_{j=1}^{\bar{N}^e_{B_v}} D^{e\zeta^2}_{\beta j} + 2\frac{dI^e_{\bar{y}\bar{y}(\beta)}}{d\zeta}\sum_{j=1}^{\bar{N}^e_{B_v}} D^{e\zeta^3}_{\beta j} + I^e_{\bar{y}\bar{y}(\beta)}\sum_{j=1}^{\bar{N}^e_{B_v}} D^{e\zeta^4}_{\beta j}\right]\tilde{V}^e_{Sj}$$
$$-\rho^e\omega^2\left\{\left[A^e_{(\beta)}\delta_{\beta j} - \frac{1}{(l^e)^2}\left(\frac{dI^e_{\bar{y}\bar{y}(\beta)}}{d\zeta}\sum_{j=1}^{\bar{N}^e_{B_v}} D^{e\zeta}_{\beta j} + I^e_{\bar{y}\bar{y}(\beta)}\sum_{j=1}^{\bar{N}^e_{B_v}} D^{e\zeta^2}_{\beta j}\right)\right]\tilde{V}^e_{Sj}\right.$$
$$\left.-x^e_{S(\beta)}A_{(\beta)}\delta_{\beta l}\tilde{\Theta}^e_{\bar{z}l}\right\} = 0, \quad \beta = 2, ..., \hat{N}^e_{B_v} - 1 \tag{7.8}$$

The discrete equation of the third of Eqs. (7.4) at a discrete point γ can be expressed as

$$-\frac{E^e}{(l^e)^2}\left(\frac{dA^e_{(\gamma)}}{d\zeta}\sum_{k=1}^{\bar{N}^e_A} D^{e\zeta}_{\gamma k} + A^e_{(\gamma)}\sum_{k=1}^{\bar{N}^e_A} D^{e\zeta^2}_{\gamma k}\right)\tilde{W}^e_k - \rho^e\omega^2 A^e_{(\gamma)}\delta_{\gamma k}\tilde{W}^e_k$$
$$= 0, \quad \gamma = 2, ..., \hat{N}^e_A - 1 \tag{7.9}$$

The discrete equation of the fourth of Eqs. (7.4) at a discrete point δ can be expressed as

$$\left[\frac{E^e}{(l^e)^4}\left(\frac{d^2I^{(s)e}_{\bar{\omega}\bar{\omega}(\delta)}}{d\zeta^2}\sum_{l=1}^{\bar{N}^e_T} D^{e\zeta^2}_{\delta l} + 2\frac{dI^{(s)e}_{\bar{\omega}\bar{\omega}(\delta)}}{d\zeta}\sum_{l=1}^{\bar{N}^e_T} D^{e\zeta^3}_{\delta l} + I^{(s)e}_{\bar{\omega}\bar{\omega}(\delta)}\sum_{l=1}^{\bar{N}^e_T} D^{e\zeta^4}_{\delta l}\right)\right.$$
$$\left.\frac{G^e}{(l^e)^2}\left(\frac{dJ^e_{(\delta)}}{d\zeta}\sum_{l=1}^{\bar{N}^e_T} D^{e\zeta}_{\delta l} + J^e_{(\delta)}\sum_{l=1}^{\bar{N}^e_T} D^{e\zeta^2}_{\delta l}\right)\right]\tilde{\Theta}^e_{\bar{z}l} + \rho^e\omega^2\left\{-\bar{y}^e_{S(\delta)}A_{(\delta)}\delta_{(\delta)i}\tilde{U}^e_{Si}\right.$$
$$+\bar{x}^e_{S(\delta)}A_{(\delta)}\delta_{\delta j}\tilde{V}^e_{Sj} - \left[I^e_{p(\delta)}\delta_{\delta l} + \frac{1}{(l^e)^2}\left(\frac{dI^{(s)e}_{\bar{\omega}\bar{\omega}(\delta)}}{d\zeta}\sum_{l=1}^{\bar{N}^e_T} D^{e\zeta}_{\delta l}\right.\right.$$
$$\left.\left.\left.+I^{(s)e}_{\bar{\omega}\bar{\omega}(\delta)}\sum_{l=1}^{\bar{N}^e_T} D^{e\zeta^2}_{\delta l}\right)\right]\tilde{\Theta}^e_{\bar{z}l}\right\} = 0, \quad \delta = 2, ..., \hat{N}^e_T - 1 \tag{7.10}$$

In Eqs. (7.7) to (7.10), the derivatives of section constants at the related discrete points can also be calculated by the DQ. The formulas for the computation are listed in Eqs. (6.9) except for the computation of the derivatives of $I^{(s)e}_{\bar{\omega}\bar{\omega}}$. Let $\tilde{N}^e_{I3}$ denote the number of nodes for defining the discretization of the derivatives of $I^{(s)e}_{\bar{\omega}\bar{\omega}}$. Then the derivatives at the related discrete points can be expressed by the following DQ discretization equations

$$\frac{dI^{(s)e}_{\bar{\omega}\bar{\omega}\delta}}{d\zeta} = \sum_{\bar{l}=1}^{\tilde{N}^e_{I3}} \bar{D}^{e\zeta}_{\delta\bar{l}} I^{(s)e}_{\bar{\omega}\bar{\omega}\bar{l}}, \quad \frac{d^2 I^{(s)e}_{\bar{\omega}\bar{\omega}\delta}}{d\zeta^2} = \sum_{\bar{l}=1}^{\tilde{N}^e_{I3}} \bar{D}^{e\zeta^2}_{\delta\bar{l}} I^{(s)e}_{\bar{\omega}\bar{\omega}\bar{l}} \tag{7.11}$$

Using Eqs. (7.7) to (7.10), the following discrete local element eigenvalue equation can be constructed

$$([\bar{\kappa}^e] - \omega^2[\bar{m}^e])\{\bar{\vartheta}^e\} = \{0\} \tag{7.12}$$

where $[\bar{\kappa}^e]$ is the local element stiffness coefficient matrix, $[\bar{m}^e]$ is the local element mass coefficient matrix

$$\begin{aligned}\{\bar{\vartheta}^e\} = \lfloor\, & \bar{U}^e_{\bar{S}1} \;\; \bar{V}^e_{\bar{S}1} \;\; \bar{W}^e_1 \;\; \bar{\Theta}^e_{\bar{x}S1} \;\; \bar{\Theta}^e_{\bar{y}S1} \;\; \bar{\Theta}^e_{\bar{z}1} \;\; \bar{\beta}^e_{\bar{z}1} \;\; 0 \;\; \ldots \;\; 0 \;\; \bar{U}^e_{\bar{S}N^e_{Bu}} \\ & \bar{V}^e_{\bar{S}N^e_{Bv}} \;\; \bar{W}^e_{N^e_A} \;\; \bar{\Theta}^e_{\bar{x}SN^e_{Bv}} \;\; \bar{\Theta}^e_{\bar{y}SN^e_{Bu}} \;\; \bar{\Theta}^e_{\bar{z}N^e_T} \;\; \bar{\beta}^e_{\bar{z}N^e_T} \rfloor^T \end{aligned} \tag{7.13}$$

the local element modal displacement vector. In the above equation, $\bar{\Theta}^e_{\bar{x}S}$, $\bar{\Theta}^e_{\bar{y}S}$ and $\bar{\beta}^e_{\bar{z}}$ represent the modal rotation angles $-\frac{d\bar{V}^e_{\bar{S}}}{d\bar{z}} = -\frac{1}{l^e}\frac{d\bar{V}^e_{\bar{S}}}{d\zeta}$, $\frac{d\bar{U}^e_{\bar{S}}}{d\bar{z}} = \frac{1}{l^e}\frac{d\bar{U}^e_{\bar{S}}}{d\zeta}$ and the modal angle of twist per unit length $\frac{d\bar{\Theta}^e_{\bar{z}}}{d\bar{z}} = \frac{1}{l^e}\frac{d\bar{\Theta}^e_{\bar{z}}}{d\zeta}$, respectively. If the element is axially rigid, $\bar{W}^e_i$ in $\{\bar{\vartheta}^e\}$ and the related columns in $[\bar{\kappa}^e]$ can also be eliminated. If the element is flexurally rigid in a specific direction, the lateral modal displacements and their derivatives in $\{\bar{\vartheta}^e\}$ and the related columns in $[\bar{\kappa}^e]$ can also be eliminated. Furthermore, if the element is torsional rigid, the modal angles of twist and the modal angles of twist per unit length in $\{\bar{\vartheta}^e\}$ and the related columns in $[\bar{\kappa}^e]$ can be eliminated. It should be noted that the local element mass coefficient matrix has to be similarly treated.

7.2.2 Coordinate Transformations

In order to solve frame problems, the discrete element fundamental relations of all discrete elements are assembled into the overall discrete eigenvalue system. All elements connected to a joint must have the same position, which is the joint node, to define the global modal displacements at that position and express the modal displacements of the element boundary nodes as the global modal displacements of that position. Thus the coordinate transformations have to be carried out. Consider a point O on the cross section. Since the cross section is rigid in its own plane, the local modal displacements at O can be related to the local modal displacements at the shear center S

$$\{\bar{\varphi}_O\} = [\bar{t}_{SO}]\{\bar{\varphi}\} \tag{7.14}$$

where

$$\{\bar{\varphi}_O\} = \lfloor \; \bar{U}_O \quad \bar{V}_O \quad \bar{W}_O \quad \bar{\Theta}_{\bar{x}O} \quad \bar{\Theta}_{\bar{y}O} \quad \bar{\Theta}_{\bar{z}O} \quad \bar{\beta}_{\bar{z}O} \; \rfloor^T \tag{7.15}$$

is the local modal displacement vector of point O, and

$$\{\bar{\varphi}\} = \lfloor \; \bar{U}_S \quad \bar{V}_S \quad \bar{W} \quad \bar{\Theta}_{\bar{x}S} \quad \bar{\Theta}_{\bar{y}S} \quad \bar{\Theta}_{\bar{z}} \quad \bar{\beta}_{\bar{z}} \; \rfloor^T \tag{7.16}$$

and

$$[\bar{t}_{SO}] = \begin{bmatrix} 1 & 0 & 0 & 0 & 0 & -\bar{y}_{SO} & 0 \\ 0 & 1 & 0 & 0 & 0 & \bar{x}_{SO} & 0 \\ 0 & 0 & 1 & \bar{y}_O & -\bar{x}_O & 0 & \tilde{\omega}_O \\ 0 & 0 & 0 & 1 & 0 & 0 & -\bar{y}_{SO} \\ 0 & 0 & 0 & 0 & 1 & 0 & \bar{x}_{SO} \\ 0 & 0 & 0 & 0 & 0 & 1 & 0 \\ 0 & 0 & 0 & 0 & 0 & 0 & 1 \end{bmatrix} \tag{7.17}$$

with $\tilde{\omega}_O = \bar{\omega}(\bar{x}_O, \bar{y}_O) - \bar{x}_O\bar{y}_S + \bar{y}_O\bar{x}_S$. If the rigid joint has a finite size, O will be at the intersection of the beam and the rigid joint. Then the relations between the local modal displacements of O and the local modal displacements of a point R on the rigid joint have to be constructed. R is the joint node used to define the transition conditions. If the rigid joint does not have a finite size, O coincides with R. Let $\bar{x}_R$, $\bar{y}_R$ and $\bar{z}_R$ denote the coordinates of R. The relations between the local modal displacements of point O and the local modal displacements $\bar{U}_R$, $\bar{V}_R$, $\bar{W}_R$, $\bar{\Theta}_{\bar{x}R}$, $\bar{\Theta}_{\bar{y}R}$, $\bar{\Theta}_{\bar{z}R}$ and $\bar{\beta}_{\bar{z}R}$ of point R are defined by

$$\{\bar{\varphi}_A\} = [\bar{t}_{RO}]\{\bar{\varphi}_R\} \tag{7.18}$$

where

$$\{\bar{\varphi}_R\} = \lfloor \; \bar{U}_R \quad \bar{V}_R \quad \bar{W}_R \quad \bar{\Theta}_{\bar{x}R} \quad \bar{\Theta}_{\bar{y}R} \quad \bar{\Theta}_{\bar{z}R} \quad \bar{\beta}_{\bar{z}R} \; \rfloor^T \tag{7.19}$$

and

$$[\bar{t}_{RO}] = \begin{bmatrix} 1 & 0 & 0 & 0 & -\bar{z}_{OR} & \bar{y}_{OR} & 0 \\ 0 & 1 & 0 & \bar{z}_{OR} & 0 & -\bar{x}_{OR} & 0 \\ 0 & 0 & 1 & -\bar{y}_{OR} & \bar{x}_{OR} & 0 & 0 \\ 0 & 0 & 0 & 1 & 0 & 0 & 0 \\ 0 & 0 & 0 & 0 & 1 & 0 & 0 \\ 0 & 0 & 0 & 0 & 0 & 1 & 0 \\ 0 & 0 & 0 & 0 & 0 & 0 & 1 \end{bmatrix} \tag{7.20}$$

Using Eqs. (7.14) and (7.19), the local modal displacements of S can be related to the local modal displacements of R through the following matrix equation

$$\{\bar{\varphi}\} = [\bar{t}_t]\,\{\bar{\varphi}_R\} \tag{7.21}$$

where

$$[\bar{t}_t] = [\bar{t}_{SO}]^{-1}[\bar{t}_{RO}] \tag{7.22}$$

is the translational transformation matrix.

In the overall discrete eigenvalue system, modal displacement components at joints are in directions of global coordinates, while the modal displacement components at the interior nodes can be in the directions of global or local coordinates. In the present numerical simulation, only the modal displacement components at joints are transformed to express them as the global joint modal displacements. Therefore the local element modal displacement vector $\{\bar{\vartheta}^e\}$ have to be transformed to obtain a local-global element modal displacement vector. The modal angle of twist per unit length of the beam, $\bar{\beta}_{\bar{z}}$, can also be expressed by the global components of the modal angle of twist per unit length, β_x, β_y and β_z, through the following transformation equation

$$\lfloor t^e_{r3} \rfloor \{\beta\} = \bar{\beta}_{\bar{z}} \tag{7.23}$$

where $\lfloor t^e_{r3} \rfloor$ is a row vector defined by the third row of the rotational transformation matrix $[t^e_r]$ expressed by Eq. (5.9) and $\{\beta\} = \lfloor \beta_x \; \beta_y \; \beta_z \rfloor^T$. Then the relations between the global modal displacements and the local modal displacements of the joint node R can be expressed by the following equation

$$[\bar{t}_R]\,\{\varphi_R\} = \{\bar{\varphi}_R\} \tag{7.24}$$

where

$$[\bar{t}_R] = \begin{bmatrix} [t^e_r] & & \\ & [t^e_r] & \\ & & \lfloor t^e_{r3} \rfloor \end{bmatrix} \tag{7.25}$$

is the joint transformation matrix, and

$$\{\varphi_R\} = \lfloor\; U_R \quad V_R \quad W_R \quad \Theta_{xR} \quad \Theta_{yR} \quad \Theta_{zR} \quad \beta_x \quad \beta_y \quad \beta_z \;\rfloor^T \tag{7.26}$$

is the global modal displacement vector of joint node R. Using Eqs. (7.21) and (7.24), the transformation relation between the local modal displacement vector of S and the global modal displacement vector of the joint node R can be expressed by the following equation

$$\{\bar{\varphi}\} = [\bar{t}^e]\,\{\varphi_R\} \tag{7.27}$$

where

$$[\bar{t}^e] = [\bar{t}_t][\bar{t}_R] \tag{7.28}$$

A transformation relation between the element modal displacement vector in global coordinate system and the element modal displacement vector in local coordinate system can be obtained through the use of $[\bar{t}^e]$

$$[\bar{T}^e]\{\vartheta^e\} = \{\bar{\vartheta}^e\} \tag{7.29}$$

where $\{\bar{\vartheta}^e\}$ is the local element modal displacement vector,

$$\{\vartheta^e\} = \lfloor U^e_{R1}\ V^e_{R1}\ W^e_{R1}\ \Theta^e_{xR1}\ \Theta^e_{yR1}\ \Theta^e_{zR1}\ \beta^e_{xR1}\ \beta^e_{yR1}\ \beta^e_{zR1}\ \dots\ U^e_{RN^e}$$
$$V^e_{RN^e}\ W^e_{RN^e}\ \Theta^e_{xRN^e}\ \Theta^e_{yRN^e}\ \Theta^e_{zRN^e}\ \beta^e_{xRN^e}\ \beta^e_{yRN^e}\ \beta^e_{zRN^e}\rfloor^T \tag{7.30}$$

the local-global element modal displacement vector, and $[\bar{T}^e]$ the element transformation matrix. $[\bar{T}^e]$ can be expressed by using $[\bar{t}^e]$ to replace $[t^e]$ in Eq. (6.30). Employing Eq. (7.29) in Eq. (7.12), the following matrix equation can be obtained

$$([\bar{\kappa}^e] - \omega^2[\bar{m}^e])[\bar{T}^e]\{\vartheta^e\} = \{0\} \tag{7.31}$$

The above equation can be rewritten as

$$([\kappa^e] - \omega^2[m^e])\{\vartheta^e\} = \{0\}, \tag{7.32}$$

where

$$[\kappa^e] = [\bar{\kappa}^e][\bar{T}^e]; \text{and} [m^e] = [\bar{m}^e][\bar{T}^e] \tag{7.33}$$

are transformed local-global element stiffness coefficient matrix, and local-global element mass coefficient matrix, respectively. Equation (7.32) is the discrete local-global element eigenvalue equation in which the local-global element modal displacement vector is defined on both of local and global coordinate systems.

7.2.3 Discrete Element Boundary Forces

Figure 7.3 shows the internal forces at the two element boundary nodes and an interior point. For calculating the internal forces of a discrete element at a specific point, certain local derivatives of modal displacements at that point must be discretized. The discrete equations can be expressed by the local-global element modal displacement vector. Using the mapping technique, the mth order derivative of $\bar{U}_S$ with respect to $\bar{z}^e$ at point α can be expressed as

$$\frac{d^m \bar{U}^e_{S\alpha}}{d(\bar{z}^e)^m} = \frac{1}{(l^e)^m} \lfloor D^{e\varsigma^m}_{\bar{u}\alpha} \rfloor \{\bar{\vartheta}^e\} \tag{7.34}$$

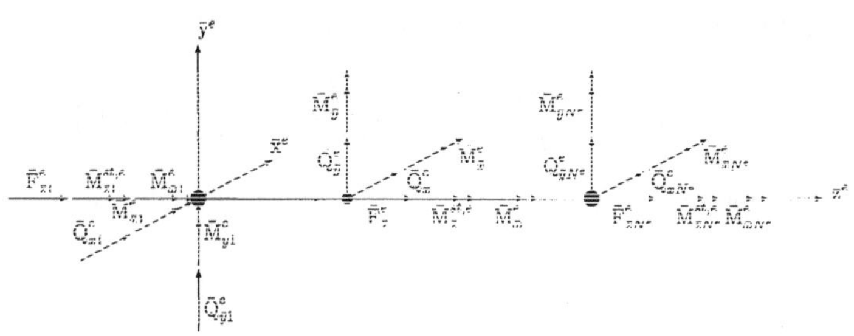

Fig. 7.3. Internal forces at the two element boundary nodes and an interior point

where $\lfloor D_{\bar{u}\alpha}^{e\varsigma^m} \rfloor = \lfloor D_{\alpha 1}^{e\varsigma^m} \; 0 \; 0 \ldots 0 \; l^e D_{\alpha \bar{N}_{Bu}^e}^{e\varsigma^m} \; 0 \; 0 \rfloor$. The introduction of Eq. (7.29) into Eq. (7.34) leads to the following equation

$$\frac{d^m \bar{U}_{S\alpha}^e}{d(\bar{z}^e)^m} = \frac{1}{(l^e)^m} \lfloor D_{\bar{u}\alpha}^{e\varsigma^m} \rfloor [\bar{T}^e]\{\vartheta^e\} \tag{7.35}$$

The mth order derivatives of $\bar{V}_S^e$, $\bar{W}^e$ and $\bar{\Theta}_{\bar{z}}^e$ with respect to $\bar{z}^e$ at discrete points β, γ and δ can be similarly obtained

$$\frac{d^m \bar{V}_{S\beta}^e}{d(\bar{z}^e)^m} = \frac{1}{(l^e)^m} \lfloor D_{\bar{v}\beta}^{e\varsigma^m} \rfloor [\bar{T}^e]\{\vartheta^e\} \tag{7.36}$$

$$\frac{d^m \bar{W}_{\gamma}^e}{d(\bar{z}^e)^m} = \frac{1}{(l^e)^m} \lfloor D_{\bar{w}\gamma}^{e\varsigma^m} \rfloor [\bar{T}^e]\{\vartheta^e\} \tag{7.37}$$

$$\frac{d^m \bar{\Theta}_{\bar{z}\delta}^e}{d(\bar{z}^e)^m} = \frac{1}{(l^e)^m} \lfloor D_{\bar{\theta}_{\bar{z}}\delta}^{e\varsigma^m} \rfloor [T^e]\{\vartheta^e\} \tag{7.38}$$

where

$$\lfloor D_{\bar{v}\beta}^{e\varsigma^m} \rfloor = \lfloor 0 \quad D_{\beta 1}^{e\varsigma^m} \quad 0 \quad \ldots \quad -l^e D_{\alpha \bar{N}_{Bv}^e}^{e\varsigma^m} \quad 0 \quad 0 \quad 0 \rfloor \tag{7.39}$$

$$\lfloor D_{\bar{w}\gamma}^{e\varsigma^m} \rfloor = \lfloor 0 \quad 0 \quad D_{\gamma 1}^{e\varsigma^m} \quad \ldots \quad D_{\gamma \bar{N}_A^e}^{e\varsigma^m} \quad 0 \quad 0 \quad 0 \quad 0 \rfloor \tag{7.40}$$

$$\lfloor D_{\bar{\theta}_{\bar{z}}\delta}^{e\varsigma^m} \rfloor = \lfloor 0 \quad 0 \quad 0 \quad 0 \quad 0 \quad D_{\delta 1}^{e\varsigma^m} \quad \ldots \quad 0 \quad 0 \quad D_{\delta(\bar{N}_T^e - 1)}^{e\varsigma^m} \quad l^e D_{\delta \bar{N}_T^e}^{e\varsigma^m} \rfloor \tag{7.41}$$

The internal forces at a discrete point can be calculated by using the local-global element modal displacement vector. Following the procedures for calculating the internal forces used in the static deflection analysis model, the axial force at the element boundary node I^e without considering a body attached to the node can be expressed by

$$\bar{F}_{\bar{z}I^e}^e = E^e A_{I^e}^e \frac{d\bar{W}_{I^e}^e}{d\bar{z}^e} = \frac{E^e A_{I^e}^e}{(l^e)} \lfloor D_{\bar{w}I^e}^{e\varsigma} \rfloor [\bar{T}^e]\{\vartheta^e\} \tag{7.42}$$

The bending moment $\bar{M}_{\bar{x}I^e}^e$ at the element boundary node I^e can be expressed by

$$\bar{M}_{\bar{x}I^e}^e = -E^e I_{\bar{y}\bar{y}I^e}^e \frac{d^2 \bar{V}_{SI^e}^e}{d(\bar{z}^e)^2} = -\frac{E^e I_{\bar{y}\bar{y}I^e}^e}{(l^e)^2} \lfloor D_{\bar{v}I^e}^{e\varsigma^2} \rfloor [\bar{T}^e]\{\vartheta^e\} \tag{7.43}$$

Then bending moment $\bar{M}_{\bar{y}I^e}^e$ at the element boundary node I^e can be expressed by

$$\bar{M}_{\bar{y}I^e}^e = E^e I_{\bar{x}\bar{x}I^e}^e \frac{d^2 \bar{U}_{SI^e}^e}{d(\bar{z}^e)^2} = \frac{E^e I_{\bar{x}\bar{x}I^e}^e}{(l^e)^2} \lfloor D_{\bar{u}I^e}^{e\varsigma^2} \rfloor [\bar{T}^e]\{\vartheta^e\} \tag{7.44}$$

The lateral forces $\bar{V}_{\bar{x}I^e}^e$ at an element boundary node I^e can be expressed by

$$
\begin{aligned}
\bar{V}^e_{\bar{x}I^e} &= -E^e \frac{d}{d\bar{z}^e}\left[I^e_{\bar{x}\bar{x}I^e}\frac{d^2\bar{U}^e_{SI^e}}{d(\bar{z}^e)^2}\right] - \rho^e I^e_{\bar{x}\bar{x}I^e}\omega^2 \frac{\partial \bar{U}^e_{SI^e}}{\partial \bar{z}^e} \\
&= -\frac{E^e}{(l^e)^3}\left(\frac{dI^e_{\bar{x}\bar{x}I^e}}{d\zeta}\lfloor D^{e\zeta^2}_{\bar{u}I^e}\rfloor + I^e_{\bar{x}\bar{x}I^e}\lfloor {}^e_{I^e}D^{e\zeta^3}_{\bar{u}I^e}\rfloor\right)[\bar{T}^e]\{\vartheta^e\} \\
&\quad -\frac{\rho^e I^e_{\bar{x}\bar{x}I^e}}{l^e}\lfloor D^{e\zeta}_{\bar{u}I^e}\rfloor[\bar{T}^e]\{\vartheta^e\}
\end{aligned}
\tag{7.45}
$$

The lateral forces $\bar{V}^e_{\bar{y}I^e}$ at an element boundary node I^e can be expressed by

$$
\begin{aligned}
\bar{V}^e_{\bar{y}I^e} &= -E^e \frac{d}{d\bar{z}^e}\left(I^e_{\bar{y}\bar{y}I^e}\frac{d^2\bar{V}^e_{SI^e}}{d(\bar{z}^e)^2}\right) - \rho^e I^e_{\bar{y}\bar{y}I^e}\omega^2 \frac{\partial \bar{V}^e_{SI^e}}{\partial \bar{z}^e} \\
&= -\frac{E^e}{(l^e)^3}\left(\frac{dI^e_{\bar{y}\bar{y}I^e}}{d\zeta}\lfloor D^{e\zeta^2}_{\bar{v}I^e}\rfloor + I^e_{\bar{y}\bar{y}I^e}\lfloor D^{e\zeta^3}_{\bar{v}I^e}\rfloor\right)[\bar{T}^e]\{\delta^e\} \\
&\quad -\frac{\rho^e I^e_{\bar{y}\bar{y}I^e}}{l^e}\lfloor D^{e\zeta}_{\bar{v}I^e}\rfloor[\bar{T}^e]\{\vartheta^e\}
\end{aligned}
\tag{7.46}
$$

The bimoment $\bar{M}^e_{\bar{\omega}I^e}$ at the element boundary node I^e can be expressed by

$$
\bar{M}^e_{\bar{\omega}I^e} = E^e I^{(s)e}_{\bar{\omega}\bar{\omega}I^e}\frac{d^2\bar{\Theta}^e_{\bar{z}I^e}}{d(\bar{z}^e)^2} = \frac{E^e I^{(s)e}_{\bar{\omega}\bar{\omega}I^e}}{(l^e)^2}\lfloor D^{e\zeta^2}_{\bar{\theta}_{\bar{z}}I^e}\rfloor[\bar{T}^e]\{\vartheta^e\}
\tag{7.47}
$$

The warping torsion moment $\bar{M}^{\omega,e}_{\bar{z}I^e}$ at the element boundary node I^e can be expressed by

$$
\begin{aligned}
\bar{M}^{\omega,e}_{\bar{z}I^e} &= -E^e \frac{d}{d\bar{z}^e}\left(I^{(s)e}_{\bar{\omega}\bar{\omega}I^e}\frac{d^2\bar{\Theta}^e_{\bar{z}I^e}}{d(\bar{z}^e)^2}\right) - \rho^e I^{(s)e}_{\bar{\omega}\bar{\omega}I^e}\omega^2 \frac{\partial \bar{\Theta}^e_{\bar{z}I^e}}{\partial \bar{z}^e} \\
&= -\frac{E^e}{(l^e)^3}\left(\frac{dI^{(s)e}_{\bar{\omega}\bar{\omega}I^e}}{d\zeta}\lfloor D^{e\zeta^2}_{\bar{\theta}_{\bar{z}}I^e}\rfloor + I^{(s)e}_{\bar{\omega}\bar{\omega}I^e}\lfloor D^{e\zeta^3}_{\bar{\theta}_{\bar{z}}I^e}\rfloor\right)[\bar{T}^e]\{\vartheta^e\} \\
&\quad -\frac{\rho^e I^{(s)e}_{\bar{\omega}\bar{\omega}I^e}}{l^e}\lfloor D^{e\zeta}_{\bar{\theta}_{\bar{z}}I^e}\rfloor[\bar{T}^e]\{\vartheta^e\}
\end{aligned}
\tag{7.48}
$$

The Saint Venant torsion moment $\bar{M}^{st,e}_{\bar{z}I^e}$ at the element boundary node I^e can be expressed by

$$
\bar{M}^{st,e}_{\bar{z}I^e} = G^e J^e_{I^e}\frac{d\bar{\Theta}^e_{\bar{z}I^e}}{d\bar{z}^e} = \frac{G^e J^e_{I^e}}{l^e}\lfloor D^{e\zeta}_{\bar{\theta}_{\bar{z}}I^e}\rfloor[T^e]\{\vartheta^e\}
\tag{7.49}
$$

The total torque $\bar{M}^e_{\bar{z}I^e}$ at the element boundary node I^e can be expressed by

$$
\bar{M}^e_{\bar{z}I^e} = \bar{M}^{\bar{\omega},e}_{\bar{z}I^e} + \bar{M}^{st,e}_{\bar{z}I^e}
\tag{7.50}
$$

7.3 Discrete Condition Equations of Joints

The DQEM requires that all condition equations at joints are satisfied. The condition equations have to be expressed as discrete forms. The compatibility conditions, which are kinematic transition conditions, of joint j can be expressed by using the modal displacement vectors in Eq. (6.41). The natural transition conditions involve the equilibriums of forces at joints.

7.3.1 Discrete Dynamic Equilibrium Equations of Joints

For the DQEM model, the dynamic equilibrium conditions of elastic restoring and inertia forces at joints also have to be satisfied. Let $\bar{V}_{\bar{x}}^{m^j}$ and $\bar{V}_{\bar{y}}^{m^j}$ denote the two lateral forces, $\bar{F}_{\bar{z}}^{m^j}$ denote the axial force, $\bar{M}_{\bar{x}}^{m^j}$ and $\bar{M}_{\bar{y}}^{m^j}$ denote the two bending moments, $\bar{M}_{\bar{\omega}}^{m^j}$ denote the bimoment, and $\bar{M}_{\bar{z}}^{m^j}$ denote the total torque, of the m^jth element at joint j. Consider that a mass $\tilde{M}^j$ with a finite volume is attached to the joint, and let $\hat{P}_x^j$, $\hat{P}_y^j$ and $\hat{P}_z^j$ denote the three translational inertia forces, and $\hat{M}_x^j$, $\hat{M}_y^j$ and $\hat{M}_z^j$ denote the three inertia moments of the mass. Using the inverse transformation of Eq. (5.9), the force vectors formed by the nodal forces at joint j for all elements connected to the joint can be transformed into the global coordinate system. Let $\{V^{m^j}\} = \lfloor V_x^{m^j}\ V_y^{m^j}\ F_z^{m^j}\ M_x^{m^j}\ M_y^{m^j}\ M_z^{m^j}\ M_{\bar{\omega}}^{m^j} \rfloor^T$ denote the globally nodal force vector of the m^jth element at joint j. Then the dynamic equilibrium conditions of joint j can be expressed as the following matrix equation:

$$\sum_{m^j=1}^{M^j} \nu^{m^j}\{V^{m^j}\} = \{\hat{P}^j\} \tag{7.51}$$

where

$$\{\hat{P}^j\} = \lfloor\ \hat{P}_x^j \quad \hat{P}_y^j \quad \hat{P}_z^j \quad \hat{M}_x^j \quad \hat{M}_y^j \quad \hat{M}_z^j \quad 0 \quad 0 \quad 0\ \rfloor^T \tag{7.52}$$

is the joint inertia force vector and the value of ν^{m^j} is expressed by Eq. (5.21). The globally nodal force vector of the m^jth element at joint j can be related to the locally nodal force vector by the following equation

$$\{V^{m^j}\} = [\bar{t}^{m^j}]^T\{\bar{Q}^{m^j}\} \tag{7.53}$$

where

$$\{\bar{Q}^{m^j}\} = \lfloor\ \bar{V}_{\bar{x}}^{m^j} \quad \bar{V}_{\bar{y}}^{m^j} \quad \bar{F}_{\bar{z}}^{m^j} \quad \bar{M}_{\bar{x}}^{m^j} \quad \bar{M}_{\bar{y}}^{m^j} \quad \bar{M}_{\bar{z}}^{m^j} \quad \bar{M}_{\bar{\omega}}^{m^j}\ \rfloor^T \tag{7.54}$$

is the locally element nodal force vector of the element at the joint, and $[t^{m^j}]$ is the joint transformation matrix.

Let (x_C, y_C, z_C) denote the global coordinates of the center C of a rigid body with the mass $\tilde{M}^j$. The mass is rigidly connected to the joint node R. Also let u_{iR}^j and θ_{jR}^j denote the translational displacements and deflection slopes, respectively, of the joint node R. The translational inertia forces can be expressed by $\check{P}_i^j = -\tilde{M}^j \frac{\partial^2}{\partial t^2}(u_{iR}^j + \epsilon_{ijk}\theta_{jR}^j x_{kRC})$, where ϵ_{ijk} is the permutation symbol and $x_{kRC} = x_{kC} - x_{kR}$. Considering the harmonic motion, the components of translational inertia forces of the vibration can be expressed by

$$\hat{P}_{xR}^j = \tilde{M}^j\omega^2\Big(U_R^j + z_{RC}\Theta_{yR}^j - y_{RC}\Theta_{zR}^j\Big),$$

$$\hat{P}^j_{yR} = \tilde{M}^j\omega^2\Big(V^j_R + x_{RC}\Theta^j_{zR} - z_{RC}\Theta^j_{xR}\Big),$$

$$\hat{P}^j_{zR} = \tilde{M}^j\omega^2\Big(W^j_R + y_{RC}\Theta^j_{xR} - x_{RC}\Theta^j_{yR}\Big) \tag{7.55}$$

Let H^j_{iC} denote the angular momentum of the mass $\tilde{M}^j$ with respect to C. Also let $\hat{I}^j_{\hat{i}\hat{j}}$ denote the inertia tensor of the mass $\tilde{M}^j$ with respect to the coordinate system $(\hat{x}, \hat{y}, \hat{z})$ which has the origin located at C and the coordinate axes $\hat{x}$, $\hat{y}$ and $\hat{z}$ oriented in x, y and z directions, respectively. The inertia moments can be expressed by $\check{M}^j_{iR} = -\frac{\partial H^j_{iC}}{\partial t} + \tilde{M}^j\epsilon_{ijk}x_{jRC}(\frac{\partial^2 x_{kR}}{\partial t^2} + \epsilon_{klm}\frac{\partial^2\theta^j_{lR}}{\partial t^2}x_{mRC})$. Considering the harmonic motion, the components of inertia moments of the vibration can be expressed by

$$\begin{aligned}\hat{M}^j_{xR} = \omega^2\Big\{&\tilde{I}_{\hat{x}\hat{x}}\Theta^j_{xR} - \tilde{I}^j_{\hat{x}\hat{y}}\Theta^j_{yR} - \tilde{I}^j_{\hat{x}\hat{z}}\Theta^j_{zR} + \tilde{M}^j\Big[y_{RC}W^j_R - z_{RC}V^j_R \\ &+(z^2_{RC} + y^2_{RC})\Theta^j_{xR} - y_{RC}x_{RC}\Theta^j_{yR} - z_{RC}x_{RC}\Theta^j_{zR}\Big]\Big\},\end{aligned}$$

$$\begin{aligned}\hat{M}^j_{yR} = \omega^2\Big\{&-\tilde{I}_{\hat{x}\hat{y}}\Theta^j_{xR} + \tilde{I}^j_{\hat{y}\hat{y}}\Theta^j_{yR} - \tilde{I}^j_{\hat{y}\hat{z}}\Theta^j_{zR} + \tilde{M}^j\Big[z_{RC}U^j_R - x_{RC}W^j_R \\ &-x_{RC}y_{RC}\Theta^j_{xR} + (x^2_{RC} + z^2_{RC})\Theta^j_{yR} - z_{RC}y_{RC}\Theta^j_{zR}\Big]\Big\},\end{aligned}$$

$$\begin{aligned}\hat{M}^j_{zR} = \omega^2\Big\{&-\tilde{I}_{\hat{x}\hat{z}}\Theta^j_{xR} - \tilde{I}^j_{\hat{y}\hat{z}}\Theta^j_{yR} + \tilde{I}^j_{\hat{z}\hat{z}}\Theta^j_{zR} + \tilde{M}^j\Big[x_{RC}V^j_R - y_{RC}U^j_R \\ &-x_{RC}z_{RC}\Theta^j_{xR} - y_{RC}z_{RC}\Theta^j_{yR} + (x^2_{RC} + y^2_{RC})\Theta^j_{zR}\Big]\Big\}\end{aligned} \tag{7.56}$$

Using Eqs. (7.42) to (7.50), and Eqs. (7.52) to (7.56), Eq. (7.51) can be rewritten as

$$\begin{aligned}&\sum_{m^j=1}^{M^j}\nu^{m^j}\Big([\bar{t}^{m^j}]^T[\bar{S}^{m^j}] - \rho^{m^j}\omega^2[\bar{t}^{m^j}_{126}][\hat{S}^{m^j}]\Big)[\bar{T}^{m^j}]\{\vartheta^{m^j}\} \\ &\qquad -\omega^2[\hat{M}^j]\{\varphi^j_R\} = 0\end{aligned} \tag{7.57}$$

where $[\bar{S}^{m^j}]$ is a matrix having seven rows with the first five rows same as those listed in Eq. (6.46) and the last two rows expressed by

$$\lfloor S_6^{m^j} \rfloor = -\frac{E^{m^j}}{(l^{m^j})^3}\left(\frac{dI^{(s)m^j}_{\bar{\omega}\bar{\omega}I^{m^j}}}{d\zeta}\lfloor D^{m^j\zeta^2}_{\bar{\theta}_{\bar{z}}I^{m^j}} \rfloor + I^{(s)m^j}_{\bar{\omega}\bar{\omega}I^{m^j}}\lfloor D^{m^j\zeta^3}_{\bar{\theta}_{\bar{z}}I^{m^j}} \rfloor\right)$$

$$+\frac{G^{m^j} J^{m^j}_{I^{m^j}}}{l^{m^j}}\lfloor D^{m^j\zeta}_{\bar{\theta}_{\bar{z}}I^{m^j}} \rfloor, \quad \lfloor S_7^{m^j} \rfloor = \frac{E^{m^j} I^{(s)m^j}_{\bar{\omega}\bar{\omega}I^{m^j}}}{(l^{m^j})^2}\lfloor D^{m^j\zeta^2}_{\bar{\theta}_{\bar{z}}I^{m^j}} \rfloor \tag{7.58}$$

$[\hat{S}^{m^j}]$ is a matrix having three rows expressed by

$$\lfloor \hat{S}_1^{m^j} \rfloor = \frac{I^{m^j}_{\bar{x}\bar{x}I^{m^j}}}{l^{m^j}}\lfloor D^{m^j\zeta}_{\bar{u}I^{m^j}} \rfloor, \quad \lfloor \hat{S}_2^{m^j} \rfloor = \frac{I^{m^j}_{\bar{y}\bar{y}I^{m^j}}}{l^{m^j}}\lfloor D^{m^j\zeta}_{\bar{v}I^{m^j}} \rfloor,$$

$$\lfloor \hat{S}_3^{m^j} \rfloor = -\frac{I^{(s)m^j}_{\bar{\omega}\bar{\omega}I^{m^j}}}{l^{m^j}}\lfloor D^{m^j\zeta}_{\bar{\theta}_{\bar{z}}I^{m^j}} \rfloor \tag{7.59}$$

and $[\bar{t}^{m^j}_{126}]$ is a matrix having three rows and is expressed by

$$[\bar{t}^{m^j}_{126}] = \begin{bmatrix} \lfloor \bar{t}_1^{m^j} \rfloor \\ \lfloor \bar{t}_2^{m^j} \rfloor \\ \lfloor \bar{t}_6^{m^j} \rfloor \end{bmatrix} \tag{7.60}$$

with $\lfloor \bar{t}_1^{m^j} \rfloor$, $\lfloor \bar{t}_2^{m^j} \rfloor$ and $\lfloor \bar{t}_6^{m^j} \rfloor$ row vectors representing the first, second and sixth rows, respectively, of $[\bar{t}^{m^j}]$, $\{\varphi^j_R\}$ the global modal displacement vector of the joint node R of joint j, and $[\hat{M}^j]$ a matrix having seven rows expressed by

$$\lfloor \hat{M}_1^j \rfloor = \lfloor\ -\tilde{M}^j \quad 0 \quad 0 \quad 0 \quad -z_{RC}\tilde{M}^j \quad y_{RC}\tilde{M}^j \quad 0 \quad 0 \quad 0\ \rfloor,$$

$$\lfloor \hat{M}_2^j \rfloor = \lfloor\ 0 \quad -\tilde{M}^j \quad 0 \quad z_{RC}\tilde{M}^j \quad 0 \quad -x_{RC}\tilde{M}^j \quad 0 \quad 0 \quad 0\ \rfloor,$$

$$\lfloor \hat{M}_3^j \rfloor = \lfloor\ 0 \quad 0 \quad -\tilde{M}^j \quad -y_{RC}\tilde{M}^j \quad x_{RC}\tilde{M}^j \quad 0 \quad 0 \quad 0 \quad 0\ \rfloor,$$

$$\lfloor \hat{M}_4^j \rfloor = \lfloor\ 0 \quad z_{RC}\tilde{M}^j \quad -y_{RC}\tilde{M}^j \quad -\tilde{I}^j_{\hat{x}\hat{x}} - (y^2_{RC} + z^2_{RC})\tilde{M}^j$$
$$\tilde{I}^j_{\hat{x}\hat{y}} + y_{RC}x_{RC}\tilde{M}^j \ \tilde{I}^j_{\hat{x}\hat{z}} + z_{RC}x_{RC}\tilde{M}^j \ 0\ 0\ 0 \rfloor,$$

$$\lfloor \hat{M}_5^j \rfloor = \lfloor\ -z_{RC}\tilde{M}^j \quad 0 \quad x_{RC}\tilde{M}^j \quad \tilde{I}^j_{\hat{x}\hat{y}} + x_{RC}y_{RC}\tilde{M}^j$$
$$-\tilde{I}^j_{\hat{y}\hat{y}} - (x^2_{RC} + z^2_{RC})\tilde{M}^j \ \tilde{I}^j_{\hat{y}\hat{z}} + z_{RC}y_{RC}\tilde{M}^j \ 0\ 0\ 0 \rfloor,$$

$$\lfloor \hat{M}_6^j \rfloor = \lfloor\ y_{RC}\tilde{M}^j \quad -x_{RC}\tilde{M}^j \quad 0 \quad \tilde{I}^j_{\hat{x}\hat{z}} + x_{RC}z_{RC}\tilde{M}^j \quad \tilde{I}^j_{\hat{x}\hat{y}} + y_{RC}z_{RC}\tilde{M}^j$$
$$-\tilde{I}^j_{\hat{z}\hat{z}} - (x^2_{RC} + y^2_{RC})\tilde{M}^j \ 0\ 0\ 0 \rfloor$$

$$\lfloor \hat{M}_7^j \rfloor = \lfloor\ 0 \quad 0 \quad 0 \quad 0 \quad 0 \quad 0 \quad 0 \quad 0 \quad 0\ \rfloor \tag{7.61}$$

7.3.2 Inclined Roller

There are nine kinematic condition equations. Among the nine condition equations of the inclined roller, the three translational condition equations need to be transformed while the six rotational conditions are the same as those of a joint. The procedure of transformation has been stated in Chapter 4 regarding the analysis of truss structures. For this frame analysis model, the kinematic condition equations can be expressed by the following matrix equation

$$\left[\dot{\tilde{t}}_r\right] \{d^j\} = \{\tilde{d}^j\} \tag{7.62}$$

where

$$\left[\dot{\tilde{t}}_r\right] = \begin{bmatrix} \left[\dot{\tilde{t}}_r\right] & [0] \\ [0]^T & [I] \end{bmatrix} \tag{7.63}$$

is the transformation matrix, $[0]$ is a 3×6 zero matrix, $[I]$ is a 6×6 unit matrix, and $\{\tilde{d}^j\} = \lfloor \tilde{U}\ \tilde{V}\ \tilde{W}\ \bar{\Theta}_x\ \bar{\Theta}_y\ \bar{\Theta}_z\ \bar{\beta}_x\ \bar{\beta}_y\ \bar{\beta}_z \rfloor^T$ is the prescribed displacement vector with the three translational components in $\tilde{x}$, $\tilde{y}$ and $\tilde{z}$ coordinate directions, and the six rotational components in x, y and z coordinate directions.

There are eight natural conditions which can be represented by the following equation

$$\left[\ddot{\tilde{t}}_r\right] \sum_{m^j=1}^{M^j} \nu^{m^j} \{V^{m^j}\} = \{\hat{P}^j\} \tag{7.64}$$

where

$$\left[\ddot{\tilde{t}}_r\right] = \begin{bmatrix} [\tilde{t}_{r12}] & [0] \\ [0]^T & [I] \end{bmatrix} \tag{7.65}$$

is a 8×8 transformation matrix with $[\tilde{t}_{r12}]^T$ the 2×3 matrix used in Eqs. (5.27) and (5.28), $[0]$ a 2×6 zero matrix and $[I]$ a 6×6 unit matrix, and $\{\hat{P}^j\} = \lfloor \hat{P}^j_{\tilde{x}}\ \hat{P}^j_{\tilde{y}}\ \hat{M}_x\ \hat{M}_y\ \hat{M}_z\ 0\ 0\ 0 \rfloor^T$. The eight natural conditions can be expressed by the element nodal displacement vectors at the joint on the inclined surface

$$\left[\ddot{\tilde{t}}_r\right] \sum_{m^j=1}^{M^j} \nu^{m^j} \left([\bar{t}^{m^j}]^T [\bar{S}^{m^j}] - \rho^{m^j} \omega^2 [\bar{t}^{m^j}_{126}][\hat{S}^{m^j}] \right) [\bar{T}^{m^j}]\{\vartheta^{m^j}\}$$
$$-\omega^2 [\hat{M}^j]\{\varphi^j_R\} = 0 \tag{7.66}$$

7.4 Assemblage

With the discrete joint compatibility conditions in mind, then by assembling all discrete local-global element eigenvalue equations represented by Eq. (7.32) for all elements, dynamic equilibrium condition equations (7.57) and (7.66), for the general joints and inclined rollers, respectively, and the prescribed joint displacement conditions, the overall eigenvalue equation system represented by Eq. (3.32) can be obtained. Like FEM, the assemblage is based on an element by element procedure. The discrete local-global element eigenvalue equation (7.32) for all elements, and discrete element boundary forces existing in Eqs. (7.57) and (7.66) defined at the two element boundary nodes and expressed by modal displacement parameters, are directly assembled to the overall eigenvalue equation system.

Another approach can be used to assemble all discrete fundamental relations. This approach includes the fourteen discrete equations for defining the element boundary forces in the local element eigenvalue equation to form another matrix equation. In this new matrix equation, each of the first seven component equations represents one equation for defining a boundary force corresponding to each individual degree of freedom assigned to node 1 while each of the last seven component equations represents one equation for defining a boundary force corresponding to each individual degree of freedom assigned to the other element boundary node. The component equations of the local element eigenvalue equation (7.12) are placed at the remaining rows. This matrix equation is expressed as

$$\left(\left[\hat{k}^e\right] - \omega^2 \left[\hat{m}^e\right]\right) \{\hat{\vartheta}^e\} = \{\hat{\gamma}^e\} \tag{7.67}$$

where the first and last seven rows of $[\hat{k}^e]$ and $[\hat{m}^e]$ are coefficients for calculating the fourteen element boundary forces caused by modal displacements and local element mass matrix, respectively, and

$$\begin{aligned}\{\hat{\gamma}^e\} = \lfloor & -\bar{V}^e_{\bar{x}1} \;\; -\bar{V}^e_{\bar{y}1} \;\; -\bar{F}^e_{\bar{z}1} \;\; -\bar{M}^e_{\bar{x}1} \;\; -\bar{M}^e_{\bar{y}1} \;\; -\bar{M}^e_{\bar{z}1} \;\; -\bar{M}^e_{\bar{\omega}1} \;\; 0 \;\ldots\; 0 \;\; \bar{V}^e_{\bar{x}N^e} \\ & \bar{V}^e_{\bar{y}N^e} \;\; \bar{F}^e_{\bar{z}N^e} \;\; \bar{M}^e_{\bar{x}N^e} \;\; \bar{M}^e_{\bar{y}N^e} \;\; \bar{M}^e_{\bar{z}N^e} \;\; \bar{M}^e_{\bar{\omega}N^e} \;\rfloor^T\end{aligned} \tag{7.68}$$

is a force vector in which the first and last seven elements are element boundary forces, caused by modal displacements, at the two element boundary nodes, while all other elements equal zero. Using the element transformation matrix $[T^e]$ to Eq. (7.67), the following matrix equation can be obtained

$$\left([\tilde{k}^e] - \omega^2[\tilde{m}^e]\right) \{\vartheta^e\} = \{\tilde{\gamma}^e\} \tag{7.69}$$

where

$$[\tilde{k}^e] = [T^e]^T[\hat{k}^e][T^e] \tag{7.70}$$

and

$$\{\tilde{\gamma}^e\} = [T^e]^T\{\hat{\gamma}^e\} \tag{7.71}$$

In Eq. (7.69) both of $\{\vartheta^e\}$ and $\{\tilde{\gamma}^e\}$ are defined on the local-global coordinate system. Then, the overall eigenvalue equation system represented by Eq. (3.32) can be obtained by assembling the transformed local-global element eigenvalue equation (7.69) for all elements and by considering the joint equilibrium equations, (7.51) and (7.64) for the general joints and inclined rollers, respectively.

7.5 Problems

In all problems analyzed, the rotary inertia of the beam member is neglected.

For the first two sample problems solved, DOF per element is the same for all elements. The DOF of flexural deformation, $\bar{N}^e_{B_u}$, and the DOF of axial deformation, N^e_A, are the same. Defining $\Delta\zeta = 1./(\bar{N}^e_{B_u} - 1)$, the interior discrete points for defining the element-based discrete eigenvalue equations of flexural vibration are located at $\zeta = (p-1)\Delta\zeta$, $p = 3, ..., \bar{N}^e_{B_u} - 2$ while the interior discrete points for defining the element-based discrete eigenvalue equations are located at $\zeta = (p-1)\Delta\zeta$, $p = 2, ..., \bar{N}^e_{B_u} - 1$.

The first problem solved involves the lateral vibration of a fixed-roller supported ASTM Standard Strong steel pipe having a concentrated mass at the mid-span. The nominal diameter of the pipe is 10 in with the outside diameter 10.75 in and inside diameter 10.02 in. The area of cross section is 11.9 in^2, while the moment of inertia of the cross section is 161 in^4. The length of the pipe is 40 ft. The weight of the concentrated mass and the weight of the pipe per feet are 1000 lb and 40.48 lb, respectively. The Young's modulus of the pipe is $E = 2.9 \times 10^7$ psi. In carrying out the DQEM analysis, two elements were used to model the pipe with the mid-span being the inter-element boundary. The rotary inertia of the concentrated mass is neglected. The convergence can be assured by the increase of DOF per element. In this analysis, Hermite EDQ model with equally spaced nodes was used to define the element basis discretization. Numerical results of the first five natural frequencies are summarized and listed in Table 7.1.

Table 7.1. Natural frequencies of a fixed-roller supported beam having a concentrated mass at mid-span (cycles/sec)

DOF per element	ω_1	ω_2	ω_3	ω_4	ω_5
6	$.1297143\times10^2$	$.4137066\times10^2$	$.8333984\times10^2$	$.1295109\times10^3$	$.3552924\times10^3$
8	$.1349680\times10^2$	$.4722486\times10^2$	$.9010459\times10^2$	$.1739265\times10^3$	$.2362307\times10^3$
10	$.1347260\times10^2$	$.4560830\times10^2$	$.9261325\times10^2$	$.1628337\times10^3$	
12	$.1347292\times10^2$	$.4562152\times10^2$	$.9229664\times10^2$	$.1629238\times10^3$	$.2381861\times10^3$
14	$.1347287\times10^2$	$.4560231\times10^2$	$.9235354\times10^2$	$.1629862\times10^3$	$.2418579\times10^3$

The second problem solved involves the in-plane vibration of a 2-D frame shown in Fig. 7.4. The frame is composed of three members represented by three elements using the Lagrange DQ model with equally spaced nodes to discretize the axial vibration and the $C^1 - C^0 - C^1$ EDQ model generated by using the equivalent Lagrange DQ model with the two auxiliary nodes inside the physical EDQ model to discretize the flexural vibration. The results

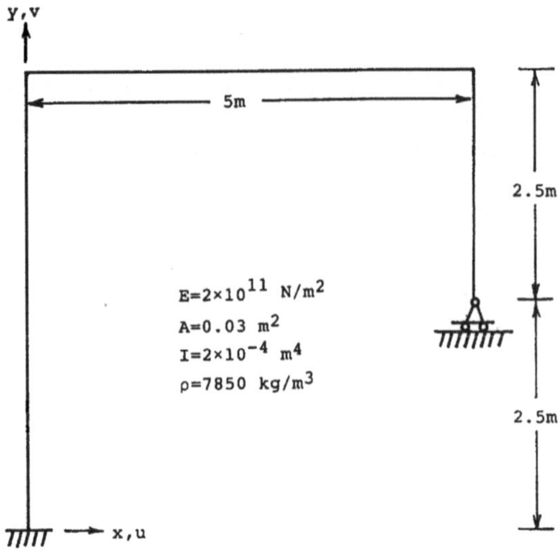

Fig. 7.4. A rigid frame

Table 7.2. Natural frequencies of a rigid frame (cycles/sec)

Order of appr.	ω_1	ω_2	ω_3	ω_4	ω_5
4	$.3804234\times10^2$	$.1637141\times10^3$	$.3021523\times10^3$	$.3310787\times10^3$	$.1718983\times10^4$
6	$.3538218\times10^2$	$.1383518\times10^3$	$.2551181\times10^3$	$.3680264\times10^3$	$.5911810\times10^3$
8	$.3501835\times10^2$	$.1365928\times10^3$	$.2609614\times10^3$	$.3626262\times10^3$	$.8123097\times10^3$
10	$.3468862\times10^2$	$.1349798\times10^3$	$.2636522\times10^3$	$.3613131\times10^3$	$.7282262\times10^3$
12	$.3445046\times10^2$	$.1340322\times10^3$	$.2652008\times10^3$	$.3607489\times10^3$	$.7352545\times10^3$
14	$.3426740\times10^2$	$.1333968\times10^3$	$.2661616\times10^3$	$.3604070\times10^3$	$.7352840\times10^3$

can converge by gradually increasing the order of approximation. Numerical results of the first five natural frequencies are summarized and listed in Table 7.2.

The last problem solved involves the in-plane vibration of a frame having an inclined roller, which is shown in Fig. 7.5. In this analysis, Chebyshev DQ model is used to define the axial discretization while the Chebyshev $C^1-C^0-C^1$ EDQ model is used to define the flexural discretization. Numerical results of the first five natural frequencies are summarized and listed in Table 7.3.

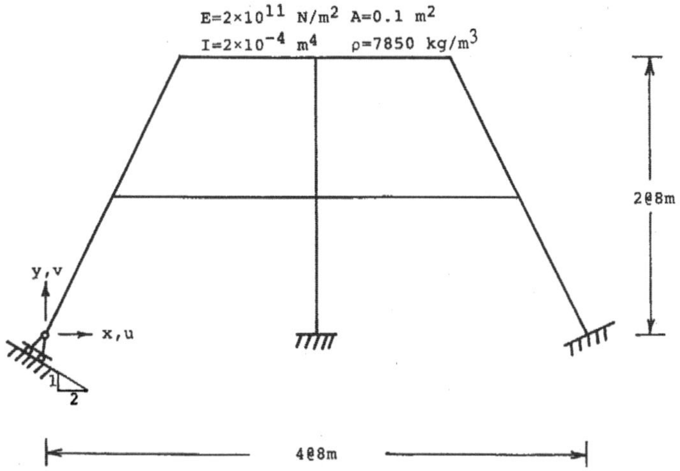

Fig. 7.5. A frame having an inclined roller

Table 7.3. Natural frequencies of a frame having an inclined roller (cycles/sec)

Order of appr.	ω_1	ω_2	ω_3	ω_4	ω_5
5	$.5475179\times10^1$	$.9100309\times10^1$	$.2007844\times10^2$	$.2402838\times10^2$	$.3027174\times10^2$
7	$.5038403\times10^1$	$.7486477\times10^1$	$.1580610\times10^2$	$.2505351\times10^2$	$.2951792\times10^2$
9	$.4933217\times10^1$	$.7477772\times10^1$	$.1560026\times10^2$	$.2486025\times10^2$	$.2924089\times10^2$
11	$.4868941\times10^1$	$.7447270\times10^1$	$.1529314\times10^2$	$.2484656\times10^2$	$.2924195\times10^2$
13	$.4830217\times10^1$	$.7429235\times10^1$	$.1511627\times10^2$	$.2483721\times10^2$	$.2924107\times10^2$
15	$.4805049\times10^1$	$.7417450\times10^1$	$.1500265\times10^2$	$.2483068\times10^2$	$.2924067\times10^2$

8

DQEM Analysis of Timoshenko Beam Structures

The Timoshenko beam theory is suitable for modelling moderately thick beam members in solving structural mechanics problems. Static deflection and vibration analysis models using DQEM are developed. The related numerical procedures are stated.

8.1 Static Deflection of Timoshenko Beam

8.1.1 Fundamental Relations

The equations of nonprismatic Timoshenko beam theory are first summarized. Consider that the beam resting on a Winkler elastic foundation, shown in Fig. 8.1, is shear deformable. The two displacement parameters employed are the transverse displacement w and the bending rotation ψ. Assume that the beam is homogeneous and isotropic. The differential equilibrium equations

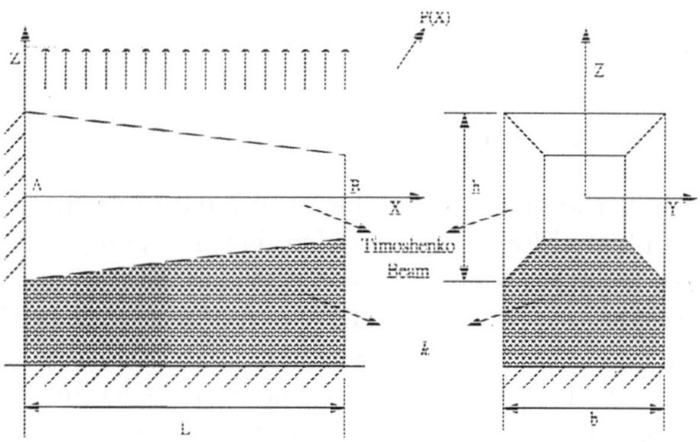

Fig. 8.1. A nonprismatic Timoshenko beam resting on a Winkler elastic foundation

of shear deformable Timoshenko beam resting on an elastic foundation are expressed by [95]

$$G\frac{d}{dx}\left[k_sA\left(\frac{dw}{dx}+\psi\right)\right]+kw=-p \tag{8.1}$$

and

$$E\frac{d}{dx}\left(I\frac{d\psi}{dx}\right)-k_sGA\left(\frac{dw}{dx}+\psi\right)=0 \tag{8.2}$$

where k is the foundation constant, I is moment of inertia of the cross-sectional area and k_s is shear correction coefficient. The stress resultants of bending moment and shear force are

$$M=EI\frac{d\psi}{dx} \tag{8.3}$$

and

$$V=k_sGA\left(\psi+\frac{dw}{dx}\right), \tag{8.4}$$

respectively. The kinematic boundary conditions are

$$w=\bar{w} \tag{8.5}$$

and

$$\psi=\bar{\psi} \tag{8.6}$$

where $\bar{w}$ and $\bar{\psi}$ are prescribed transverse displacement and bending rotation, respectively. Let $\bar{V}$ and $\bar{M}$ denote the transverse force and moment, respectively, applied on the natural boundary. The natural boundary conditions can be expressed by

$$k_sGA\left(\psi+\frac{dw}{dx}\right)=\bar{V} \tag{8.7}$$

and

$$EI\frac{d\psi}{dx}=\bar{M} \tag{8.8}$$

8.1.2 DQEM Formulation

The DQ which uses the function variable at element nodes to define the DQ discretization is adopted for the formulation. For analyzing the nonprismatic shear deformable beam problems, two-node element can be used for elements having no distributed external load. But the number of element node must be at least three for elements having distributed external cause. Assume that shear correction coefficient, shear modulus and Young's modulus are constant in an element. Then by using Eq. (3.6) and DQ, Eqs. (8.7) and (8.8) can be discretized which show to have the following forms:

$$\left\{\frac{k_s^e G^e}{(l^e)^2}\left[\sum_{\bar{\beta}=1}^{N^e} D_{(\alpha)\bar{\beta}}^{e\xi} A_{\bar{\beta}}^e \sum_{\beta=1}^{N^e} D_{\alpha\beta}^{e\xi} + A_{(\alpha)}^e \sum_{\beta=1}^{N^e} D_{\alpha\beta}^{e\xi^2}\right] + k_{(\alpha)}^e \delta_{\alpha\beta}\right\} w_\beta^e$$
$$+\frac{k_s^e G^e A_{(\alpha)}^e}{l^e} \sum_{\beta=1}^{N^e} D_{\alpha\beta}^{e\xi} \psi_\beta^e + \frac{k_s^e G^e}{l^e} \sum_{\bar{\beta}=1}^{N^e} D_{(\alpha)\bar{\beta}}^{e\xi} A_{\bar{\beta}}^e \psi_\alpha^e = 0,$$
$$\alpha = 2, 3, ..., N^e - 1 \tag{8.9}$$

$$-\frac{k_s^e G^e A_{(\alpha)}^e}{l^e} \sum_{\beta=1}^{N^e} D_{\alpha\beta}^{e\xi} w_\beta^e - k_s^e G^e A_{(\alpha)}^e \psi_\alpha^e + \frac{E^e}{(l^e)^2}\left[\sum_{\bar{\beta}=1}^{N^e} D_{(\alpha)\bar{\beta}}^{e\xi} I_{\bar{\beta}}^e \sum_{\beta=1}^{N^e} D_{\alpha\beta}^{e\xi}\right.$$
$$\left. + I_{(\alpha)}^e \sum_{\beta=1}^{N^e} D_{\alpha\beta}^{e\xi^2}\right] \psi_\beta^e = 0, \quad \alpha = 2, 3, ..., N^e - 1 \tag{8.10}$$

The bending moment and shear force at an element node α in the element can be obtained by using Eqs. (8.3) and (8.4)

$$M_\alpha^e = \frac{E^e I_{(\alpha)}^e}{l^e} \sum_{\beta=1}^{N^e} D_{\alpha\beta}^{e\xi} \psi_\beta^e \tag{8.11}$$

$$V_\alpha^e = k_s^e G^e A_{(\alpha)}^e \left(\psi_\alpha^e + \frac{1}{l^e} \sum_{\beta=1}^{N^e} D_{\alpha\beta}^{e\xi} w_\beta^e\right) \tag{8.12}$$

kinematic transition conditions are continuities of transverse displacement and bending rotation. Let $x = x^{i,i+1}$ denote the inter-element boundary of two adjacent elements i and $i+1$. The discrete kinematic transition conditions are

$$w_{N^i}^i = w_1^{i+1} \tag{8.13}$$

and

$$\psi_{N^i}^i = \psi_1^{i+1} \tag{8.14}$$

Natural transition conditions are equilibriums of discrete element boundary forces and external loads at the inter-element boundary of two adjacent elements i and $i+1$. Figure 8.2 shows the forces on the inter-element boundary in which $M^{i,i+1}$, $P^{i,i+1}$ and $\bar{k}^{i,i+1}$ are moment, transversely applied load and constant of the concentrated spring located at the inter-element boundary. The equilibrium of lateral forces is expressed by

$$k_s^i G^i A_{N^i}^i \left(\psi_{N^i}^i + \frac{dw_{N^i}^i}{dx}\right) - k_s^{i+1} G^{i+1} A_1^{i+1} \left(\psi_1^{i+1} + \frac{dw_1^{i+1}}{dx}\right)$$
$$- \bar{k}^{i,i+1} w^{i,i+1} = P^{i,i+1} \tag{8.15}$$

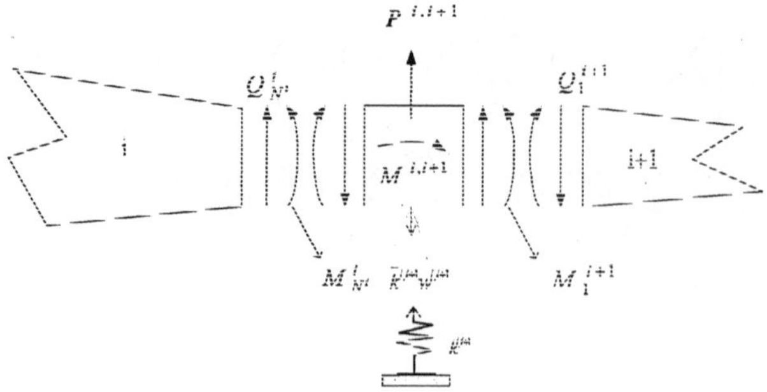

Fig. 8.2. Forces on the inter-element boundary

The introduction of mapping transformation and DQ discretization into the above equation leads to

$$k_s^i G^i A_{N^i}^i \left(\psi_{N^i}^i + \frac{1}{l^i} \sum_{\alpha=1}^{N^i} D_{N^i\alpha}^{i\xi} w_\alpha^i \right) - k_s^{i+1} G^{i+1} A_1^{i+1} \left(\psi_1^{i+1} + \frac{1}{l^{i+1}} \sum_{\beta=1}^{N^{i+1}} D_{1\beta}^{(i+1)\xi} w_\beta^{i+1} \right) - \bar{k}^{i,i+1} w_{N^i}^{i,i+1} = P^{i,i+1} \tag{8.16}$$

The equilibrium of moments is expressed by

$$-E^i I_{N^i}^i \frac{d\psi_{N^i}^i}{dx^i} + E^{i+1} I_1^{i+1} \frac{d\psi_1^{i+1}}{dx^{i+1}} = M^{i,i+1} \tag{8.17}$$

The above equation can be similarly discretized

$$-\frac{E^i I_{N^i}^i}{l^i} \sum_{\alpha=1}^{N^i} D_{N^i\alpha}^{i\xi} \psi_\alpha^i + \frac{E^{i+1} I_1^{i+1}}{l^{i+1}} \sum_{\beta=1}^{N^{i+1}} D_{1\beta}^{(i+1)\xi} \psi_\beta^{i+1} = M^{i,i+1} \tag{8.18}$$

For the pin connected inter-element boundary, Eqs. (8.14) and (8.18) are replaced by the following two equations

$$\frac{E^i I_{N^i}^i}{l^i} \sum_{\beta=1}^{N^i} D_{N^i\beta}^{i\xi} \psi_\beta^i = 0 \tag{8.19}$$

and

$$\frac{E^{i+1} I_1^{i+1}}{l^{i+1}} \sum_{\beta=1}^{N^{i+1}} D_{1\beta}^{(i+1)\xi} \psi_\beta^{i+1} = 0 \tag{8.20}$$

If forced displacements are imposed on an inter-element boundary, the discrete natural transition conditions (8.16) and (8.18) are replaced by the condition equations of forced displacement $w^i_{N^i} = \bar{w}^{i,i+1}$ and displacement gradient $\frac{1}{l^i}\sum_{\beta=1}^{N^i} D^{i\xi}_{N^i\beta} w^i_\beta = \frac{d\bar{w}^{i,i+1}}{dx}$ where $\bar{w}^{i,i+1}$ and $\frac{d\bar{w}^{i,i+1}}{dx}$ are prescribed values.

Letting element m be an element having one or more nodes on the kinematic boundary, the kinematic boundary conditions (8.13) and (8.14) can be rewritten as

$$w^m_{I^m} = \bar{w}^m_{I^m}, \quad I^m = 1 \quad \text{or} \quad N^m \tag{8.21}$$

and

$$\psi^m_{I^m} = \bar{\psi}^m_{I^m}, \quad I^m = 1 \quad \text{or} \quad N^m \tag{8.22}$$

Also letting element n be an element having one or more nodes on the natural boundary, the discrete natural boundary conditions can be defined using Eqs. (8.7) and (8.8)

$$k^n_s G^n A^n_{I^n}\left(\psi^n_{I^n} + \frac{1}{l^n}\sum_{\beta=1}^{N^n} D^{n\xi}_{I^n\beta} w^n_\beta\right) = \nu^n \bar{V}, \quad I^n = 1 \quad \text{or} \quad N^n \tag{8.23}$$

$$\frac{E^n I^n_{I^n}}{l^n}\sum_{\beta=1}^{N^n} D^{n\xi}_{I^n\beta}\psi^n_\beta = -\nu^n \bar{M}, \quad I^n = 1 \quad \text{or} \quad N^n \tag{8.24}$$

where ν^n is a sign indicator equal to 1 for the right boundary and -1 for the left boundary.

8.1.3 Assemblage

With the kinematic transition conditions (8.13) and (8.14) defined at the inter-element boundary in mind, upon assembling the discrete element equilibrium equations (8.9) and (8.10) for elements having more than two nodes, discrete natural boundary conditions (8.15) and (8.16) for all inter-element boundaries, and discrete natural boundary conditions (8.23) and (8.24) for boundary elements with natural boundary, an overall linear algebraic system can be obtained. It is the overall stiffness equation which is represented by Eq. (3.18). The overall stiffness equation can be solved by Considering the kinematic boundary conditions (8.21) and (8.22) for boundary elements with kinematic boundary.

Like FEM, the assemblage is based on an element by element procedure. When assembling the discrete equations of element e, the discrete element equilibrium equations (8.9) and (8.10), and the four discrete element boundary forces, expressed by displacement parameters, at the two element boundary nodes are directly assembled to the overall discrete equation system. Consequently, an element basis explicit matrix equation, containing discrete element equilibrium equations and discrete element boundary forces placed at the first and last two rows, is not necessary to be formed in the assembling process. This element basis explicit matrix equation is an element stiffness equation which can be expressed by Eq. (3.20) in which $[k^e]$ is a $2N^e \times 2N^e$ element stiffness matrix,

$$\{\delta^e\} = \lfloor\ w_1^e \quad \psi_1^e \quad \ldots \quad w_{N^e}^e \quad \psi_{N^e}^e\ \rfloor^T \tag{8.25}$$

is the element displacement vector, and $\{r^e\}$ is the element load vector. The element load vector can be expressed by

$$\begin{aligned}\{r^e\} &= \lfloor -V_1^e - \bar{R}^{e-1,e}\ M_1^e\ -p_2^e\ 0\ \ldots\ -p_{N^e-1}^e\ 0\ V_{N^e}^e\ -M_{N^e}^e\ \rfloor^T, \text{ for} \\ &\quad \text{an interior element with } \bar{R}^{e-1,e} = \bar{k}^{e-1,e} w^{e-1,e} \text{ the spring force,} \\ &= \lfloor -V_1^1 - \bar{R}^1\ \ M_1^1\ \ -p_2^1\ \ 0\ \ \ldots\ \ -p_{N^1-1}^1\ \ 0\ \ V_{N^1}^1\ \ -M_{N^1}^1\ \rfloor^T, \\ &\quad \text{for the first or left boundary element,} \\ &= \lfloor -V_1^s - \bar{R}^s\ M_1^s\ -p_2^s\ 0\ \ldots\ -p_{N^s-1}^s\ 0\ V_{N^s}^s - \bar{R}^s\ -M_{N^s}^s\ \rfloor^T, \text{ for} \\ &\quad \text{the last or right boundary element}\end{aligned} \tag{8.26}$$

As the element stiffness equation contains discrete resultant forces and spring forces at the two element boundary nodes, equilibriums of resultant forces, spring forces and external forces at the inter-element boundary of two adjacent elements and the natural boundary are exactly satisfied in the assembling process. Consequently, the DQEM is different from FEM which needs to form the element stiffness equation, and which neglects the exact equilibriums.

8.1.4 Problems

A cantilever Timoshenko beam composed of two prismatic elements resting on a Winkler foundation and subjected to a concentrated tip load was solved. The structure is shown in Fig. 8.3. The values of Young's modulus, shear modulus, shear correction coefficient and foundation constant are: $E = 2.66666667\ Pa$,

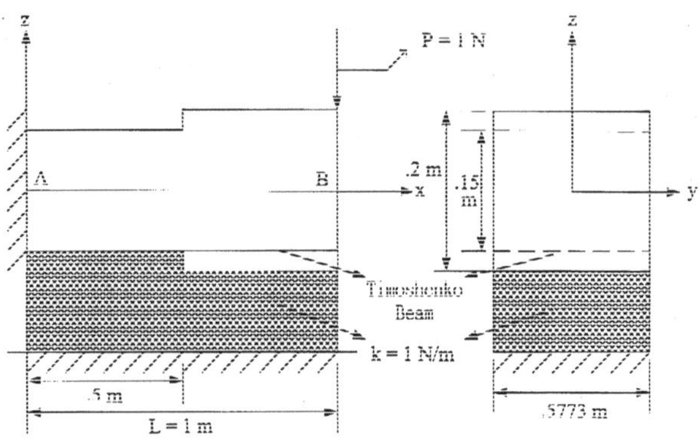

Fig. 8.3. A cantilever Timoshenko beam composed of two prismatic elements resting on a Winkler foundation

$G = 1\ Pa$, $k_s = .66666667$ and $k = 1\ N/m$. In the analysis, Chebyshev DQ model is adopted for the element basis discretization. Chebyshev polynomials are used to implicitly calculate the weighting coefficients. In an element, the position of DQ nodes are defined by the roots of Chebyshev polynomials. Numerical results obtained are summarized and listed in Table 8.1. It shows that the results can converge by either increasing the order of approximation or the number of elements used to model the beam structure.

Table 8.1. Results of the beam composed of two prismatic elements resting on an elastic foundation and subjected to a concentrated tip load

Nodes per element grid	Number of elements	Displacement at B (mm)	Bending moment at A ($N{\cdot}m$)	Shear force at A (N)
3	2	$.1000193\times10^{1}$	$.8326154\times10^{-2}$	$-.2630099\times10^{0}$
	4	$.8962702\times10^{1}$	$.7011425\times10^{-2}$	$-.5999550\times10^{-1}$
	6	$.8838056\times10^{1}$	$.5646578\times10^{-2}$	$-.3119224\times10^{-1}$
5	2	$.7898007\times10^{1}$	$.2872152\times10^{-2}$	$-.2738131\times10^{-1}$
	4	$.8669395\times10^{1}$	$.4164640\times10^{-2}$	$-.1506932\times10^{-1}$
	6	$.8745051\times10^{1}$	$.4201717\times10^{-2}$	$-.1400644\times10^{-1}$
7	2	$.8745051\times10^{1}$	$.4137852\times10^{-2}$	$-.1330204\times10^{-1}$
	4	$.8772212\times10^{1}$	$.4205389\times10^{-2}$	$-.1372138\times10^{-1}$
	6	$.8772605\times10^{1}$	$.4206730\times10^{-2}$	$-.1372650\times10^{-1}$
9	2	$.8772902\times10^{1}$	$.4207227\times10^{-2}$	$-.1369951\times10^{-1}$
	4	$.8772644\times10^{1}$	$.4206849\times10^{-2}$	$-.1372682\times10^{-1}$
	6	$.8772642\times10^{1}$	$.4206852\times10^{-2}$	$-.1372692\times10^{-1}$

Another problem solved involves the static deflection of a nonprismatic cantilever Timoshenko beam resting on a Winkler foundation. The beam is subjected to a concentrated tip load and shown in Fig. 8.4. With the fixed end A being the origin of the coordinate system, the variation of depth is $d(z) = d_0(1 - z/L + z^2/2L^2)$ with $d_0 = .2\ m$ and $L = 1\ m$. The values of Young's modulus, shear modulus, shear correction coefficient and foundation constant are: $E = 2.6\ Pa$, $G = 1\ Pa$, $k_s = .85$ and $k = 1\ N/m$. In carrying out the DQEM analysis, Lagrange DQ model is used for the element basis discretization. Elements and nodes in an element are equally spaced. The DQEM results are summarized and listed in Table 8.2. It also shows excellent convergence.

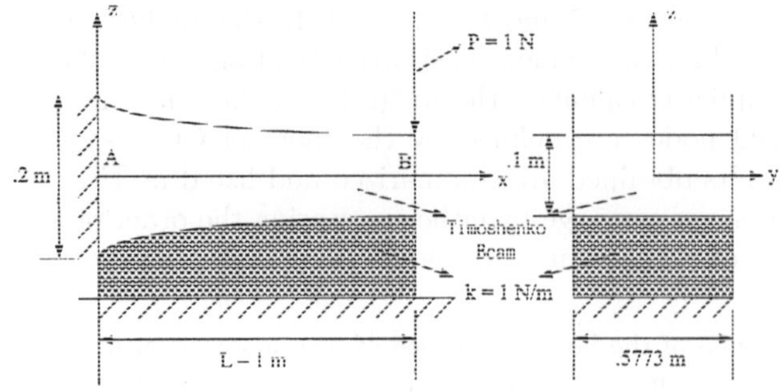

Fig. 8.4. A nonprismatic cantilever Timoshenko beam resting on a Winkler foundation

Table 8.2. Results of a nonprismatic beam resting on an elastic foundation and subjected to a concentrated tip load

Nodes per element grid	Number of elements	Displacement at B (mm)	Bending moment at A ($N{\cdot}m$)	Shear force at A (N)
3	2	$.15669494\times10^{2}$	$.25558355\times10^{-1}$	$-.42480499\times10^{0}$
	4	$.13490726\times10^{2}$	$.12921292\times10^{-1}$	$-.45070993\times10^{-1}$
	6	$.12972890\times10^{2}$	$.73025184\times10^{-2}$	$-.13758632\times10^{-1}$
5	2	$.96386025\times10^{1}$	$.19531531\times10^{-1}$	$-.17100329\times10^{-1}$
	4	$.11430845\times10^{2}$	$.34378907\times10^{-2}$	$-.76044995\times10^{-2}$
	6	$.12115994\times10^{2}$	$.34054717\times10^{-2}$	$-.47640084\times10^{-2}$
7	2	$.11223759\times10^{2}$	$.18152135\times10^{-2}$	$-.69916636\times10^{-2}$
	4	$.12337228\times10^{2}$	$.33698395\times10^{-2}$	$-.41576248\times10^{-2}$
	6	$.12426370\times10^{2}$	$.33836369\times10^{-2}$	$-.39137329\times10^{-2}$
9	2	$.1222707\times10^{2}$	$.31169229\times10^{-2}$	$-.44340489\times10^{-2}$
	4	$.12435530\times10^{2}$	$.33842902\times10^{-2}$	$-.38963901\times10^{-2}$
	6	$.12440173\times10^{2}$	$.33849592\times10^{-2}$	$-.38848593\times10^{-2}$

8.2 Free Vibration of Timoshenko Beam

By using the modal displacement parameters W and Ψ to replace w and ψ, respectively, $\rho A\omega^2 W$ to replace the distributed external load p, and considering the effect of rotary inertia, the static equilibrium equations (8.1) and (8.2) can be converted into the following differential eigenvalue equations of dynamic equilibrium

$$\frac{d}{dx}\left[k_s GA\left(\frac{dW}{dx}+\Psi\right)\right]+kW+\rho A\omega^2 W=0 \tag{8.27}$$

$$\frac{d}{dx}\left(EI\frac{d\Psi}{dx}\right) - k_s GA\left(\frac{dW}{dx} + \Psi\right) + \rho I\omega^2\Psi = 0 \tag{8.28}$$

Assume that a body having the mass $\tilde{M}^n$ and moment of inertia $\tilde{I}^n$ is attached to the natural boundary. Then by using the inertia force $-\nu^n\tilde{M}^n\omega^2 W$ to replace $\bar{V}$ in Eq. (8.7), and the inertia moment $\nu^n\tilde{I}\omega^2\Psi$ to replace $\bar{M}$ in Eq. (8.8), the corresponding natural boundary conditions can be expressed by

$$k_s GA\left(\Psi + \frac{dW}{dx}\right) - \nu^n\bar{k}^n W + \nu^n\tilde{M}^n\omega^2 W = 0 \tag{8.29}$$

and

$$EI\frac{d\Psi}{dx} - \nu^n\tilde{I}^n\Psi = 0 \tag{8.30}$$

where ν^n is the sign indicator equal to 1 for the right boundary and -1 for the left boundary.

8.2.1 DQEM Formulation

Assume that the mass density ρ is constant in an element. Equations (8.27) and (8.28) at a node α in an element e can be discretized

$$\begin{aligned}&\left\{\frac{k_s^e G^e}{(l^e)^2}\left[\sum_{\bar{\beta}=1}^{N^e} D_{(\alpha)\bar{\beta}}^{e\xi} A_{\bar{\beta}}^e \sum_{\beta=1}^{N^e} D_{\alpha\beta}^{e\xi} + A_{(\alpha)}^e \sum_{\beta=1}^{N^e} D_{\alpha\beta}^{e\xi^2}\right] + k_{(\alpha)}^e \delta_{\alpha\beta}\right\} W_\beta^e \\ &+ \frac{k_s^e G^e A_{(\alpha)}^e}{l^e}\sum_{\beta=1}^{N^e} D_{\alpha\beta}^{e\xi}\Psi_\beta^e + \frac{k_s^e G^e}{l^e}\sum_{\bar{\beta}=1}^{N^e} D_{(\alpha)\bar{\beta}}^{e\xi} A_{\bar{\beta}}^e \Psi_\alpha^e + \rho^e A_{(\alpha)}^e \omega^2 W_\alpha \\ &= 0, \quad \alpha = 2, 3, ..., N^e - 1\end{aligned} \tag{8.31}$$

$$\begin{aligned}&-\frac{k_s^e G^e A_{(\alpha)}^e}{l^e}\sum_{\beta=1}^{N^e} D_{\alpha\beta}^{e\xi} W_\beta^e - k_s^e G^e A_{(\alpha)}^e \Psi_\alpha^e + \frac{E^e}{(l^e)^2}\left[\sum_{\bar{\beta}=1}^{N^e} D_{(\alpha)\bar{\beta}}^{e\xi} I_{\bar{\beta}}^e \sum_{\beta=1}^{N^e} D_{\alpha\beta}^{e\xi}\right. \\ &\left. + I_{(\alpha)}^e \sum_{\beta=1}^{N^e} D_{\alpha\beta}^{e\xi^2}\right]\Psi_\beta^e + \rho^e I_{(\alpha)}^e \omega^2 \Psi_\alpha = 0, \quad \alpha = 2, 3, ..., N^e - 1\end{aligned} \tag{8.32}$$

Assume that a body having the mass $\tilde{M}^{i,i+1}$ and moment of inertia $\tilde{I}^{i,i+1}$ is attached to the inter-element boundary of two adjacent element i and $i+1$. By using the inertia force $-\tilde{M}^{i,i+1}\omega^2 W_{N^i}^{i,i+1}$ to replace $P^{i,i+1}$ in Eq. (8.16), the following discrete dynamic equilibrium equation of lateral forces at the inter-element boundary can be obtained

$$k_s^i G^i A_{N^i}^i\left(\Psi_{N^i}^i + \frac{1}{l^i}\sum_{\alpha=1}^{N^i} D_{N^i\alpha}^{i\xi} W_\alpha^i\right)$$

$$
\begin{aligned}
& - k_s^{i+1} G^{i+1} A_1^{i+1} \left(\Psi_1^{i+1} + \frac{1}{l^{i+1}} \sum_{\beta=1}^{N^{i+1}} D_{1\beta}^{(i+1)\xi} W_\beta^{i+1} \right) \\
& - \bar{k}^{i,i+1} W_{N^i}^i + \tilde{M}^{i,i+1} \omega^2 W_{N^i}^i = 0
\end{aligned} \tag{8.33}
$$

Also by using the inertia moment $\tilde{I}^{i,i+1}\omega^2\Psi_{N^i}^i$ to replace $M^{i,i+1}$ in Eq. (8.18), the following discrete dynamic equilibrium equation of moments can be obtained

$$
\begin{aligned}
& -\frac{E^i I_{N^i}^i}{l^i} \sum_{\alpha=1}^{N^i} D_{N^i\alpha}^{i\xi} \Psi_\alpha^i - \frac{E^{i+1} I_1^{i+1}}{l^{i+1}} \sum_{\beta=1}^{N^{i+1}} D_{1\beta}^{(i+1)\xi} \Psi_\beta^{i+1} - \tilde{I}^{i,i+1} \omega^2 W^{i,i+1} \\
& = 0
\end{aligned} \tag{8.34}
$$

Using Eqs. (8.33) and (8.34), the following discrete natural boundary conditions can be obtained

$$
\begin{aligned}
& k_s^n G^n A_{I^n}^n \left(\Psi_{I^n}^n + \frac{1}{l^n} \sum_{\beta=1}^{N^n} D_{I^n\beta}^{n\xi} W_\beta^n \right) - \nu^n \bar{k}^n W_{I^n}^n + \nu^n \tilde{M}^n \omega^2 W_{I^n}^n = 0, \\
& I^n = 1 \quad \text{or} \quad N^n
\end{aligned} \tag{8.35}
$$

$$
\frac{E^n I_{I^n}^n}{l^n} \sum_{\beta=1}^{N^n} D_{I^n\beta}^{n\xi} \Psi_\beta^n - \nu^n \tilde{I}^n \omega^2 \Psi_{I^n}^n = 0, \quad I^n = 1 \quad \text{or} \quad N^n \tag{8.36}
$$

8.2.2 Assemblage

With the kinematic transition conditions (8.13) and (8.14) in mind, upon assembling the discrete element eigenvalue equations (8.31) and (8.32) for elements having more than two nodes, discrete natural transition conditions (8.33) and (8.34), and discrete natural boundary conditions (8.35) and (8.36), an overall discrete eigenvalue equation system of nonprismatic Timoshenko beam problems can be obtained. The overall eigenvalue equation considering the discrete kinematic boundary conditions is represented by Eq. (3.32). If the rotary inertia is neglected, modal displacement parameters Ψ_αs at all element interior nodes can be eliminated before solving the eigenvalue system. If no mass is attached to an inter-element boundary or natural boundary, the modal displacement parameters associated with it can also be eliminated. Considering the nonexistence of inertia forces for some component equations existing in the overall discrete eigenvalue equation system, the procedures stated in Subsection 3.2.3 for reducing the solution system and efficiently solving the eigenvalue problem can be used to calculate the eigenpairs of a Timoshenko beam structure resting on a Winkler foundation.

8.2.3 Problems

The free vibration of a uniform cantilever Timoshenko beam was solved. The non-dimensional rotary inertia parameter of the beam defined as $r = \gamma/L$ with γ being the radius of gyration of the cross section is selected to be 0.05, while the non-dimensional shear flexibility parameter defined as $s = (EI/k_sAGL^2)^{1/2}$ is selected to be 0.1. A representative beam has the following material and geometrical properties: rectangular cross section with the depth of the cross section $h = .17320508\ m$, the width $b = .57735027\ m$, the beam length $L = 1\ m$, the shear correction coefficient $k_s = .66666667$, mass density $\rho = .66666667 \times 10^{-4}\ kg/m^3$, Young's modulus $E = 2.66666667$ Pa and shear modulus $G = 1\ Pa$. Let b_i denote the analytical solutions of the natural frequencies of Euler-Bernoulli beam. Numerical results and analytical solutions of the ratios of the first three natural frequencies of the Timoshenko beam to b_i are summarized and listed in Table 8.3 [104]. It shows that the results can converge fast to analytical solutions by either increasing the number of nodes per element or the number of elements.

Table 8.3. The first three natural frequencies of a prismatic cantilever Timoshenko beam considering the effect of rotary inertia

Nodes per element	Number of elements	ω_1/b_1	ω_2/b_2	ω_3/b_3
3	4	1.303559	1.182778	1.060439
	6	1.130902	1.002834	.8846225
5	2	.9544300	.7495082	.6532353
	4	.9711311	.8370633	.7052438
	6	.9720688	.8431605	.7195133
7	2	.9723885	.8475282	.7305820
	4	.9716780	.8447689	.7237495
	6	.9723245	.8447462	.7234423
9	2	.9720762	.8446972	.7231421
	4	.9731479	.8447698	.7234108
	6	.9710939	.8446963	.7234039
Analytical sol.		.9723	.8447	.7234

Free vibration of the cantilever beam composed of two prismatic elements and shown in Fig. 8.3 was solved. It was first analyzed without resting the composed beam on an Winkler foundation. The effect of rotary inertia is considered. Two elements adopting Chebyshev DQ model to define the element basis discretization are used to model the structure. Numerical results of the first four natural frequencies are summarized and listed in Table 8.4. It also

Table 8.4. The first four natural frequencies of a shear deformable cantilever beam composed of two prismatic elements considering the effect of rotary inertia (*cycles/sec*)

Nodes per element	ω_1	ω_2	ω_3	ω_4
3	$.5863078\times10^1$	$.3798482\times10^2$	$.1599888\times10^3$	$.2037513\times10^3$
5	$.2768917\times10^1$	$.1584239\times10^2$	$.4127034\times10^2$	$.6109054\times10^2$
7	$.2813875\times10^1$	$.1749714\times10^2$	$.4578586\times10^2$	$.7700861\times10^2$
9	$.2813415\times10^1$	$.1745975\times10^2$	$.4547572\times10^2$	$.7405805\times10^2$
11	$.2813242\times10^1$	$.1746005\times10^2$	$.4548344\times10^2$	$.7418299\times10^2$

shows that the convergence properties are excellent. The problem was reanalyzed by resting the composed beam on an Winkler foundation with the foundation constant $k = 1\ N/m$. Numerical results obtained by using the same DQEM modelling as the previous analysis are summarized and listed in Table 8.5. It shows the same convergence tendency. It also shows that the values of natural frequencies of lower modes increase more than the higher modes by resting the composed beam on an Winkler foundation.

Table 8.5. The first four natural frequencies of a shear deformable cantilever beam composed of two prismatic elements resting on a Winkler foundation and considering the effect of rotary inertia (*cycles/sec*)

Nodes per element	ω_1	ω_2	ω_3	ω_4
3	$.3627203\times10^2$	$.5688747\times10^2$	$.1310366\times10^3$	$.2529273\times10^3$
5	$.3591770\times10^2$	$.4057903\times10^2$	$.5616760\times10^2$	$.7204069\times10^2$
7	$.3590355\times10^2$	$.4129053\times10^2$	$.5948750\times10^2$	$.8575786\times10^2$
9	$.3590296\times10^2$	$.4127607\times10^2$	$.4547572\times10^2$	$.8323640\times10^2$

The last problem solved involves the free vibration of a nonprismatic cantilever beam resting on a Winkler foundation which is shown in Fig. 8.4. In the analysis, the effect of rotary inertia is considered. The same DQEM modelling as that is used for the static deflection analysis was adopted. Numerical results of the first three natural frequencies are summarized and listed in Table 8.6. It also shows that DQEM has excellent convergence properties.

Table 8.6. The first three natural frequencies of a shear deformable nonprismatic cantilever beam resting on an elastic foundation and considering the effect of rotary inertia (*cycles/sec*)

Nodes per element	Number of elements	ω_1	ω_2	ω_3
3	2	$.4443144\times10^2$	$.5689754\times10^2$	$.2001291\times10^3$
	4	$.4448764\times10^2$	$.4971019\times10^2$	$.7629094\times10^2$
	6	$.4393554\times10^2$	$.4877331\times10^2$	$.6706347\times10^2$
5	2	$.4273247\times10^2$	$.4749201\times10^2$	$.5705289\times10^2$
	4	$.4321339\times10^2$	$.4817797\times10^2$	$.5842086\times10^2$
	6	$.4324357\times10^2$	$.4823142\times10^2$	$.5910570\times10^2$
7	2	$.4327764\times10^2$	$.4835176\times10^2$	$.5925708\times10^2$
	4	$.4325222\times10^2$	$.4824901\times10^2$	$.5931240\times10^2$
	6	$.4325158\times10^2$	$.4824598\times10^2$	$.5930851\times10^2$
9	2	$.4325257\times10^2$	$.4825859\times10^2$	$.5929692\times10^2$
	4	$.4325147\times10^2$	$.4824596\times10^2$	$.5930803\times10^2$
	6	$.4325126\times10^2$	$.4824601\times10^2$	$.5930836\times10^2$

Table 5.6. The first three natural frequencies of a [illegible] cantilever beam resting on [illegible] elastic foundation and considering the effect of [illegible] rotary inertia [illegible]

Node [illegible]	Number of [illegible]	[illegible]
[illegible]	[illegible]	[illegible]

9

DQEM Analysis of Curved Beam Structures

DQEM is effective for solving out-of-plane deflection and in-plane deflection of curved beam structures. The developed DQEM algorithms neglect the effect of transverse shear deformation. The EDQ is used to the DQEM element discretization.

9.1 Out-of-Plane Deflection Analysis

9.1.1 Fundamental Relations

The fundamental equations of the deflection of out-of-plane nonprismatic curved beam structures are first summarized [105]. Consider the out-of-plane deflection of a circular beam shown in Fig. 9.1, the coordinate system $Oxyz$ is a Cartesian one with the origin O at the centroid of the cross section, the x- and y-axes coinciding with the principal axes of the cross section and the z-axis coinciding with the tangent to the center line at O. In Fig. 9.1, v is the displacement of centroid O in the direction of y-axis, β the angle of twist of the cross section about the z-axis, M_x the moment about the x-axis, M_z the

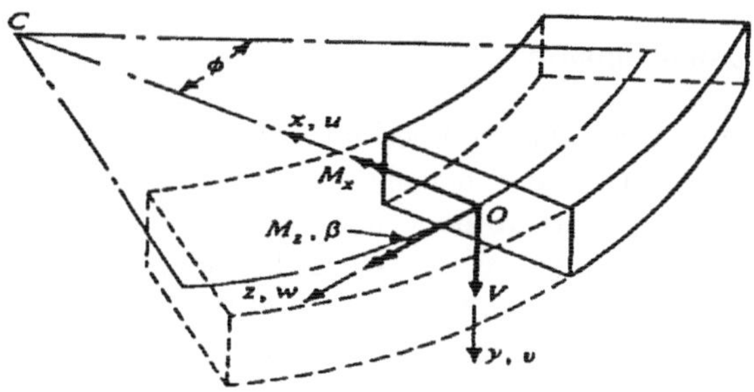

Fig. 9.1. Coordinates of the curved beam

moment about the z-axis, V the shear force in the direction of the y-axis and ϕ the angle measured from the reference end. Let EI and GJ be the flexural and torsional rigidities, respectively. The moments M_x and M_z, and shear force V for the nonprismatic curved beam can be expressed as

$$V = \frac{1}{R^2}\left\{\frac{d}{d\phi}\left[EI\left(\beta - \frac{1}{R}\frac{d^2v}{d\phi^2}\right)\right] + GJ\left(\frac{d\beta}{d\phi} + \frac{1}{R}\frac{dv}{d\phi}\right)\right\},$$

$$M_x = \frac{EI}{R}\left(\beta - \frac{1}{R}\frac{d^2v}{d\phi^2}\right), \quad M_z = \frac{GJ}{R}\left(\frac{d\beta}{d\phi} + \frac{1}{R}\frac{dv}{d\phi}\right) \tag{9.1}$$

Let $p(\phi)$ denote the distributed load in the y-direction per unit angle. Also let $m_z(\phi)$ denote the distributed torque per unit angle. The equilibrium equations of the curved beam are expressed as

$$\frac{1}{R^3}\frac{d}{d\phi}\left\{\frac{d}{d\phi}\left[EI\left(\beta - \frac{1}{R}\frac{d^2v}{d\phi^2}\right)\right] + GJ\left(\frac{d\beta}{d\phi} + \frac{1}{R}\frac{dv}{d\phi}\right)\right\} = -p,$$

$$\frac{1}{R^2}\frac{d}{d\phi}\left[GJ\left(\frac{d\beta}{d\phi} + \frac{1}{R}\frac{dv}{d\phi}\right)\right] - \frac{EI}{R^2}\left(\beta - \frac{1}{R}\frac{d^2v}{d\phi^2}\right) = -m_z \tag{9.2}$$

The kinematic boundary conditions at the kinematic boundary are

$$v = \bar{v}, \quad \beta = \bar{\beta} \tag{9.3}$$

where $\bar{v}$ and $\bar{\beta}$ are prescribed values. The natural boundary conditions at the natural boundary are

$$\frac{1}{R^2}\left[\frac{EI}{R}\frac{d^3v}{d\phi^3} - \frac{1}{R}\frac{d(EI)}{d\phi}\frac{d^2v}{d\phi^2} + \frac{GJ}{R}\frac{dv}{d\phi} + (EI + GJ)\frac{d\beta}{d\phi} + \frac{d(EI)}{d\phi}\beta\right] = \bar{V},$$

$$\frac{EI}{R}\left(\beta - \frac{1}{R}\frac{d^2v}{d\phi^2}\right) = \bar{M}_x, \quad \frac{GJ}{R}\left(\frac{d\beta}{d\phi} + \frac{1}{R}\frac{dv}{d\phi}\right) = \bar{M}_z \tag{9.4}$$

where $\bar{M}_x$, $\bar{M}_z$ and $\bar{V}$ are prescribed values.

9.1.2 DQEM Formulation

Since the second of Eqs. (9.2) is a second order differential equation, without using a certain technique to calculate the two equivalent nodal loads for the distributed torque of a two-node linear element and include them into the natural transition conditions or natural boundary conditions, the order of approximate angle of twist must at least be two, and the element needs at least one discrete point for defining the discrete element torsional equilibrium equation. The DQEM linear element is equivalent to the FEM linear element. Since the first of Eqs.(9.2) is a fourth order differential equation, without using a certain technique to calculate the four nodal loads for the distributed loads of a two-node cubic element and include them into the natural transition conditions or natural boundary conditions, the order of approximate lateral displacement must at least be four and the element needs at least one discrete point for defining one discrete element lateral equilibrium equation. The DQEM cubic element is equivalent to the FEM cubic Hermite element. Discrete points for defining discrete element equilibrium equations can be either in the interior of the element or on the element boundary.

In the present analysis model, only interior discrete points are used to define the discrete element equilibrium equations. In the numerical simulation, the nodes used to define the flexural discretizations and the nodes used to define the torsional discretization can be different. In addition to the DOF assigned to interior nodes, three DOF of the lateral displacement v, rotation about the x-axis and the angle of twist are assigned to each of the two element boundary nodes. Consequently, the second and last columns of the weighting coefficient matrices for the flexural discretization defined by using the natural coordinate, ξ, must be modified by multiplying each element in the two columns by $-R\Phi^e$ with Φ^e the range of open angle of the element. Let N_B^e and N_T^e denote the numbers of nodes for defining the flexural discretization and the torsional discretization, respectively, $\bar{N}_B^e$ and $\bar{N}_T^e$ denote the corresponding element degrees of freedom. Also let $\hat{N}_B^e$ denote the number of the corresponding interior discrete points for defining the v-related discrete element equilibrium equations plus the two element boundary nodes, and $D_{B\alpha r}^{e\xi^m}$ denote the corresponding weighting coefficients for the mth order derivative with respect to ξ. And the number of the corresponding interior discrete points for defining the β-related discrete element equilibrium equations plus the two element boundary nodes is denoted by $\hat{N}_T^e$ with the corresponding weighting coefficients for the mth order derivative with respect to ξ being denoted by $D_{T\beta s}^{e\xi^m}$. Then the numbers of discrete points need to be used to discretize the element lateral and torsional equilibrium equations are $\hat{N}_B^e - 2$ and $\bar{N}_T^e - 2$, respectively. Let $\bar{\vartheta}_i(\xi)$ denote the interpolation functions. $\beta^e(\xi)$ in an element can be expressed as $\beta^e(\xi) = \bar{\vartheta}_i(\xi)\tilde{\beta}_i^e$. The introduction of mapping and EDQ into the first of Eqs. (9.2) leads to the following discrete element equilibrium equation at a discrete point α in the element e

$$
\begin{aligned}
&-\frac{1}{R^3}\left\{\frac{(EI)^e_{(\alpha)}}{R(\Phi^e)^4}\sum_{r=1}^{\bar{N}_B^e} D_{B\alpha r}^{e\xi^4} + \frac{2}{R(\Phi^e)^3}\frac{d(EI)^e_{(\alpha)}}{d\phi^e}\sum_{r=1}^{\bar{N}_B^e} D_{B\alpha r}^{e\xi^3}\right.\\
&\left.+\frac{1}{R(\Phi^e)^2}\left[\frac{d^2(EI)^e_{(\alpha)}}{d(\phi^e)^2} - (GJ)^e_{(\alpha)}\right]\sum_{r=1}^{\bar{N}_B^e} D_{B\alpha r}^{e\xi^2} - \frac{1}{R\Phi^e}\frac{d(GJ)^e_{(\alpha)}}{d\phi^e}\sum_{r=1}^{\bar{N}_B^e} D_{B\alpha r}^{e\xi}\right\}\tilde{v}_r^e\\
&+\frac{1}{R^3}\left\{\frac{(EI+GJ)^e_{(\alpha)}}{(\Phi^e)^2}\sum_{s=1}^{\bar{N}_T^e} D_{T\alpha s}^{e\xi^2} + \frac{1}{\Phi^e}\left[2\frac{d(EI)^e_{(\alpha)}}{d\phi^e} + \frac{d(GJ)^e_{(\alpha)}}{d\phi^e}\right]\sum_{s=1}^{\bar{N}_T^e} D_{T\alpha s}^{e\xi}\right.
\end{aligned}
$$

$$+\bar{\vartheta}_s(\xi_\alpha)\frac{d^2(EI)^e_{(s)}}{d(\phi^e)^2}\Bigg\}\tilde{\beta}^e_s = -p^e_\alpha, \quad \alpha = 2, ..., \hat{N}^e_B - 1 \tag{9.5}$$

If the discrete point α is a node for discretizing the angle of twist β, $\bar{\vartheta}_s(\xi_\alpha)$ represents the Kronecker delta $\delta_{\alpha s}$. Similarly, the discrete element equilibrium equation of the second of Eqs. (9.2) at a discrete point β in the element e is expressed as

$$\begin{aligned}
&\frac{1}{R^2}\left[\frac{(EI+GJ)^e_{(\beta)}}{(\Phi^e)^2}\sum_{r=1}^{\bar{N}^e_B} D^{e\xi^2}_{B\beta r} + \frac{1}{R\Phi^e}\frac{d(GJ)^e_{(\beta)}}{d\phi^e}\sum_{r=1}^{\bar{N}^e_B} D^{e\xi}_{B\beta r}\right]\tilde{v}^e_r \\
&+\frac{1}{R^2}\left[\frac{(GJ)^e_{(\beta)}}{(\Phi^e)^2}\sum_{s=1}^{\bar{N}^e_T} D^{e\xi^2}_{T\beta s} + \frac{1}{\Phi^e}\frac{d(GJ)^e_{(\beta)}}{d\phi^e}\sum_{s=1}^{\bar{N}^e_T} D^{e\xi}_{T\beta s} - \bar{\vartheta}_s(\xi_\beta)\frac{(EI)^e_{(s)}}{R^2}\right]\tilde{\beta}^e_s \\
&= -m^e_{z\beta}, \quad \beta = 2, ..., \hat{N}^e_T - 1
\end{aligned} \tag{9.6}$$

In Eqs. (9.5) and (9.6), if α and β are not coincident, the matrices of v related weighting coefficients $D^{e\xi^m}_{B\alpha r}$ and $D^{e\xi^m}_{B\beta r}$ are different and they need to be calculated, separately. The same calculations are also necessary for the weighting coefficients $D^{e\xi^m}_{T\alpha r}$ and $D^{e\xi^m}_{T\beta r}$.

The moments and shear force at a discrete point γ in an element can be expressed by

$$\begin{aligned}
V^e_\gamma = &-\frac{1}{R^3\Phi^e}\left[\frac{(EI)^e_{(\gamma)}}{(\Phi^e)^2}\sum_{r=1}^{\bar{N}^e_B} D^{e\xi^3}_{B\gamma r} + \frac{1}{\Phi^e}\frac{d(EI)^e_{(\gamma)}}{d\phi^e}\sum_{r=1}^{\bar{N}^e_B} D^{e\xi^2}_{B\gamma r}\right. \\
&\left.-(GJ)^e_{(\gamma)}\sum_{r=1}^{\bar{N}^e_B} D^{e\xi}_{B\gamma r}\right]\tilde{v}^e_r + \frac{1}{R^2}\left[\frac{(EI+GJ)^e_{(\gamma)}}{\Phi^e}\sum_{s=1}^{\bar{N}^e_T} D^{e\xi}_{T\gamma s}\right. \\
&\left.+\bar{\vartheta}_s(\xi_\gamma)\frac{d(EI)^e_{(s)}}{d\phi^e}\right]\tilde{\beta}^e_s,
\end{aligned}$$

$$M^e_{x\gamma} = \frac{(EI)^e_{(\gamma)}}{R}\left[\bar{\vartheta}_r(\xi_\gamma)\tilde{\beta}^e_r - \frac{1}{R(\Phi^e)^2}\sum_{r=1}^{\bar{N}^e_B} D^{e\xi^2}_{B\gamma r}\tilde{v}^e_r\right],$$

$$M^e_{z\gamma} = \frac{(GJ)^e_{(\gamma)}}{R\Phi^e}\left[\sum_{s=1}^{\bar{N}^e_T} D^{e\xi}_{T\gamma s}\tilde{\beta}^e_s + \frac{1}{R}\sum_{r=1}^{\bar{N}^e_B} D^{e\xi}_{B\gamma r}\tilde{v}^e_r\right] \tag{9.7}$$

The transition conditions of two adjacent elements are kinematic transition conditions and natural transition conditions. Kinematic transition conditions are continuities of transverse displacements, first derivatives of transverse displacements and angles of twist of cross section. The discrete kinematic transition conditions at the inter-element boundary $(i, i+1)$ of two adjacent elements i and $i+1$ are expressed as

$$\tilde{v}^i_{\bar{N}^i_B-1} - \tilde{v}^{i+1}_1 = 0, \quad \tilde{v}^i_{\bar{N}^i_B} - \tilde{v}^{i+1}_2 = 0, \quad \tilde{\beta}^i_{\bar{N}^i_T} - \tilde{\beta}^{i+1}_1 = 0 \tag{9.8}$$

These conditions are automatically satisfied when assembling discrete fundamental equations. The natural transition conditions are equilibriums of forces on the inter-element boundary. Let $P^{i,i+1}$, $M^{i,i+1}_x$ and $M^{i,i+1}_z$ denote the concentrated force in the y direction and moments directed toward x and z, respectively, applied on the inter-element boundary of element i and element $i+1$. Then, the equilibrium of lateral forces on the inter-element boundary can be defined

$$V^i_{\hat{N}^i_B} - V^{i+1}_1 = P^{i,i+1} \tag{9.9}$$

The substitution of the first of Eqs. (9.7) into the above equilibrium equation leads to the following explicit discrete equation

$$\begin{aligned}
&\frac{1}{R^3}\left[-\frac{(EI)^i_{\hat{N}^i_B}}{(\Phi^i)^3}\sum_{\alpha=1}^{\bar{N}^i_B} D^{i\xi^3}_{B\hat{N}^i_B\alpha} - \frac{1}{(\Phi^i)^2}\frac{d(EI)^i_{\hat{N}^i_B}}{d\phi^i}\sum_{\alpha=1}^{\bar{N}^i_B} D^{i\xi^2}_{B\hat{N}^i_B\alpha} + \frac{(GJ)^i_{\hat{N}^i_B}}{\Phi^i}\right.\\
&\left.\times\sum_{\alpha=1}^{\bar{N}^i_B} D^{i\xi}_{B\hat{N}^i_B\alpha}\right]\tilde{v}^i_\alpha + \frac{1}{R^2}\left[\frac{(EI+GJ)^i_{\hat{N}^i_T}}{\Phi^i}\sum_{\beta=1}^{\bar{N}^i_T} D^{i\xi}_{T\hat{N}^i_T\beta} + \delta_{\bar{N}^i_T\beta}\frac{d(EI)^i_{\hat{N}^i_T}}{d\phi^i}\right]\tilde{\beta}^i_\beta\\
&+\frac{1}{R^3}\left[\frac{(EI)^{i+1}_1}{(\Phi^{i+1})^3}\sum_{\alpha=1}^{\bar{N}^{i+1}_B} D^{(i+1)\xi^3}_{B1\alpha} + \frac{1}{(\Phi^{i+1})^2}\frac{d(EI)^{i+1}_1}{d\phi^{i+1}}\sum_{\alpha=1}^{\bar{N}^{i+1}_B} D^{(i+1)\xi^2}_{B1\alpha}\right.\\
&\left.-\frac{(GJ)^{i+1}_1}{\Phi^i}\sum_{\alpha=1}^{\bar{N}^{i+1}_B} D^{(i+1)\xi}_{B1\alpha}\right]\tilde{v}^{i+1}_\alpha - \frac{1}{R^2}\left[\frac{(EI+GJ)^{i+1}_1}{\Phi^{i+1}}\sum_{\beta=1}^{\bar{N}^{i+1}_T} D^{(i+1)\xi}_{T1\beta}\right.\\
&\left.+\delta^{i+1}_{1\beta}\frac{d(EI)^{i+1}_1}{d\phi^{i+1}}\right]\tilde{\beta}^{i+1}_\beta = P^{i,i+1}
\end{aligned} \tag{9.10}$$

The equilibrium of moments about the x-axis on the inter-element boundary can also be defined

$$M^i_{x\hat{N}^i_B} - M^{i+1}_{x1} = M^{i,i+1}_x \tag{9.11}$$

The substitution of the second of Eqs. (9.7) into the above equilibrium equation leads to the following explicit discrete equation

$$\begin{aligned}
&-\frac{(EI)^i_{\hat{N}^i_B}}{(R\Phi^i)^2}\sum_{\alpha=1}^{\bar{N}^i_B} D^{i\xi^2}_{B\hat{N}^i_B\alpha}\tilde{v}^i_\alpha + \frac{(EI)^i_{\hat{N}^i_T}}{R}\tilde{\beta}^i_{\bar{N}^i_T} + \frac{(EI)^{i+1}_1}{(R\Phi^{i+1})^2}\sum_{\alpha=1}^{\bar{N}^{i+1}_B} D^{(i+1)\xi^2}_{B1\alpha}\tilde{v}^{i+1}_\alpha\\
&-\frac{(EI)^{i+1}_1}{R}\tilde{\beta}^{i+1}_1 = M^{i,i+1}_x
\end{aligned} \tag{9.12}$$

The equilibrium of torques on the inter-element boundary can also be defined

$$M^i_{z\hat{N}^i_T} - M^{i+1}_{z1} = M^{i,i+1}_z \tag{9.13}$$

The substitution of the third of Eqs. (9.7) into the above equilibrium equation leads to the following explicit discrete equation

$$\frac{(GJ)^i_{\hat{N}^i_T}}{R\Phi^i}\left[\frac{1}{R}\sum_{\alpha=1}^{\bar{N}^i_B} D^{i\xi}_{B\hat{N}^i_B\alpha}\tilde{v}^i_\alpha + \sum_{\beta=1}^{\bar{N}^i_T} D^{i\xi}_{T\hat{N}^i_T\beta}\tilde{\beta}^i_\beta\right]$$

$$-\frac{(GJ)^{i+1}_1}{R\Phi^{i+1}}\left[\frac{1}{R}\sum_{\alpha=1}^{\bar{N}^{i+1}_B} D^{(i+1)\xi}_{B1\alpha}\tilde{v}^{i+1}_\alpha + \sum_{\beta=1}^{\bar{N}^{i+1}_T} D^{(i+1)\xi}_{T1\beta}\tilde{\beta}^{i+1}_\beta\right] = M^{i,i+1}_z \quad (9.14)$$

Letting element m be an element consisting of the kinematic boundary, and $\underline{I}^m$ and $\bar{I}^m$ equal to 1 for the kinematic boundary with the node number of the node at the kinematic boundary equal to 1, and $\bar{N}^m_B - 1$ and $\bar{N}^m_T$, respectively, for the kinematic boundary with the node number of the node at the kinematic boundary equal to N^m, the kinematic boundary conditions can be rewritten as

$$\tilde{v}^m_{\underline{I}^m} = \bar{v}^m_{\underline{I}^m}, \quad \tilde{v}^m_{\underline{I}^m+1} = \bar{v}^m_{\underline{I}^m+1}, \quad \tilde{\beta}^m_{\bar{I}^m} = \bar{\beta}^m_{\bar{I}^m} \quad (9.15)$$

where $\bar{v}^m_{\underline{I}^m}$, $\bar{v}^m_{\underline{I}^m+1}$ and $\bar{\beta}^m_{\bar{I}^m}$ are prescribed transverse displacement, bending rotation and angle of twist, respectively. Let element n be an element having one or more nodes on the natural boundary with $\bar{V}^n$ the concentrated forces in y direction, $\bar{M}^n_x$ and $\bar{M}^n_z$ the moments directed toward the coordinate directions x and z, respectively, applied on the natural boundary, and $\breve{I}^n$, $\underline{I}^n$ and $\bar{I}^n$ equal to 1 for the natural boundary with the node number of the node at the natural boundary equal to 1, and $\hat{N}^n_v$, $\bar{N}^n_v - 1$ and $\hat{N}^n_\beta$, respectively, for the natural boundary with the node number of the node at the natural boundary equal to N^n. Also let ν^n denote an indicator defined by the node number of element n. ν^n is defined as: $\nu^n = -1$, if the node number of the node at the natural boundary equal to 1; $\nu^n = +1$, if the node number of the node equal to N^n. Then the discrete natural boundary conditions can be obtained from Eqs. (9.10), (9.12) and (9.14)

$$\frac{1}{R^3}\left[\frac{(EI)^n_{\underline{I}^n}}{(\Phi^n)^3}\sum_{\beta=1}^{\bar{N}^n_B} D^{n\xi^3}_{B\underline{I}^n\beta} - \frac{1}{(\Phi^n)^2}\frac{d(EI)^n_{\underline{I}^n}}{d\phi^n}\sum_{\beta=1}^{\bar{N}^n_B} D^{n\xi^2}_{B\underline{I}^n\beta} + \frac{(GJ)^n_{\underline{I}^n}}{\Phi^n}\sum_{\beta=1}^{\bar{N}^n_B} D^{n\xi}_{B\underline{I}^n\beta}\right]$$

$$\tilde{v}^n_\beta + \frac{1}{R^2}\left[\frac{(EI+GJ)^n_{\underline{I}^n}}{\Phi^n}\sum_{\beta=1}^{\bar{N}^n_T} D^{n\xi}_{T\underline{I}^n\beta} + \frac{d(EI)^n_{\underline{I}^n}}{d\phi^n}\right]\tilde{\beta}^n_\beta = \nu^n\bar{V}^n_{\breve{I}^n},$$

$$-\frac{(EI)^n_{\underline{I}^n}}{R^2(\Phi^n)}\sum_{\beta=1}^{\bar{N}^n_B} D^{n\xi^2}_{B\underline{I}^n\beta}\tilde{v}^n_\beta + \frac{1}{R}(EI)^n_{\underline{I}^n}\tilde{\beta}^n_{\bar{I}^n} = \nu^n\bar{M}^n_x,$$

$$\frac{(GJ)^n_{\bar{I}^n}}{R\Phi^n}\left[\frac{1}{R}\sum_{\beta=1}^{\bar{N}^n_B} D^{n\xi}_{B\bar{I}^n\beta}\tilde{v}^n_\beta + \sum_{\beta=1}^{\bar{N}^n_T} D^{n\xi}_{T\bar{I}^n\beta}\tilde{\beta}^n_\beta\right] = \nu^n\bar{M}^n_z \quad (9.16)$$

9.1.3 Assemblage

With the kinematic transition conditions in mind, upon assembling the discrete element equilibrium equations (9.5) and (9.6) for elements having more than two nodes, discrete natural transition conditions, and discrete natural boundary conditions, an overall discrete equilibrium/transition/boundary equation can be obtained. It is the overall stiffness equation represented by Eq. (3.18). Consider the kinematic boundary conditions and solve the overall discrete equilibrium/transition/boundary equation, transverse displacements and bending rotations at all nodes can be obtained.

Like FEM, the assemblage is based on an element by element procedure. When assembling the discrete equations of element e, the discrete element equilibrium equations (9.5) and (9.6), and the six discrete element boundary forces of moments and shear forces, expressed by displacements, at the two element boundary nodes are directly assembled to the overall discrete equation system. Consequently, an element basis explicit matrix equation, containing the discrete element equilibrium equations and the discrete element boundary forces placed at the first and last three rows, is not necessary to be formed in the assemblage process. This element basis explicit matrix equation is an element stiffness equation which can be expressed by Eq. (3.20) with $[k^e]$ a $(\bar{N}_B^e + \bar{N}_T^e) \times (\bar{N}_B^e + \bar{N}_T^e)$ element stiffness matrix,

$$\{\delta^e\} = \lfloor \; \tilde{v}_1^e \quad \tilde{v}_2^e \quad \tilde{\beta}_1^e \quad \ldots \quad \tilde{v}_{\bar{N}_B^e-1}^e \quad \tilde{v}_{\bar{N}_B^e}^e \quad \tilde{\beta}_{\bar{N}_T^e}^e \; \rfloor^T \tag{9.17}$$

is the element displacement vector, and

$$\begin{aligned}\{r^e\} = \lfloor \; -V_1^e \quad -M_{x1}^e \quad -M_{z1}^e \quad -p_1^e \quad -m_{z1}^e \quad \ldots \\ -p_{\hat{N}_B^e-1}^e \quad -m_{z(\hat{N}_T^e-1)}^e \quad V_{\hat{N}_B^e}^e \quad M_{x\hat{N}_B^e}^e \quad M_{z\hat{N}_T^e}^e \; \rfloor^T\end{aligned} \tag{9.18}$$

is the element load vector. In Eq. (9.17), $\tilde{v}_1^e$ and $\tilde{v}_{\bar{N}_B^e-1}^e$ represent the lateral displacements at first and last element nodes, respectively, $\tilde{v}_2^e$ and $\tilde{v}_{\bar{N}_B^e}^e$ represent the deflection slope, $-\frac{1}{R}\frac{dv^e}{d\phi^e} = -\frac{1}{R\Phi^e}\frac{dv^e}{d\xi}$, or θ_x^e, at the first and last element nodes, respectively, and $\tilde{\beta}_1^e$ and $\tilde{\beta}_{\bar{N}_T^e}^e$ represent the angle of twist at the first and last element nodes, respectively. As the element stiffness equation contains discrete resultant forces at the two element boundary nodes, equilibriums of resultant forces and external forces at the inter-element boundary of two adjacent elements and the natural boundary are exactly satisfied in the assemblage process. Consequently, the DQEM model is different from the FEM model which needs to form the element stiffness equation, and which neglects the exact equilibriums.

9.1.4 Problems

In solving the problems, the elements are equally spaced. The DOF of flexural deformation, $\bar{N}_B^e$, and the DOF of torsional deformation, $\bar{N}_T^e$, are the same. Defining $\Delta\xi = 1./(\bar{N}_B^e - 1)$, the interior discrete points for defining the element-based transverse equilibrium equations are located at $\xi = (p-1)\Delta\xi$, $p = 3, ..., \bar{N}_B^e - 2$ while the interior discrete points for defining the element-based torsional equilibrium equations are located at $\xi = (p-1)\Delta\xi$, $p = 2, ..., \bar{N}_B^e - 1$. Only one DOF representing the lateral displacement and one DOF representing the angle of twist are assigned to an interior node. Explicit Lagrange DQ weighting coefficients are used for the torsional discretization. They are also used to generate the $C^1 - C^0 - C^1$ EDQ model with the node points of the equivalent Lagrange DQ model equally spaced and having two auxiliary node points. The resulting weighting coefficients are used for the flexural discretization.

The first problem solved involves a clamped prismatic circular bow girder made of reinforced concrete with the Young's modulus $E = 3 \times 10^6$ psi and subjected to a uniformly distributed load of 0.25 $tons/ft$ [106]. The cross section of the curved beam is a rectangle having the width 20 in and the depth 12 in. The value of radius R equals 10 ft while the range of the open angle between the two ends is 120^0. The ratio of the torsional stiffness to the flexural stiffness is 0.5. The analyses by both the p refinement procedure of increasing the number of DOF per element and the h refinement procedure of increasing the number of elements are carried out, separately. Numerical results of displacement parameters and stress resultants at some positions are summarized and listed in Table 9.1. They are compared with exact solutions [107]. It shows that the results converge fast to exact solutions by either increasing the DOF per element or the number of elements. It also shows that

Table 9.1. Results of a clamped prismatic circular bow girder subjected to a uniformly distributed load

DOF per element	No. of elements	Displacement at $\phi = 60°$ (in)	Angle of twist at $\phi = 60°$ (rad)	Bending moment at $\phi = 60°$ $(ft - tons)$
10	2	$.2557343\times10^{-1}$	$.4168421\times10^{-3}$	$.3892901\times10^{1}$
	4	$.2481342\times10^{-1}$	$.3868535\times10^{-3}$	$.3759391\times10^{1}$
	6	$.2456347\times10^{-1}$	$.3561457\times10^{-3}$	$.3659023\times10^{1}$
14	2	$.2420569\times10^{-1}$	$.3544746\times10^{-3}$	$.3639998\times10^{1}$
	4	$.2417680\times10^{-1}$	$.3482164\times10^{-3}$	$.3625651\times10^{1}$
	6	$.2416554\times10^{-1}$	$.3469548\times10^{-3}$	$.3628673\times10^{1}$
18	2	$.2416061\times10^{-1}$	$.3463458\times10^{-3}$	$.3629501\times10^{1}$
	6	$.2415964\times10^{-1}$	$.3468756\times10^{-3}$	$.3629923\times10^{1}$
22	2	$.2415543\times10^{-1}$	$.3468755\times10^{-3}$	$.3629821\times10^{1}$
	4	$.2415467\times10^{-1}$	$.3468737\times10^{-3}$	$.3629938\times10^{1}$
	6	$.2415424\times10^{-1}$	$.3468676\times10^{-3}$	$.3629956\times10^{1}$
Exact sol.		$.2415429\times10^{-1}$	$.3468679\times10^{-3}$	$.3629987\times10^{1}$

the solution procedure of increasing the DOF per element is more efficient than the solution procedure of increasing the number of elements. Let e_r denote the convergence indicator which is the relative error of the angle of twist at $\phi = 60°$, obtained by the DQEM, with respect to the exact solution.

The convergence rates can be seen in Fig. 9.2 which shows that the two refinement procedures possess high convergence rates. It also shows that the convergence rate of the p refinement procedure is higher than that of the h refinement procedure.

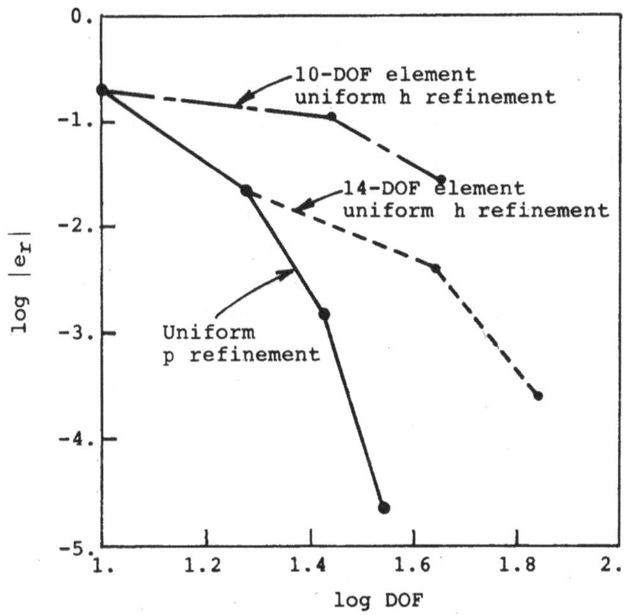

Fig. 9.2. Convergence of the angle of tist at $\phi = 60°$

The second problem solved involves a clamped nonprismatic curved box beam composed of three segments with the Young's modulus $E = 3 \times 10^6$ *psi* and subjected to a uniformly distributed load of 0.25 *tons*/*ft*. The structure is shown in Fig. 9.3.

The value of radius R equals 10 *ft*. Each segment has the open angle of 40°. The middle segment is a prismatic thin-walled square box with the side length 24 *in* and the wall thickness 1 *in*. The two side segments also have the thin-walled square cross section with the wall thickness 1 *in*. They are nonprismatic segments with the side length increasing linearly from 24 *in* at the intersection to 48 *in* at the support end. In carrying out the numerical computations, three elements are used to model the structure with each element representing a segment. Results listed in Table 9.2 show that the convergence is assured by

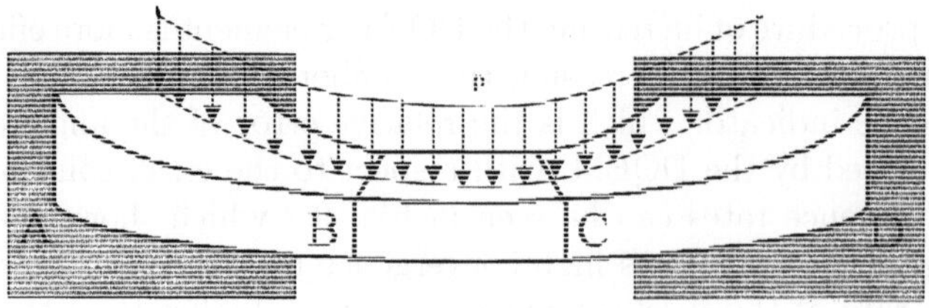

Fig. 9.3. A nonprismatic circular beam girder subjected to a uniformly distributed load

Table 9.2. Results of a nonprismatic circular beam girder subjected to a uniformly distributed load

DOF per element	No. of elements	Displacement at $\phi = 60°$ (in)	Angle of twist at $\phi = 60°$ (rad)	Bending moment at $\phi = 60°$ $(ft - tons)$
10	3	$.4357344 \times 10^1$	$-.2568429 \times 10^1$	$-.1389251 \times 10^1$
14	3	$.4249578 \times 10^1$	$-.2344485 \times 10^1$	$-.1363998 \times 10^1$
18	3	$.4236414 \times 10^1$	$-.2223731 \times 10^1$	$-.1369501 \times 10^1$
22	3	$.4238689 \times 10^1$	$-.2224340 \times 10^1$	$-.1362911 \times 10^1$
26	3	$.4238711 \times 10^1$	$-.2224421 \times 10^1$	$-.1362751 \times 10^1$
30	3	$.4238715 \times 10^1$	$-.2224428 \times 10^1$	$-.1362749 \times 10^1$

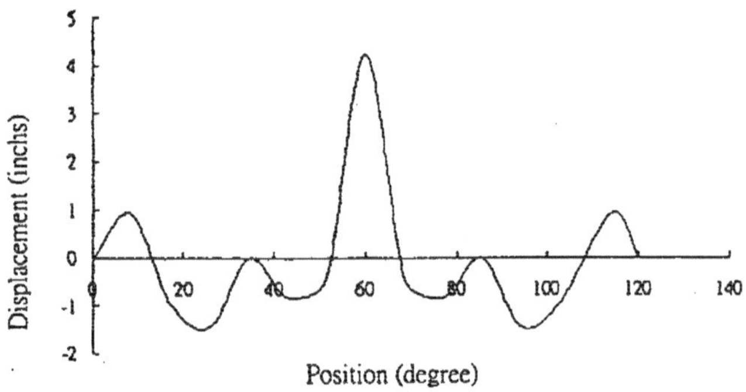

Fig. 9.4. Distribution of lateral displacement

the increase of the number of DOF per element. The distributions of converged lateral displacement and stress resultants are shown in Figs. 9.4 to 9.7.

9.2 In-Plane Deflection Analysis

The results are difficult to converge by using some FEM models to analyze the in-plane deflection of curved beam structures, due to the locking. Many

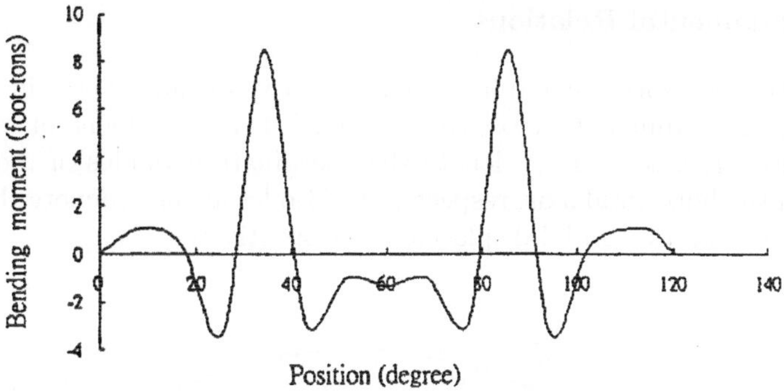

Fig. 9.5. Distribution of bending moment

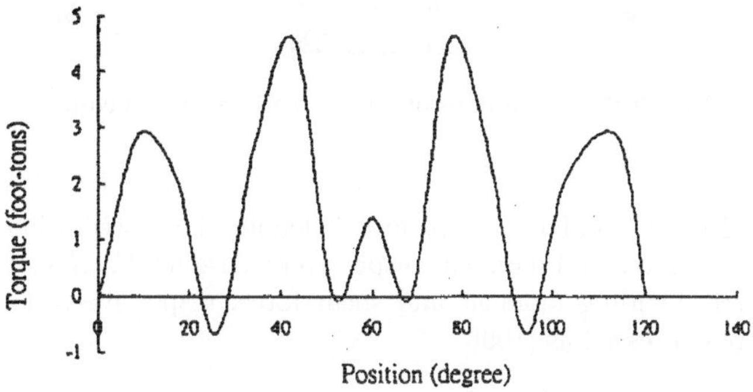

Fig. 9.6. Distribution of torque

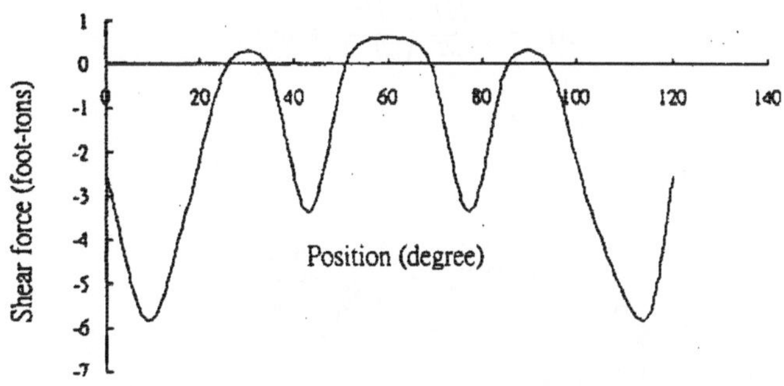

Fig. 9.7. Distribution of lateral shear

techniques including the reduced integration, partial approximation, mixed formulation,..., etc. have been used to overcome the locking problem and solve the curved beams [108]. DQEM analysis model for solving this type of structural problems is introduced.

9.2.1 Fundamental Relations

Consider that an isotropic and homogeneous curved beam, shown in Fig. 9.8, with Young's modulus, E, cross-section area, A, and moment of inertia of cross-section, I. Let r and ϕ denote the coordinate variables of radius and angle with the horizontal axis, respectively. Also let w and v denote the radial displacement and tangential displacement, respectively.

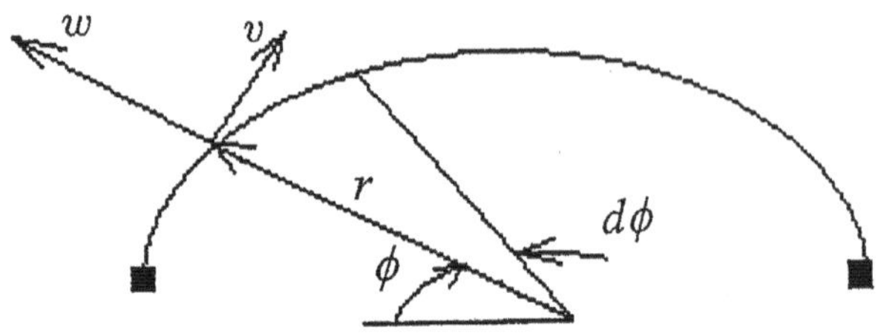

Fig. 9.8. Coordinates of an arbitrarily curved beam

Refer to Fig. 9.9 and let p_r, p_t and t denote the distribution of radial force, tangential force and moment, respectively. Also let F, M and V denote the axial force, bending moment and shear force, respectively. These stress resultants are expressed as [109]

$$F = \frac{EA}{r}\left[\left(\frac{dv}{d\phi} + w\right) + \frac{I}{r^2 A}\left(\frac{d^2 w}{d\phi^2} + w\right)\right],$$

$$M = -\frac{EI}{r^2}\left(\frac{d^2 w}{d\phi^2} + w\right),$$

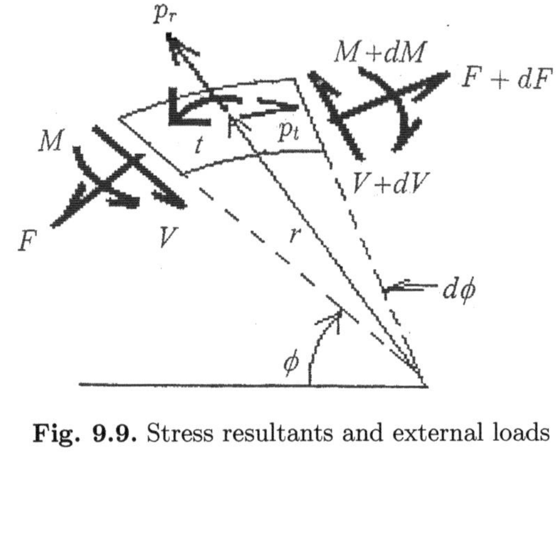

Fig. 9.9. Stress resultants and external loads

$$
\begin{aligned}
V &= \frac{1}{r}\frac{dM}{d\phi} - t \\
&= -EI\left[\frac{1}{r^3}\left(\frac{d^3w}{d\phi^3} + \frac{dw}{d\phi}\right) - \frac{2}{r^4}\frac{dr}{d\phi}\left(\frac{d^2w}{d\phi^2} + w\right)\right]
\end{aligned}
\tag{9.19}
$$

Assume that EA and EI are constant in a DQEM element. The equilibrium equation in the tangential direction at any point in a DQEM element is expressed as

$$
\frac{1}{r}\left(\frac{dF}{d\phi} + V\right) = -p_t
$$

or

$$
\begin{aligned}
&-\frac{EI}{r^5}\frac{dr}{d\phi}\frac{d^2w}{d\phi^2} + \frac{EA}{r^2}\frac{dw}{d\phi} - \left(\frac{EA}{r^3} + \frac{EI}{r^5}\right)\frac{dr}{d\phi}w + \frac{EA}{r^2}\frac{d^2v}{d\phi^2} - \frac{EA}{r^3}\frac{dr}{d\phi}\frac{dv}{d\phi} \\
&= -p_t
\end{aligned}
\tag{9.20}
$$

The equilibrium equation in the radial direction is expressed as

$$
\frac{1}{r}\left(\frac{dv}{d\phi} - F\right) = -p_r
$$

or

$$
\begin{aligned}
&-\frac{EI}{r^4}\frac{d^4w}{d\phi^4} + \frac{5EI}{r^5}\frac{dr}{d\phi}\frac{d^3w}{d\phi^3} + EI\left[\frac{2}{r^5}\frac{d^2r}{d\phi^2} - \frac{8}{r^6}\left(\frac{dr}{d\phi}\right)^2 - \frac{1}{r^4}\right]\frac{d^2w}{d\phi^2} \\
&+ \frac{5EI}{r^5}\frac{dr}{d\phi}\frac{dw}{d\phi} + \left\{EI\left[\frac{2}{r^5}\frac{d^2r}{d\phi^2} - \frac{8}{r^6}\left(\frac{dr}{d\phi}\right)^2 - \frac{1}{r^4}\right] - \frac{EA}{r^2}\right\}w - \frac{EA}{r^2}\frac{dv}{d\phi} \\
&= -p_r
\end{aligned}
\tag{9.21}
$$

The kinematic boundary conditions are expressed as

$$
w = \bar{w}, \qquad \frac{1}{r}\frac{dw}{d\phi} = \frac{1}{r}\frac{d\bar{w}}{d\phi}, \qquad v = \bar{v}
\tag{9.22}
$$

In the above equations, $\frac{1}{r}\left(\frac{d\bar{w}}{d\phi} - v\right)$ represents the rotation of cross section. Assume that $\bar{F}$, $\bar{M}$ and $\bar{V}$ are prescribed tangential force, moment and radial force, respectively. The natural boundary conditions are expressed as

$$
\frac{EA}{r}\left[\left(\frac{dv}{d\phi} + w\right) + \frac{I}{r^2A}\left(\frac{d^2w}{d\phi^2} + w\right)\right] = \bar{F}, \quad -\frac{EI}{r^2}\left(\frac{d^2w}{d\phi^2} + w\right) = \bar{M},
$$

$$
-EI\left[\frac{1}{r^3}\left(\frac{d^3w}{d\phi^3} + \frac{dw}{d\phi}\right) - \frac{2}{r^4}\frac{dr}{d\phi}\left(\frac{d^2w}{d\phi^2} + w\right)\right] = \bar{V}
\tag{9.23}
$$

9.2.2 DQEM Formulation

Since the orders of w and v related differentiations existing in the fundamental relations are four and two, respectively, the orders of the corresponding approximate displacement components can be three and one, respectively, for an element without distributed load. However, the orders of the corresponding approximate displacement components must at least be four and two, respectively, and each of the two equilibrium equations needs at least one discrete point for defining one discrete element equilibrium equation. The discrete points for defining discrete equilibrium equations can be either in the interior of the element or on the element boundary.

In the present DQEM curved beam analysis model, only the degrees of freedom representing w, $\frac{dw}{rd\phi}$ and v, which are necessary for automatically setting the kinematic transition conditions and kinematic boundary conditions are assigned to the element boundary nodes. Consequently, the second and last columns of the weighting coefficient matrices for the flexural discretization defined by using the natural coordinate, ξ, must be modified by multiplying each element in the two columns by $r\Phi^e$. Only interior discrete points are used to define the discrete element equilibrium equations. Let N_w^e denote the number of nodes for defining the w related discretization, $\bar{N}_w^e$ denote the number of the corresponding element DOF, $\tilde{w}_i^e$ denote the w related element basis discrete displacement parameters, and $D_{w\alpha i}^{e\xi^q}$ denote the corresponding qth order weighting coefficients. Also let N_v^e denote the number of nodes for defining the v related discretization, $\bar{N}_v^e$ denote the number of the corresponding element DOF, $\tilde{v}_i^e$ denote the v related element basis discrete displacement parameters, and $D_{v\alpha i}^{e\xi^q}$ denote the corresponding weighting coefficients. When defining the discrete fundamental relations, the number of discrete points at which the w related discrete element equilibrium equations in the radial direction are defined is $\bar{N}_w^e - 4$, while the number of discrete points at which the v related discrete element equilibrium equations in the tangential direction are defined is $\bar{N}_v^e - 2$. The DOF assigned to an interior node for defining the w or v related discretization can be flexible. If only the DOF representing v is assigned to the interior nodes for defining the v related discretization and the Lagrange interpolation functions are used to calculate the weighting coefficients, Lagrange DQ model is adopted. If only the DOF representing w and $\frac{dw}{d\phi}$ are assigned to the interior nodes for defining the w related discretization and Hermite interpolation functions are used to calculate the weighting coefficients, Hermite EDQ model is adopted.

When discretizing the fundamental relations, if a discrete point at which a discrete fundamental relation needs to be defined is not an element node, the interpolation is necessary for expressing w existing in the fundamental relation. Let $\vartheta_i(\xi)$ denote the interpolation functions. Then, $w^e(\xi)$ in an element can be expressed as $w^e(\xi) = \vartheta_i(\xi)\tilde{w}_i^e$. If both of p_t and p_r are not zero, each of Eqs. (20) and (21) needs at least one interior discrete point for defining the discrete element equilibrium equations. Let $\hat{N}_w^e$ and $\hat{N}_v^e$ denote the numbers of the corresponding interior discrete points for defining the discrete element equilibrium equations related to w and v, respectively, plus the two element nodes. Then the following two relations hold: $\hat{N}_w^e = \bar{N}_w^e - 2$ and $\hat{N}_v^e = \bar{N}_v^e$. The introduction of mapping into Eq. (9.20) and the use of EDQ discretization at an interior discrete point α in an element e lead to the following discrete equation

$$\left\{ \frac{(EI)^e}{(r_{(\alpha)}^e)^5(\Phi^e)^2} \frac{dr_{(\alpha)}^e}{d\phi} \sum_{i=1}^{\bar{N}_w^e} D_{w\alpha i}^{e\xi^2} + \frac{(EA)^e}{(r_{(\alpha)}^e)^2\Phi^e} \sum_{i=1}^{\bar{N}_w^e} D_{w\alpha i}^{e\xi} - \left[\frac{(EA)^e}{(r_{(\alpha)}^e)^3} + \frac{(EI)^e}{(r_{(\alpha)}^e)^5} \right] \right.$$
$$\left. \times \frac{dr_{(\alpha)}^e}{d\phi} \vartheta_i(\xi_\alpha) \right\} \tilde{w}_i^e + \left[\frac{(EA)^e}{(r_{(\alpha)}^e)^2(\Phi^e)^2} \sum_{j=1}^{\bar{N}_v^e} D_{v\alpha j}^{e\xi^2} - \frac{(EA)^e}{(r_{(\alpha)}^e)^3\Phi^e} \frac{dr_{(\alpha)}^e}{d\phi} \sum_{j=1}^{\bar{N}_v^e} D_{v\alpha j}^{e\xi} \right] \tilde{v}_j^e$$
$$= -p_{t\alpha}^e, \quad \alpha = 2, ..., \hat{N}_v^e - 1 \tag{9.24}$$

If the discrete point α is a node for discretizing w, $\vartheta_i(\xi_\alpha)$ represents the Kronecker delta $\delta_{\alpha i}$. Similarly, the introduction of mapping into Eq. (9.20) and the use of EDQ discretization at a discrete point β lead to the following discrete equation

$$\left\{ -\frac{(EI)^e}{(r_{(\beta)}^e)^4(\Phi^e)^4} \sum_{i=1}^{\bar{N}_w^e} D_{w\beta i}^{e\xi^4} + \frac{5(EI)^e}{(r_{(\beta)}^e)^5(\Phi^e)^3} \frac{dr_{(\beta)}^e}{d\phi} \sum_{i=1}^{\bar{N}_w^e} D_{w\beta i}^{e\xi^3} \right.$$
$$+ \frac{(EI)^e}{(r_{(\beta)}^e)^4(\Phi^e)^2} \left[\frac{2}{r_{(\beta)}^e} \frac{d^2 r_{(\beta)}^e}{d\phi^2} - \frac{8}{(r_{(\beta)}^e)^2} \left(\frac{dr_{(\beta)}^e}{d\phi} \right)^2 - 1 \right] \sum_{i=1}^{\bar{N}_w^e} D_{w\beta i}^{e\xi^2}$$
$$+ \frac{5(EI)^e}{(r_{(\beta)}^e)^5\Phi^e} \frac{dr_{(\beta)}^e}{d\phi} \sum_{i=1}^{\bar{N}_w^e} D_{w\beta i}^{e\xi} + \left[\frac{(EI)^e}{(r_{(\beta)}^e)^4} \left(\frac{2}{r_{(\beta)}^e} \frac{d^2 r_{(\beta)}^e}{d\phi^2} - \frac{8}{(r_{(\beta)}^e)^2} \left(\frac{dr_{(\beta)}^e}{d\phi} \right)^2 - 1 \right) \right.$$
$$\left. \left. - \frac{(EA)^e}{(r_{(\beta)}^e)^2} \right] \vartheta_i(\xi_\beta) \right\} \tilde{w}_i^e - \frac{(EA)^e}{(r_{(\beta)}^e)^2\Phi^e} \sum_{j=1}^{\bar{N}_v^e} D_{v\beta j}^{e\xi} \tilde{v}_j^e = -p_{r\beta}^e,$$
$$\beta = 2, ..., \hat{N}_w^e - 1 \tag{9.25}$$

The stress resultants F_α^e, M_α^e and V_α^e at a discrete point α of an element e can be expressed by

$$F_\alpha^e = \left[\frac{(EI)^e}{(r_{(\alpha)}^e)^3(\Phi^e)^2} \sum_{i=1}^{\bar{N}_w^e} D_{w\alpha i}^{e\xi^2} + \left(\frac{(EA)^e}{r_{(\alpha)}^e} + \frac{(EI)^e}{(r_{(\alpha)}^e)^3} \right) \vartheta_i(\xi_\alpha) \right] \tilde{w}_i^e$$
$$+ \frac{(EA)^e}{r_{(\alpha)}^e \Phi^e} \sum_{j=1}^{\bar{N}_v^e} D_{v\alpha j}^{e\xi} \tilde{v}_j^e,$$

$$M_\alpha^e = -\frac{(EI)^e}{(r_{(\alpha)}^e)^2} \left[\frac{1}{(\Phi^e)^2} \sum_{i=1}^{\bar{N}_w^e} D_{w\alpha i}^{e\xi^2} + \vartheta_i(\xi_\alpha) \right] \tilde{w}_i^e,$$

$$V_\alpha^e = \frac{(EI)^e}{(r_{(\alpha)}^e)^3}\left[-\frac{1}{(\Phi^e)^3}\sum_{i=1}^{\bar{N}_w^e} D_{w\alpha i}^{e\xi^3} + \frac{2}{r_{(\alpha)}^e(\Phi^e)^2}\frac{dr_{(\alpha)}^e}{d\phi}\sum_{i=1}^{\bar{N}_w^e} D_{w\alpha i}^{e\xi^2} \right.$$

$$\left. -\frac{1}{\Phi^e}\sum_{i=1}^{\bar{N}_w^e} D_{w\alpha i}^{e\xi} + \frac{2}{r_\alpha^e}\frac{dr_{(\alpha)}^e}{d\phi}\vartheta_i(\xi_\alpha)\right]\tilde{w}_i^e \tag{9.26}$$

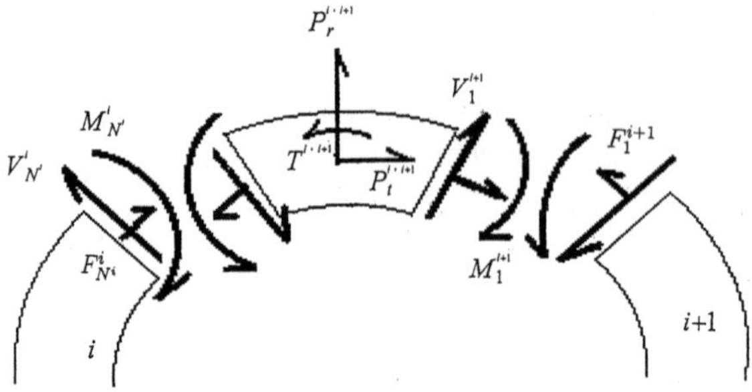

Fig. 9.10. Forces at the inter-element of two adjacent elements i and $i+1$

The compatibility and conformability conditions at the inter-element boundary of two adjacent elements are automatically satisfied. The equilibrium conditions of stress resultants and external forces at the inter-element boundary or natural boundary also need to be satisfied. Each equilibrium condition is either a natural transition condition or a natural boundary condition. Assume that $P_r^{i,i+1}$, $P_t^{i,i+1}$ and $T^{i,i+1}$ shown in Fig. 9.10 are external radial force, tangential force and moment, respectively, applied on the inter-element boundary $i, i+1$. The equilibrium condition of moments is expressed as

$$M_1^{i+1} - M_{\hat{N}_w^i}^i = T^{i,i+1} \tag{9.27}$$

Introducing the second of Eqs. (9.26) into the above equation, the following explicit discrete equilibrium condition of moments can be obtained:

$$\frac{(EI)^i}{(r_{\hat{N}_w^i}^i)^2}\left[\frac{1}{(\Phi^i)^2}\sum_{i=1}^{\bar{N}_w^i} D_{w\hat{N}_w^i i}^{i\xi^2} + \delta_{(\bar{N}_w^i-1)i}\right]\tilde{w}_i^i$$

$$-\frac{(EI)^{i+1}}{(r_1^{i+1})^2}\left[\frac{1}{(\Phi^{i+1})^2}\sum_{i=1}^{\bar{N}_w^{i+1}} D_{w1i}^{(i+1)\xi^2} + \delta_{1i}\right]\tilde{w}_i^{i+1} = T^{i,i+1} \tag{9.28}$$

where δ_{1i} and $\delta_{(\bar{N}_w^i-1)i}$ are first and $(\bar{N}_w^i-1)$th rows of Kronecker delta $\delta_{\alpha i}$. The equilibrium condition of forces in the radial direction is expressed as

$$V_{\hat{N}_w^i}^i - V_1^{i+1} = P_r^{i,i+1} \tag{9.29}$$

The introduction of the third of Eqs. (9.26) into the above equation leads to the following explicit discrete equation

$$\frac{(EI)^i}{(r^i_{\hat{N}^i_w})^3}\left[-\frac{1}{(\Phi^i)^3}\sum_{i=1}^{\bar{N}^i_w} D^{i\xi^3}_{w\hat{N}^i_w i} + \frac{2}{r^i_{\hat{N}^i_w}(\Phi^i)^2}\frac{dr^i_{\hat{N}^i_w}}{d\phi}\sum_{i=1}^{\bar{N}^i_w} D^{i\xi^2}_{w\hat{N}^i_w i}\right.$$
$$\left.-\frac{1}{\Phi^i}\sum_{i=1}^{\bar{N}^i_w} D^{i\xi}_{w\hat{N}^i_w i} + \frac{2}{r^i_{\hat{N}^i_w}}\frac{dr^i_{\hat{N}^i_w}}{d\phi}\delta_{(\bar{N}^i_w-1)i}\right]\tilde{w}^i_i$$
$$+\frac{(EI)^{i+1}}{(r^{i+1}_1)^3}\left[\frac{1}{(\Phi^{i+1})^3}\sum_{i=1}^{\bar{N}^{i+1}_w} D^{(i+1)\xi^3}_{w1i} - \frac{2}{r^{i+1}_1(\Phi^{i+1})^2}\frac{dr^{i+1}_1}{d\phi}\sum_{i=1}^{\bar{N}^{i+1}_w} D^{(i+1)\xi^2}_{w1i}\right.$$
$$\left.+\frac{1}{\Phi^{i+1}}\sum_{i=1}^{\bar{N}^{i+1}_w} D^{(i+1)\xi}_{w1i} - \frac{2}{r^{i+1}_1}\frac{dr^{i+1}_1}{d\phi}\delta_{1i}\right]\tilde{w}^{i+1}_i = P^{i,i+1}_r \tag{9.30}$$

The equilibrium condition of forces in the tangential direction is expressed as

$$F^i_{\hat{N}^i_v} - F^{i+1}_1 = P^{i,i+1}_t \tag{9.31}$$

The introduction of the first of Eqs. (9.26) into the above equation leads to the following explicit discrete equation

$$\left[\frac{(EI)^i}{(r^i_{\hat{N}^i_v})^3(\Phi^i)^2}\sum_{i=1}^{\bar{N}^i_w} D^{i\xi^2}_{w\hat{N}^i_w i} + \left(\frac{(EA)^i}{r^i_{\hat{N}^i_w}} + \frac{(EI)^i}{(r^i_{\hat{N}^i_v})^3}\right)\delta_{(\bar{N}^i_w-1)i}\right]\tilde{w}^i_i$$
$$+\frac{(EA)^i}{r^i_{\hat{N}^i_v}\Phi^i}\sum_{j=1}^{\bar{N}^i_v} D^{i\xi}_{v\hat{N}^i_v j}\tilde{v}^i_j - \left[\frac{(EI)^{i+1}}{(r^{i+1}_1)^3(\Phi^{i+1})^2}\sum_{i=1}^{\bar{N}^{i+1}_w} D^{(i+1)\xi^2}_{w1i} + \left(\frac{(EA)^{i+1}}{r^{i+1}_1}\right.\right.$$
$$\left.\left.+\frac{(EI)^{i+1}}{(r^{i+1}_1)^3}\right)\delta_{1i}\right]\tilde{w}^{i+1}_i - \frac{(EA)^{i+1}}{r^{i+1}_1\Phi^{i+1}}\sum_{j=1}^{\bar{N}^{i+1}_v} D^{(i+1)\xi}_{v1j}\tilde{v}^{i+1}_j = P^{i,i+1}_t \tag{9.32}$$

Letting element m be an element consisting of the kinematic boundary, and $\underline{I}^m$ and $\bar{I}^m$ equal to 1 for the kinematic boundary with the node number of the node at the kinematic boundary equal to 1, and $\bar{N}^m_w - 1$ and $\bar{N}^m_v$, respectively, for the kinematic boundary with the node number of the node at the kinematic boundary equal to N^m, the discrete kinematic boundary conditions are expressed by

$$\tilde{w}^m_{\underline{I}^m} = \bar{w}^m_{\underline{I}^m}, \quad \tilde{w}^m_{\underline{I}^m+1} = \bar{\theta}^m_{\underline{I}^m+1}, \quad \tilde{v}^m_{\bar{I}^m} = \bar{v}^m_{\bar{I}^m} \tag{9.33}$$

Where $\bar{w}^m_{I^m}$, $\bar{w}^m_{I^m+1}$ and $\bar{v}^m_{\bar{I}^m}$ are prescribed radial displacement, bending rotation and tangential displacement. Letting element n be an element consisting of the natural boundary, and $\breve{I}^n$, $\underline{I}^n$ and $\bar{I}^n$ equal to 1 for the natural boundary with the node number of the node at the natural boundary equal to 1, and $\hat{N}_w^n$, $\bar{N}_w^n - 1$ and $\hat{N}_v^n$, respectively, for the natural boundary with the node number of the node at the natural boundary equal to N^n, the discrete natural boundary conditions can be obtained by discretizing Eqs. (9.23)

$$\left\{ \frac{(EI)^n}{(r_{\bar{I}^n}^n)^3 (\Phi^n)^2} \sum_{i=1}^{\bar{N}_w^n} D_{w\bar{I}^n i}^{n\xi^2} + \left[\frac{(EA)^n}{r_{\bar{I}^n}^n} + \frac{(EI)^n}{(r_{\bar{I}^n}^n)^3} \right] \delta_{\underline{I}^n i} \right\} \tilde{w}_i^n$$
$$+ \frac{(EA)^n}{r_{\bar{I}^n}^n \Phi^n} \sum_{j=1}^{\bar{N}_v^n} D_{v\bar{I}^n j}^{n\xi} \tilde{v}_j^n = \bar{F},$$

$$-\frac{(EI)^n}{(r_{\check{I}^n}^n)^2} \left[\frac{1}{(\Phi^n)^2} \sum_{i=1}^{\bar{N}_w^n} D_{w\check{I}^n i}^{n\xi^2} + \delta_{\underline{I}^n i} \right] \tilde{w}_i^n = \bar{M},$$

$$\frac{(EI)^n}{(r_{\check{I}^n}^n)^3} \left[-\frac{1}{(\Phi^n)^3} \sum_{i=1}^{\bar{N}_w^n} D_{w\check{I}^n i}^{n\xi^3} + \frac{2}{r_{\check{I}^n}^n (\Phi^n)^2} \frac{dr_{\check{I}^n}^n}{d\phi} \sum_{i=1}^{\bar{N}_w^n} D_{w\check{I}^n i}^{n\xi^2} \right.$$
$$\left. -\frac{1}{\Phi^n} \sum_{i=1}^{\bar{N}_w^n} D_{w\check{I}^n i}^{n\xi} + \frac{2}{r_{\check{I}^n}^n} \frac{dr_{\check{I}^n}^n}{d\phi} \delta_{\underline{I}^n i} \right] \tilde{w}_i^n = \bar{V} \tag{9.34}$$

9.2.3 Assemblage

With the kinematic transition conditions in mind, then assemble the discrete element equilibrium equations (9.24) and (9.25) for elements having more than two nodes, discrete natural transition conditions (9.28), (9.30) and (9.32), and discrete natural boundary conditions (9.34), an overall discrete equilibrium/transition/boundary equation can be obtained. It is the overall stiffness equation represented by Eq. (3.18). Consider the kinematic boundary conditions and solve the overall discrete equilibrium/transition/boundary equation, tangential displacements, radial displacements and bending rotations at all nodes can be obtained. Like FEM, the assemblage is based on an element by element procedure. When assembling the discrete equations of element e, the discrete element equilibrium equations (9.24) and (9.25), and the six discrete element boundary forces of moments and shear forces, expressed by displacements, at the two element boundary nodes are directly assembled to the overall discrete equation system. An element basis explicit matrix equation, containing the discrete element equilibrium equations and the discrete element boundary forces placed at the first and last three rows, is not necessary to be formed in the assemblage process. This element basis explicit matrix equation is an element stiffness equation which can be expressed by Eq. (3.20) with $[k^e]$ a $(\bar{N}_w^e + \bar{N}_v^e) \times (\bar{N}_w^e + \bar{N}_v^e)$ element stiffness matrix,

$$\{\delta^e\} = \lfloor \; \tilde{w}_1^e \quad \tilde{w}_2^e \quad \tilde{v}_1^e \quad \ldots \quad \tilde{w}_{\bar{N}_w^e - 1}^e \quad \tilde{w}_{\bar{N}_w^e}^e \quad \tilde{v}_{\bar{N}_v^e}^e \; \rfloor^T \tag{9.35}$$

is the element displacement vector, and

$$\{r^e\} = \lfloor -V_1^e \quad -M_1^e \quad -F_1^e \quad -r_2^e p_{r2}^e \quad -r_2^e p_{t2}^e \quad \ldots$$
$$-r^e_{\hat{N}_w^e-1} p^e_{r(\hat{N}_w^e-1)} \; -r^e_{\hat{N}_v^e-1} p^e_{r(\hat{N}_v^e-1)} \; V^e_{\hat{N}_w^e} \; M^e_{\hat{N}_w^e} \; F^e_{\hat{N}_v^e} \rfloor^T \tag{9.36}$$

is the element load vector. In Eq. (9.35), $\tilde{w}_2^e$ and $\tilde{w}^e_{\bar{N}_w^e}$ represent the values of $\frac{1}{r}\frac{dw^e}{d\phi^e} = \frac{1}{r\Phi^e}\frac{dw^e}{d\xi}$ at nodes 1 and N^e, respectively. As Eq. (3.20) contains discrete resultant forces at the two element boundary nodes, equilibriums of resultant forces and external forces at the inter-element boundary of two adjacent elements and the natural boundary are exactly satisfied in the assemblage process.

9.2.4 Problems

In solving the problem, the elements are equally spaced. The DOF of flexural deformation, $\bar{N}_w^e$, and the DOF of tangential deformation, N_v^e, are the same. Defining $\Delta\xi = 1./(\bar{N}_w^e - 1)$, the interior discrete points for defining the element-based radial equilibrium equations are located at $\xi = (p-1)\Delta\xi$, $p = 3, ..., \bar{N}_w^e - 2$ while the interior discrete points for defining the element-based extensional equilibrium equations are located at $\xi = (p-1)\Delta\xi$, $p = 2, ..., \bar{N}_w^e - 1$. Only one DOF representing the radial displacement and one DOF representing the tangential displacement are assigned to an interior node. Explicit Lagrange DQ weighting coefficients are used for the v-related discretization. They are also used to generate the $C^1 - C^0 - C^1$ EDQ model used for the w-related discretization.

The problem solved involves a simply supported prismatic circular arch, subjected to a concentrated force P and shown in Fig. 9.11, with the values of radius r, area of cross section A, moment of inertia of cross section I, Young's modulus E, and concentrated force P equal to 1. The range of open

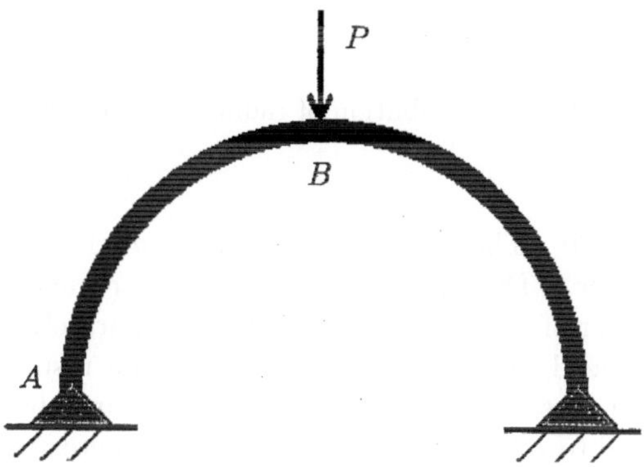

Fig. 9.11. A simply supported circular arch subjected to a concentrated force

Table 9.3. Results of a simply supported curved beam

DOF per element	Number of elements	w_B	N_A	M_B
10	2	$-.185194870\times10^{-1}$	$-.526387466\times10^{0}$	$-.185164252\times10^{0}$
	4	$-.184702325\times10^{-1}$	$-.508564260\times10^{0}$	$-.184500372\times10^{0}$
	6	$-.183169531\times10^{-1}$	$-.504656312\times10^{0}$	$-.184092580\times10^{0}$
14	2	$-.183099455\times10^{-1}$	$-.500194872\times10^{0}$	$-.183857117\times10^{0}$
	4	$-.183019268\times10^{-1}$	$-.500005198\times10^{0}$	$-.183198376\times10^{0}$
	6	$-.182851892\times10^{-1}$	$-.500001875\times10^{0}$	$-.182913655\times10^{0}$
18	2	$-.182751443\times10^{-1}$	$-.500002427\times10^{0}$	$-.182538425\times10^{0}$
	4	$-.182654927\times10^{-1}$	$-.500001253\times10^{0}$	$-.182377047\times10^{0}$
	6	$-.182550496\times10^{-1}$	$-.500000362\times10^{0}$	$-.182373975\times10^{0}$
Analytical	solution	$-.182550423\times10^{-1}$	$-.500000000\times10^{0}$	$-.182373661\times10^{0}$

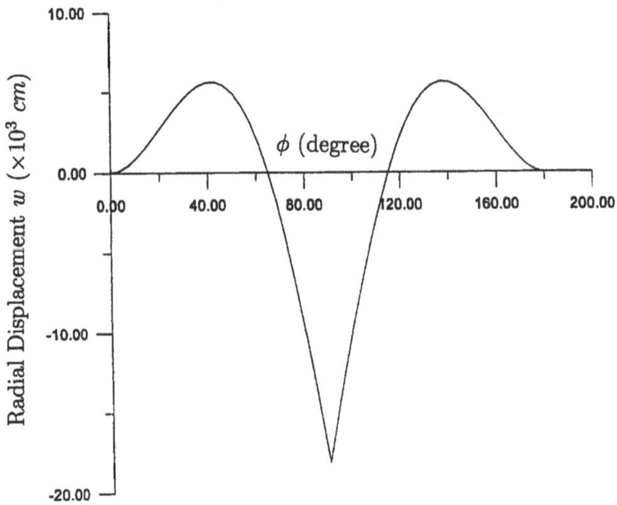

Fig. 9.12. Distribution of radial displacement

angle of the arch is 180°. The analyses by both the p refinement procedure of increasing the number of DOF per element and the h refinement procedure of increasing the number of elements are carried out, separately [110]. Numerical results are summarized and listed in Table 9.3. The results are compared with analytical solutions which can be obtained by using the complementary virtual work method to solve this statically indeterminate structural problem [111]. It shows that the results converge fast to the exact solutions by either increasing the DOF per element or the number of elements. It also shows that

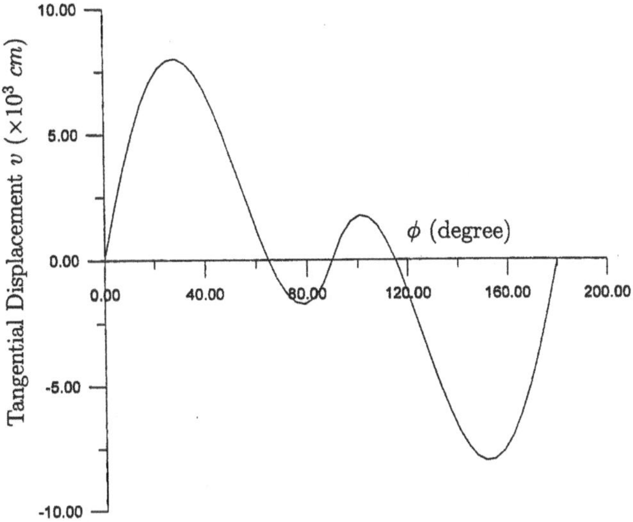

Fig. 9.13. Distribution of tangential displacement

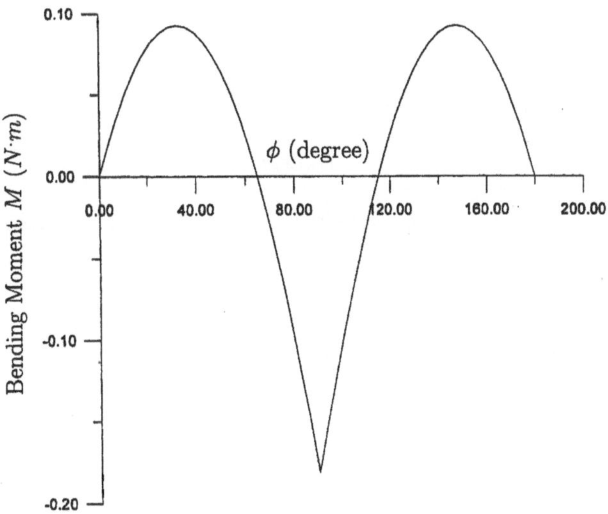

Fig. 9.14. Distribution of bending moment

the solution procedure of increasing the DOF per element is more efficient than the solution procedure of increasing the number of elements. The converged distributions of displacements and stress resultants are shown in Figs. 9.12 to 9.16.

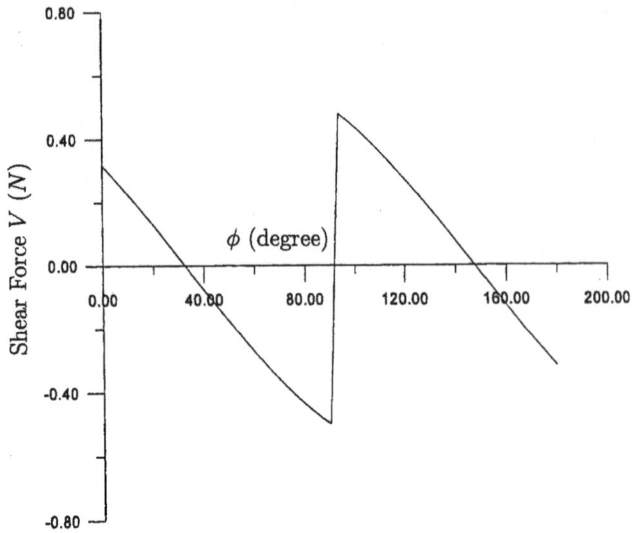

Fig. 9.15. Distribution of axial

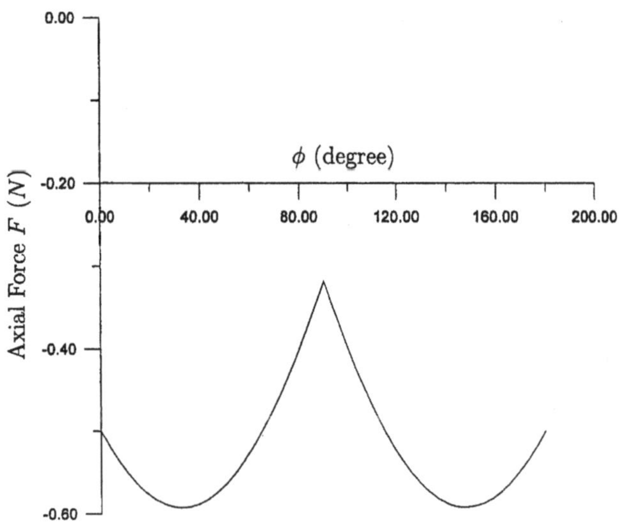

Fig. 9.16. Distribution of shear force

10

Development of DQEM Irregular Elements

The DQ or EDQ discretizations are carried out on an element basis with the element regular. However the physical element might be irregular. Consequently, mapping is necessary for carrying out the DQEM analysis of problems with irregular domain configurations. Some two-dimensional irregular elements with quadrilateral or triangular shapes are summarized. The generation of mesh and design of element grid are introduced. The calculation of outward unit normal vector on the element boundary is also necessary for the DQEM analysis. The method of using shape functions for mapping associated with the tangent relation, and the method of using secant relation are introduced.

10.1 Irregular Elements

The element configuration would change from element to element in the mesh [20,25]. By introducing an invertible transformation between a master element $\tilde{\Omega}$ of regular shape and an arbitrary physical element Ω^e it should be possible to transform the partial differential operations on Ω^e so that they hold on $\tilde{\Omega}$ [113].

The mapping of $\tilde{\Omega}$ onto Ω^e is defined by the following coordinate transformations

$$x = x(\xi, \eta), \quad y = y(\xi, \eta) \tag{10.1}$$

where x and y are physical coordinates in Ω^e and ξ and η are natural coordinates in $\tilde{\Omega}$. Then the transformations of the first derivatives of the variable function ϕ of element e are

$$\phi^e_{i,x} = \xi_{,x}\phi^e_{i,\xi} + \eta_{,x}\phi^e_{i,\eta}, \quad \phi^e_{i,y} = \xi_{,y}\phi^e_{i,\xi} + \eta_{,y}\phi^e_{i,\eta} \tag{10.2}$$

And the transformations of the second order derivatives of the variable function are

$$\phi^e_{i,xx} = \xi^2_{,x}\phi^e_{i,\xi\xi} + \eta^2_{,x}\phi^e_{i,\eta\eta} + 2\xi_{,x}\eta_{,x}\phi^e_{i,\xi\eta} + \xi_{,xx}\phi^e_{i,\xi} + \eta_{,xx}\phi^e_{i,\eta},$$

$$\phi^e_{i,xy} = \xi_{,x}\xi_{,y}\phi^e_{i,\xi\xi} + \eta_{,x}\eta_{,y}\phi^e_{i,\eta\eta} + (\xi_{,x}\eta_{,y} + \xi_{,y}\eta_{,x})\phi^e_{i,\xi\eta} + 2\xi_{,xy}\phi^e_{i,\xi} + 2\eta_{,xy}\phi^e_{i,\eta},$$

$$\phi^e_{i,yy} = \xi^2_{,y}\phi^e_{i,\xi\xi} + \eta^2_{,y}\phi^e_{i,\eta\eta} + 2\xi_{,y}\eta_{,y}\phi^e_{i,\xi\eta} + \xi_{,yy}\phi^e_{i,\xi} + \eta_{,yy}\phi^e_{i,\eta} \tag{10.3}$$

Suppose that the functions x and y are continuously differentiable with respect to ξ and η. Then the infinitesimals $d\xi$ and $d\eta$ can be transformed into dx and dy according to the following equations

$$dx = x_{,\xi}d\xi + x_{,\eta}d\eta, \quad dy = y_{,\xi}d\xi + y_{,\eta}d\eta \tag{10.4}$$

The above two equations can be expressed by the following matrix equation

$$\begin{Bmatrix} dx \\ dy \end{Bmatrix} = \begin{bmatrix} x_{,\xi} & x_{,\eta} \\ y_{,\xi} & y_{,\eta} \end{bmatrix} \begin{Bmatrix} d\xi \\ d\eta \end{Bmatrix} \tag{10.5}$$

The 2×2 matrix of partial derivatives in Eq. (10.5) is the Jacobian matrix of the transformation Eq. (10.4), and is denoted $\mathbf{J}$. The determinant $|\mathbf{J}|$ of the Jacobian matrix is

$$|\mathbf{J}| = det\mathbf{J} = x_{,\xi}y_{,\eta} - x_{,\eta}y_{,\xi} \tag{10.6}$$

An inverse transformation of Eq. (10.5) can be constructed if $|\mathbf{J}|$ is larger than zero. This inverse transformation is expressed as

$$\begin{Bmatrix} d\xi \\ d\eta \end{Bmatrix} = \mathbf{J}^{-1} \begin{Bmatrix} dx \\ dy \end{Bmatrix} = |\mathbf{J}|^{-1} \begin{bmatrix} y_{,\eta} & -x_{,\eta} \\ -y_{,\xi} & x_{,\xi} \end{bmatrix} \begin{Bmatrix} dx \\ dy \end{Bmatrix} \tag{10.7}$$

The infinitesimals dx and dy can also be transformed into $d\xi$ and $d\eta$ according to the following matrix equation

$$\begin{Bmatrix} d\xi \\ d\eta \end{Bmatrix} = \begin{bmatrix} \xi_{,x} & \xi_{,y} \\ \eta_{,x} & \eta_{,y} \end{bmatrix} \begin{Bmatrix} dx \\ dy \end{Bmatrix} \tag{10.8}$$

Equating terms in Eqs. (10.7) and (10.8), the following relationships can be obtained

$$\xi_{,x} = |\mathbf{J}|^{-1} y_{,\eta}, \quad \xi_{,y} = -|\mathbf{J}|^{-1} x_{,\eta}, \quad \eta_{,x} = -|\mathbf{J}|^{-1} y_{,\xi}, \quad \eta_{,y} = |\mathbf{J}|^{-1} x_{,\xi} \tag{10.9}$$

Using Eq. (10.9), the following two equations can be obtained

$$\frac{\partial}{\partial x} = |\mathbf{J}|^{-1}\left(y_{,\eta}\frac{\partial}{\partial \xi} - y_{,\xi}\frac{\partial}{\partial \eta}\right), \quad \frac{\partial}{\partial y} = |\mathbf{J}|^{-1}\left(-x_{,\eta}\frac{\partial}{\partial \xi} + x_{,\xi}\frac{\partial}{\partial \eta}\right) \tag{10.10}$$

And the use of Eq. (10.6) leads to the following two equations

$$|\mathbf{J}|_{,\xi} = x_{,\xi}y_{,\xi\eta} - y_{,\xi}x_{,\xi\eta} + y_{,\eta}x_{,\xi\xi} - x_{,\eta}y_{,\xi\xi},$$

$$| \mathbf{J} |_{,\eta} = y_{,\eta}x_{,\xi\eta} - x_{,\eta}y_{,\xi\eta} + x_{,\xi}y_{,\eta\eta} - y_{,\xi}x_{,\eta\eta} \tag{10.11}$$

Using Eq. (10.10), the following second derivatives of the natural coordinates with respect to physical coordinates can be obtained

$$\begin{aligned}
\xi_{,xx} &= | \mathbf{J} |^{-2} (y_{,\eta}y_{,\xi\eta} - y_{,\xi}y_{,\eta\eta}) - | \mathbf{J} |^{-3} (y_{,\eta}^2 \, | \mathbf{J} |_{,\xi} - y_{,\xi}y_{,\eta} \, | \mathbf{J} |_{,\eta}), \\
\xi_{,xy} &= | \mathbf{J} |^{-2} (x_{,\xi}y_{,\eta\eta} - x_{,\eta}y_{,\xi\eta}) + | \mathbf{J} |^{-3} (x_{,\eta}y_{,\eta} \, | \mathbf{J} |_{,\xi} - x_{,\xi}y_{,\eta} \, | \mathbf{J} |_{,\eta}), \\
\xi_{,yy} &= | \mathbf{J} |^{-2} (x_{,\eta}x_{,\xi\eta} - x_{,\xi}x_{,\eta\eta}) - | \mathbf{J} |^{-3} (x_{,\eta}^2 \, | \mathbf{J} |_{,\xi} - x_{,\xi}x_{,\eta} \, | \mathbf{J} |_{,\eta}), \\
\eta_{,xx} &= | \mathbf{J} |^{-2} (y_{,\xi}y_{,\xi\eta} - y_{,\eta}y_{,\xi\xi}) + | \mathbf{J} |^{-3} (y_{,\xi}y_{,\eta} \, | \mathbf{J} |_{,\xi} - y_{,\xi}^2 \, | \mathbf{J} |_{,\eta}), \\
\eta_{,xy} &= | \mathbf{J} |^{-2} (y_{,\eta}x_{,\xi\xi} - y_{,\xi}x_{,\xi\eta}) - | \mathbf{J} |^{-3} (x_{,\xi}y_{,\eta} \, | \mathbf{J} |_{,\xi} - x_{,\xi}y_{,\xi} \, | \mathbf{J} |_{,\eta}), \\
\eta_{,yy} &= | \mathbf{J} |^{-2} (x_{,\xi}x_{,\xi\eta} - x_{,\eta}x_{,\xi\xi}) + | \mathbf{J} |^{-3} (x_{,\xi}x_{,\eta} \, | \mathbf{J} |_{,\xi} - x_{,\xi}^2 \, | \mathbf{J} |_{,\eta})
\end{aligned} \tag{10.12}$$

The outlined mapping transformations are generic which hold good for adopting any kinds of appropriate analytical functions. Thus various domain configurations and mapping techniques can be adopted [113]. The simulation for transformation adopting polynomials is carried out [25–27]. The transformation relations are expressed by

$$x_{\bar{i}} = N_\gamma(\xi_r)\tilde{x}_{\bar{i}\gamma}, \quad \gamma = 1, 2, ..., N_c \tag{10.13}$$

where $\tilde{x}_{\bar{i}\gamma}$ are $x_{\bar{i}}$ and their possible partial derivatives with respective to ξ_r at nodes used to define the transformations, $N_\gamma(\xi_r)$ are the corresponding shape functions and N_c is the total degrees of freedom. Using Eq. (10.13), the first order partial derivatives of the physical coordinates with respect to the natural coordinates can be obtained.

$$x_{\bar{i},\xi_{\bar{j}}} = N_{\gamma,\xi_{\bar{j}}}(\xi_r)\tilde{x}_{\bar{i}\gamma} \tag{10.14}$$

And the second order partial derivatives of the physical coordinates with respect to the natural coordinates are

$$x_{\bar{i},\xi_{\bar{j}}\xi_{\bar{k}}} = N_{\gamma,\xi_{\bar{j}}\xi_{\bar{k}}}(\xi_r)\tilde{x}_{\bar{i}\gamma} \tag{10.15}$$

The shape functions of serendipity C^0 triangular elements, serendipity triangular elements with incomplete first order derivatives, serendipity C^0 quadrilateral elements and serendipity Hermitian quadrilateral elements which are necessary for carrying out the mapping transformation and developing irregular DQEM elements are summarized [114–115].

10.1.1 Serendipity C^0 Triangular Elements

In constructing the mapping transformation, the master triangular element in the natural space may be an arbitrary linear triangle. For convenience, the rectangular unit triangle is adopted. Let the natural coordinates be L_1 and L_2. A representative serendipity rectangular unit C^0 triangular element is shown in Fig. 10.1. Assume that the physical coordinates are nth order on side $1 - n + 1$, lth order on side $n + 1 - n + l + 1$ and mth order on side $n + l + 1 - 1$ in terms of the natural coordinates. The shape functions can be expressed by

$$
\begin{aligned}
N_i = {} & \frac{L_1}{2(1-L_2)} \Psi^n_{n-i+1}(1-L_2)\Psi^n_{i-1}(L_2) \\
& + \frac{L_2}{2(1-L_1)} \Psi^n_{n-i+1}(L_1)\Psi^n_{i-1}(1-L_1), \quad 2 \le i \le n,
\end{aligned}
$$

$$
\begin{aligned}
N_i = {} & \frac{L_2}{2(L_1+L_2)} \Psi^l_{l+n+1-i}(L_1+L_2)\Psi^l_{i-n-1}(1-L_1-L_2) \\
& + \frac{1-L_1-L_2}{2(1-L_2)} \Psi^l_{l+n+1-i}(L_2)\Psi^l_{i-n-1}(1-L_2), \quad n+2 \le i \le n+l,
\end{aligned}
$$

$$
\begin{aligned}
N_i = {} & \frac{1-L_1-L_2}{2(1-L_1)} \Psi^m_{l+m+n+1-i}(1-L_1)\Psi^m_{i-l-n-1}(L_1) \\
& + \frac{L_1}{2(L_1+L_2)} \Psi^m_{l+m+n+1-i}(1-L_1-L_2)\Psi^m_{i-l-n-1}(L_1+L_2), \\
& l+n+2 \le i \le l+m+n,
\end{aligned}
$$

$$
\begin{aligned}
N_1 = {} & \frac{1-L_1-L_2}{2(1-L_1)} \Psi^m_m(L_1) + \frac{L_1}{2(L_1+L_2)} \Psi^m_m(L_1+L_2) \\
& + \frac{L_1}{2(1-L_2)} \Psi^n_n(1-L_2) + \frac{L_2}{2(1-L_1)} \Psi^n_n(L_1) - \frac{L_1}{2},
\end{aligned}
$$

$$
\begin{aligned}
N_{n+1} = {} & \frac{L_1}{2(1-L_2)} \Psi^n_n(L_2) + \frac{L_2}{2(1-L_1)} \Psi^n_n(1-L_1) \\
& + \frac{L_2}{2(L_1+L_2)} \Psi^l_l(L_1+L_2) + \frac{1-L_1-L_2}{2(1-L_2)} \Psi^l_l(L_2) - \frac{L_2}{2},
\end{aligned}
$$

$$
\begin{aligned}
N_{l+n+1} = {} & \frac{L_2}{2(L_1+L_2)} \Psi^l_l(1-L_1-L_2) + \frac{1-L_1-L_2}{2(1-L_2)} \Psi^l_l(1-L_2) \\
& + \frac{1-L_1-L_2}{2(1-L_1)} \Psi^m_m(1-L_1) \\
& + \frac{L_1}{2(L_1+L_2)} \Psi^m_m(1-L_1-L_2) - \frac{1-L_1-L_2}{2}
\end{aligned}
\tag{10.16}
$$

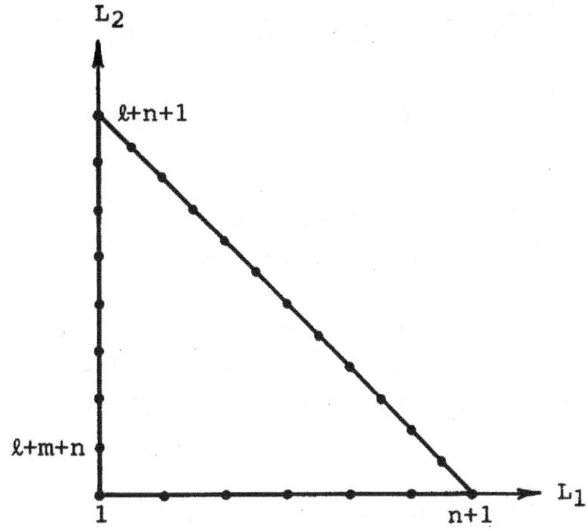

Fig. 10.1. The serendipity rectangular unit C^0 triangular element

where

$$\begin{aligned} \Psi_{\bar{p}}^{\bar{n}}(L) &= \textstyle\prod_{k=1}^{\bar{p}} \frac{\bar{n}L-k+1}{k}, \quad 1 \le \bar{p} \le \bar{n} \\ &= 1, \qquad\qquad\qquad \bar{p} = 0 \end{aligned} \tag{10.17}$$

in which $\bar{n}$ is the order of one side. Assume that the three sides have the same order n and define

$$\begin{aligned} \lambda_p^n(L) &= \textstyle\prod_{k=1}^{p-1} \frac{nL-k}{p-k}, \quad 2 \le p \le n \\ &= 1, \qquad\qquad\qquad p = 1 \end{aligned} \tag{10.18}$$

Then the shape functions can be expressed by

$$\begin{aligned} N_i = \frac{n^2 L_1 L_2}{2(i-1)(n-i+1)} &\big[\lambda_{n-i+1}^n(L_1)\lambda_{i-1}^n(1-L_1) \\ &+\lambda_{n-i+1}^n(1-L_2)\lambda_{i-1}^n(L_2)\big], \quad 2 \le i \le n, \end{aligned}$$

$$\begin{aligned} N_i = \frac{n^2 L_2(1-L_1-L_2)}{2(i-n-1)(2n-i+1)} &\big[\lambda_{2n-i+1}^n(L_2)\lambda_{i-n-1}^n(1-L_2) \\ &+\lambda_{2n-i+1}^n(L_1+L_2)\lambda_{i-n-1}^n(1-L_1-L_2)\big], \quad n+2 \le i \le 2n, \end{aligned}$$

$$\begin{aligned} N_i = \frac{n^2 L_1(1-L_1-L_2)}{2(i-2n-1)(3n-i+1)} &\big[\lambda_{3n-i+1}^n(1-L_1-L_2)\lambda_{i-2n-1}^n(L_1+L_2) \\ &+\lambda_{3n-i+1}^n(1-L_1)\lambda_{i-2n-1}^n(L_1)\big], \quad 2n+2 \le i \le 3n, \end{aligned}$$

$$N_1 = \frac{L_1}{2}\big[\lambda_n^n(L_1) + \lambda_n^n(L_1+L_2) + \lambda_n^n(1-L_2) - 1\big],$$

$$N_{n+1} = \frac{L_2}{2}\left[\lambda_n^n(L_2) + \lambda_n^n(1-L_1) + \lambda_n^n(L_1+L_2) - 1\right],$$

$$N_{2n+1} = \frac{1-L_1-L_2}{2}\left[\lambda_n^n(1-L_1-L_2)+\lambda_n^n(1-L_2)+\lambda_n^n(1-L_1)-1\right] \quad (10.19)$$

10.1.2 Serendipity Triangular Element with Incomplete First Order Derivatives

Denote ζ the natural coordinate along a straight line having n nodes in the $L_1 - L_2$ plane with $\zeta_1 = 0$, $\zeta_n = 1$. Then by using the first order one-dimensional Hermitian interpolation formula, the physical coordinate variable x can be expressed by

$$x(\zeta) = f_i^n(\zeta)x_i + g_i^n(\zeta)\left[(L_{1(n)} - L_{1(1)})x_{i,L_1} + (L_{2(n)} - L_{2(1)})x_{i,L_2}\right], \quad i = 1, 2, ..., n \quad (10.20)$$

where f_i^n represents $H_i(\zeta)$ expressed by Eq. (2.77). It can also be written as

$$f_i^n = \left[1 + \psi_i^n(\zeta - \zeta_i)\right]\left[\varphi_i^n(\zeta)\right]^2, \quad g_i^n(\zeta) = (\zeta - \zeta_i)\left[\varphi_i^n(\zeta)\right]^2,$$
$$\psi_i^n = -2\prod_{k=1,k\neq i}^{n}\frac{1}{\zeta_i - \zeta_k}, \quad \varphi_i^n(\zeta) = \prod_{k=1,k\neq i}^{n}\frac{\zeta - \zeta_k}{\zeta_i - \zeta_k} \quad (10.21)$$

In the above equation $\varphi_i^n(\zeta)$ represents the $(n-1)$th order Lagrange interpolation function with respect to i, $\hat{L}_i^n(\zeta)$ expressed by Eq. (2.70), and ψ_i^n represents the first order derivative of the Lagrange interpolation function $\hat{L}_i^n(\zeta)$ with respect to ζ at ζ_i multiplied by -2. Equation (10.20) can be used to derive shape functions of serendipity triangular elements with incomplete first order derivatives. Serendipity Hermitian triangular elements with higher order derivatives can also be used for the mapping transformation.

10.1.3 Serendipity C^0 Quadrilateral Elements

In constructing the mapping transformation, the master quadrilateral element in the natural space may be an arbitrary rectangle. For convenience, the unit square is adopted. A representative serendipity unit C^0 quadrilateral element is shown in Fig. 10.2. Assume that the physical coordinates are pth order on side 1 - $p+1$, qth order on side $p+1$ - $p+q+1$, rth order on side $p+q+1$ - $p+q+r+1$ and sth order on side $p+q+r+1$ - 1 in terms of the natural coordinates. The shape functions can be expressed by

$$N_i = (1-\eta)\varphi_i^{p+1}(\xi), \quad 2 \le i \le p,$$

$$N_i = \xi\varphi_{i-p}^{q+1}(\eta), \quad p+2 \le i \le p+q,$$

$$N_i = \eta\varphi_{p+q+r+2-i}^{r+1}(\xi), \quad p+q+2 \le i \le p+q+r,$$

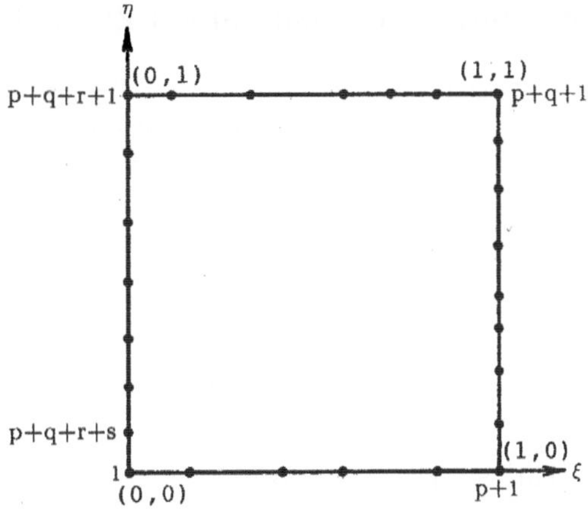

Fig. 10.2. The serendipity unit C^0 quadrilateral element

$$
\begin{aligned}
N_i &= (1-\xi)\varphi^{s+1}_{p+q+r+s+2-i}(\eta), \quad p+q+r+2 \le i \le p+q+r+s,\\
N_1 &= (1-\eta)\varphi^{p+1}_1(\xi) + (1-\xi)\varphi^{s+1}_1(\eta) - (1-\xi)(1-\eta),\\
N_{p+1} &= (1-\eta)\varphi^{p+1}_{p+1}(\xi) + \xi\varphi^{q+1}_1(\eta) - \xi(1-\eta),\\
N_{p+q+1} &= \eta\varphi^{r+1}_{r+1}(\xi) + \xi\varphi^{q+1}_{q+1} - \xi\eta,\\
N_{p+q+r+1} &= \eta\varphi^{r+1}_1(\xi) + (1-\xi)\varphi^{s+1}_{s+1} - (1-\xi)\eta
\end{aligned}
\tag{10.22}
$$

Assume that the four sides have the same order p. Then the shape functions can be expressed by

$$
\begin{aligned}
N_i &= (1-\eta)\varphi^{p+1}_i(\xi), \quad 2 \le i \le p,\\
N_i &= \xi\varphi^{p+1}_{i-p}(\eta), \quad p+2 \le i \le 2p,\\
N_i &= \eta\varphi^{p+1}_{3p+2-i}(\xi), \quad 2p+2 \le i \le 3p,\\
N_i &= (1-\xi)\varphi^{p+1}_{4p+2-i}(\eta), \quad 3p+2 \le i \le 4p,\\
N_1 &= (1-\eta)\varphi^{p+1}_1(\xi) + (1-\xi)\varphi^{p+1}_1(\eta) - (1-\xi)(1-\eta),\\
N_{p+1} &= (1-\eta)\varphi^{p+1}_{p+1}(\xi) + \xi\varphi^{p+1}_1(\eta) - \xi(1-\eta),\\
N_{2p+1} &= \eta\varphi^{p+1}_{p+1}(\xi) + \xi\varphi^{p+1}_{p+1}(\eta) - \xi\eta,\\
N_{3p+1} &= \eta\varphi^{p+1}_1(\xi) + (1-\xi)\varphi^{p+1}_{p+1}(\eta) - (1-\xi)\eta
\end{aligned}
\tag{10.23}
$$

10.1.4 Serendipity Hermitian Quadrilateral Elements

Consider that the unit square in the natural space has an $m \times n$ serendipity grid. By using the two-dimensional node identification method, the following relations hold

$$\hat{D}_\xi^s \hat{D}_\eta^t x_{\alpha\beta} = \hat{D}_\xi^s \hat{D}_\eta^t \tilde{x}_{\alpha\beta}, \quad 0 \le s \le m-1, 0 \le t \le n-1 \tag{10.24}$$

where the following relation is used

$$\hat{D}_\xi^s \hat{D}_\eta^t x_{\alpha\beta} = \frac{\partial^{(s+t)} x(\xi,\eta)}{\partial \xi^s \partial \eta^t} |_{\alpha\beta} \tag{10.25}$$

The physical coordinate x can be expressed by

$$x(\xi,\eta) = x_\xi + x_\eta - \bar{x} \tag{10.26}$$

Let $a(0,0)$, $b(1,0)$, $c(1,1)$ and $d(0,1)$ be the four corner nodes. Then

$$x_\xi = \sum_{\alpha=1}^{m} \Bigg[\sum_{s=0}^{m-1} H_\alpha^{m-1,s}(m,\xi) \hat{D}_\xi^s \times \sum_{t=0}^{n-1} \Big(H_1^{n-1,t}(2,\eta) \hat{D}_\eta^t x_\alpha |_{ab} + H_2^{n-1,t}(2,\eta) \hat{D}_\eta^t x_\alpha |_{dc} \Big) \Bigg], \tag{10.27}$$

$$x_\eta = \sum_{\beta=1}^{n} \Bigg[\sum_{t=0}^{n-1} H_\beta^{n-1,t}(n,\eta) \hat{D}_\eta^t \times \sum_{s=0}^{m-1} \Big(H_1^{m-1,s}(2,\xi) \hat{D}_\xi^s x_\beta |_{ad} + H_2^{m-1,s}(2,\xi) \hat{D}_\xi^s x_\beta |_{bc} \Big) \Bigg], \tag{10.28}$$

and

$$\begin{aligned} \bar{x} = \sum_{s=0}^{m-1} \sum_{t=0}^{n-1} \Big[& H_1^{m-1,s}(2,\xi) H_1^{n-1,t}(2,\eta) \hat{D}_\xi^s \hat{D}_\eta^t x_a \\ & + H_2^{m-1,s}(2,\xi) H_1^{n-1}(2,\eta) \hat{D}_\xi^s \hat{D}_\eta^t x_b \\ & + H_2^{m-1}(2,\xi) H_2^{n-1}(2,\eta) \hat{D}_\xi^s \hat{D}_\eta^t x_c \\ & + H_1^{m-1}(2,\xi) H_2^{n-1}(2,\eta) \hat{D}_\xi^s \hat{D}_\eta^t x_d \Big] \end{aligned} \tag{10.29}$$

where $H_\alpha^{m-1,s}$ and $H_\beta^{n-1,t}$ are Hermitian polynomials expressed by Eq. (2.77).

10.1.5 C^{n*} Elements

The C^{n*} elements are also effective for defining the mapping transformation [116]. For illustration, consider the four-node twelve-DOF quadrilateral C^{1*} element. Each of the four nodes has three DOF representing x_i, $\frac{\partial x_i}{\partial \xi}$ and $\frac{\partial y_i}{\partial \eta}$. Considering that the natural coordinates of the four element nodes 1, 2, 3 and 4 are $(\xi_1, \eta_1) = (-1, -1)$, $(\xi_2, \eta_2) = (1, -1)$, $(\xi_3, \eta_3) = (1, 1)$, and $(\xi_4, \eta_4) = (-1, 1)$, respectively, the twelve shape functions are expressed by

$$N_1 = \frac{1}{4}\left[1 - (3\xi/2 - \xi^3/2) - (3\eta/2 - \eta^3/2) + (2\xi\eta - \xi^3\eta/2 - \xi\eta^3/2)\right],$$

$$N_2 = \frac{1}{8}\left[1 - (\xi + \xi^2 - \xi^3) - \eta + (\xi\eta + \xi^2\eta - \xi^3\eta)\right],$$

$$N_3 = \frac{1}{8}\left[1 - \xi + (-\eta - \eta^2 + \eta^3) + (\xi\eta + \xi\eta^2 - \xi\eta^3)\right],$$

$$N_4 = \frac{1}{4}\left[1 + (3\xi/2 - \xi^3/2) - (3\eta/2 - \eta^3/2) - (2\xi\eta - \xi^3\eta/2 - \xi\eta^3/2)\right],$$

$$N_5 = \frac{1}{8}\left[-1 - (\xi - \xi^2 - \xi^3) + \eta + (\xi\eta - \xi^2\eta - \xi^3\eta)\right],$$

$$N_6 = \frac{1}{8}\left[1 + \xi - (\eta + \eta^2 - \eta^3) - (\xi\eta + \xi\eta^2 - \xi\eta^3)\right],$$

$$N_7 = \frac{1}{4}\left[1 + (3\xi/2 - \xi^3/2) + (3\eta/2 - \eta^3/2) + (2\xi\eta - \xi^3\eta/2 - \xi\eta^3/2)\right],$$

$$N_8 = \frac{1}{8}\left[-1 - (\xi - \xi^2 - \xi^3) - \eta - (\xi\eta - \xi^2\eta - \xi^3\eta)\right],$$

$$N_9 = \frac{1}{8}\left[-1 - \xi(\eta - \eta^2 - \eta^3) - (\xi\eta - \xi\eta^2 - \xi\eta^3)\right],$$

$$N_{10} = \frac{1}{4}\left[1 - (3\xi/2 - \xi^3/2) + (3\eta/2 - \eta^3/2) - (2\xi\eta - \xi^3\eta/2 - \xi\eta^3/2)\right],$$

$$N_{11} = \frac{1}{8}\left[1 - (\xi + \xi^2 - \xi^3) + \eta - (\xi\eta + \xi^2\eta - \xi^3\eta)\right],$$

$$N_{12} = \frac{1}{8}\left[-1 + \xi - (\eta - \eta^2 - \eta^3) + (\xi\eta - \xi\eta^2 - \xi\eta^3)\right]$$

10.2 Mesh and Element Grids

Both of h and p DQEM procedures can be used to solve a problem and get converged results. Since the p version is more efficient than the h version, in the real engineering or scientific application the concept of adaptive discretization can be adopted and the number of elements used to model the analysis domain must be as small as possible [20,25,117]. Various techniques can be used to generate the mesh and element grids. The mapping technique is used. The grid of an element can be flexible. Consider the quadrilateral element and let ξ and η denote the nondimensional one unit natural coordinates corresponding to the physical coordinates x and y, respectively. Also consider the Lagrange family grid and let N_ξ and N_η denote the numbers of levels in ξ and η directions, respectively, in the master element of a physical element. For solving a problem having irregular analysis domain boundary which has a curved line or a straight line not parallel to x and y axes, by designing the mesh in such a way that interior elements are rectangles with the element grid lines parallel to x or y axes the assembled overall algebraic system will have more zero elements in the coefficient matrix. It can reduce the computer memory and CPU time required for solving the problem. Figure 10.3 shows an efficient mesh. Mapping is also not necessary for triangular elements having three linear sides. For the solution of field problems, the concept of generating efficient mesh can be similarly used to design efficient element grid of an irregular element with the inclined or curved element side attached to the analysis domain boundary. Figure 10.4 shows an efficient element grid model of an irregular element. The form of the assembled overall coefficient matrix, when assemble a discrete governing equation at an interior node, the number of data filled in is N_ξ+N_η-1 for the efficient element grid model as compared to $N_\xi \times N_\eta$ for a fully irregular element grid model. When assemble a discrete natural boundary condition if the natural boundary is a straight line parallel to one of the coordinate axes, the adoption of efficient element grid model can also reduce the number of data filled in. Considering a boundary node at the intersection of a natural boundary line parallel to the y axis and an interior grid line parallel to the x axis, the number of data

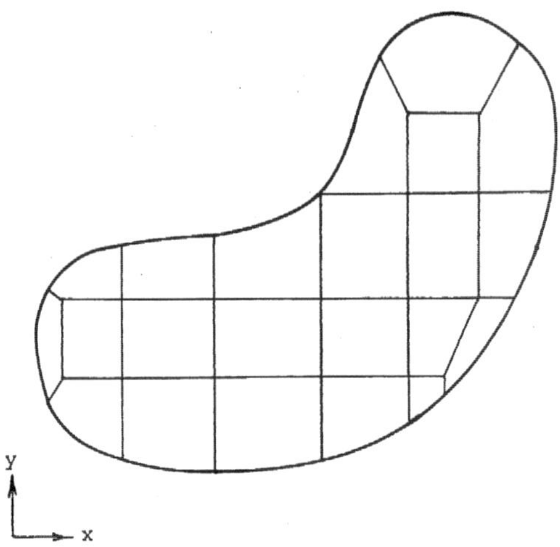

Fig. 10.3. The efficient mesh

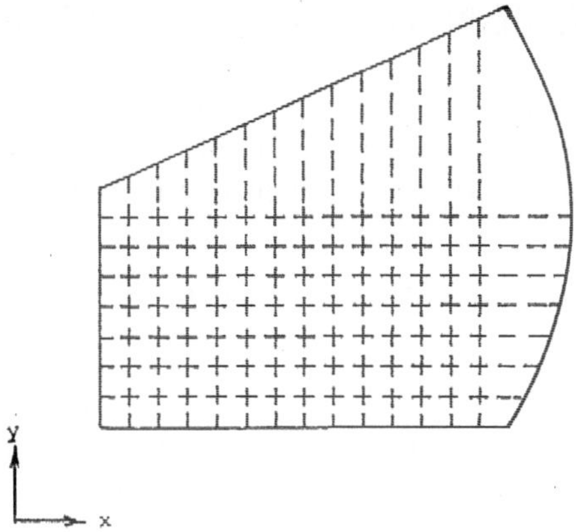

Fig. 10.4. Efficient element grid model

filled in is N_ξ for the efficient element grid model as compared to N_ξ+N_η-1 for a fully irregular element grid model. When assembling the discrete natural boundary condition at a node on an inclined or curved natural boundary line the number of data filled in is also N_ξ+N_η-1. When assembling the discrete natural transition condition the number of data filled in can also be reduced if the efficient mesh and efficient element grid model are used. It should be mentioned that the design of efficient element grid model for an element with inclined and/or curved sides attached to the analysis domain boundary might result in an extremely nonuniform distribution of element nodes. The DQEM analysis using elements having extremely nonuniform distributions of element nodes also have excellent numerical performance [27].

For two adjacent elements having different numbers of nodes on the inter-element boundary, the number of kinematic transition conditions must be larger than the number of natural transition conditions. Let n_d denote the difference between the two node numbers which equals the difference between the two numbers of transition conditions. To set up the kinematic transition conditions, the nodes on the inter-element boundary must be arranged in such a way that only n_d nodes in one element are not coincident with the nodes in the other element. The n_d extra nodes are used to define n_d extra kinematic transition conditions. In defining the n_d extra kinematic conditions, the interpolation technique must be used. In addition, the transition conditions can be easily set up by designing the grids of the two adjacent elements in such a way that both elements have the same numbers of nodes on the inter-element boundary no matter what the orders of approximations and grid configurations are. The concept of efficient element grid can be similarly used to design the grids of two adjacent elements which have this type of connection.

10.3 Outward Unit Normal Vector on Element Boundary

The direction cosines of the outward unit normal vector at a discrete point on the element boundary is necessary for defining the discrete natural transition conditions or natural boundary conditions at the discrete point.

10.3.1 Mapping for Calculating the Direction Cosines

The mapping technique can be used to calculate the direction cosines of the outward unit normal vector on the element boundary [25–27].

For illustration, consider the mapping of the bilinear element shown in Fig. 10.5. The four shape functions which define the mapping are expressed

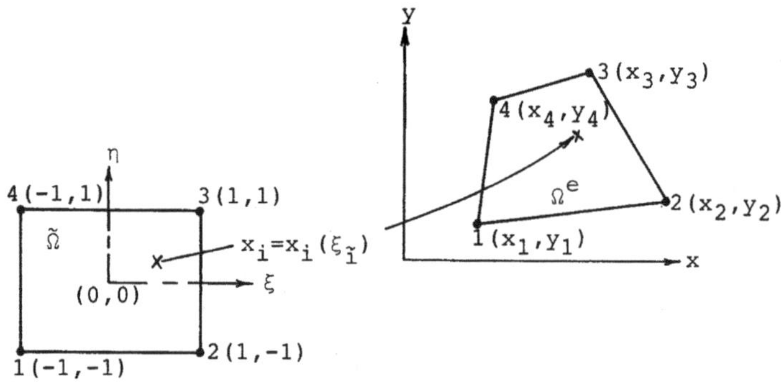

Fig. 10.5. Mapping of the bilinear element

as

$$\Psi_1 = \frac{1}{4}(1-\xi)(1-\eta), \quad \Psi_2 = \frac{1}{4}(1+\xi)(1-\eta),$$
$$\Psi_3 = \frac{1}{4}(1+\xi)(1+\eta), \quad \Psi_4 = \frac{1}{4}(1-\xi)(1+\eta) \tag{10.30}$$

Referring to Fig. 10.6, the position vector of a point on side $\eta = -1$ is

$$\begin{aligned}\mathbf{R}(\xi) &= x\mathbf{i} + y\mathbf{j} \\ &= \frac{1}{2}\left[(1-\xi)x_1 + (1+\xi)x_2\right]\mathbf{i} + \frac{1}{2}\left[(1-\xi)y_1 + (1+\xi)y_2\right]\mathbf{j}\end{aligned} \tag{10.31}$$

Then the unit tangent vector $\mathbf{t}$ can be expressed by

$$\mathbf{t} = \frac{\frac{d\mathbf{R}(\xi)}{d\xi}}{\left|\frac{d\mathbf{R}(\xi)}{d\xi}\right|} = \frac{(x_2-x_1)\mathbf{i} + (y_2-y_1)\mathbf{j}}{\left[(x_2-x_1)^2 + (y_2-y_1)^2\right]^{\frac{1}{2}}} = \bar{\alpha}\mathbf{i} + \bar{\beta}\mathbf{j} \tag{10.32}$$

Hence, the direction cosines of $\mathbf{t}$ are

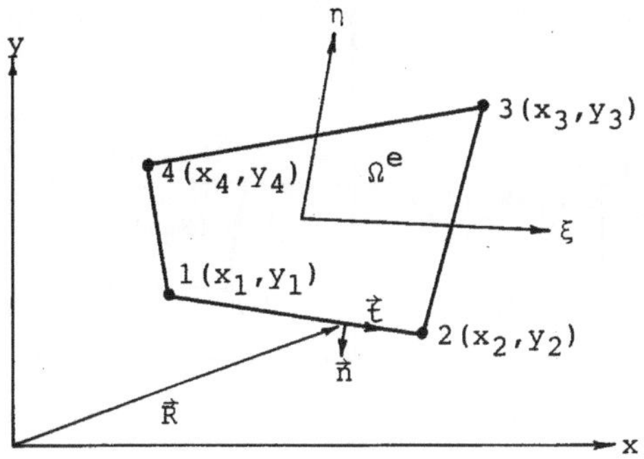

Fig. 10.6. A bilinear element

$$\bar{\alpha} = \frac{x_2 - x_1}{[(x_2 - x_1)^2 + (y_2 - y_1)^2]^{\frac{1}{2}}}, \quad \bar{\beta} = \frac{y_2 - y_1}{[(x_2 - x_1)^2 + (y_2 - y_1)^2]^{\frac{1}{2}}} \tag{10.33}$$

And the direction cosines of the outward unit normal vector $\mathbf{n} = l\mathbf{i} + m\mathbf{j}$ can be obtained

$$l = \bar{\alpha} cos\frac{\pi}{2} + \bar{\beta} sin\frac{\pi}{2} = \bar{\beta}, \quad m = -\bar{\alpha} sin\frac{\pi}{2} + \bar{\beta} cos\frac{\pi}{2} = -\bar{\alpha} \tag{10.34}$$

For side $\xi = 1$, the direction cosines of the unit tangent vector are found to be

$$\bar{\alpha} = \frac{x_3 - x_2}{[(x_3 - x_2)^2 + (y_3 - y_2)^2]^{\frac{1}{2}}}, \quad \bar{\beta} = \frac{y_3 - y_2}{[(x_3 - x_2)^2 + (y_3 - y_2)^2]^{\frac{1}{2}}} \tag{10.35}$$

And the direction cosines of the outward unit normal vector are

$$l = \bar{\beta}, \quad m = -\bar{\alpha} \tag{10.36}$$

For side $\eta = 1$, the direction cosines of the unit tangent vector are found to be

$$\bar{\alpha} = \frac{x_3 - x_4}{[(x_3 - x_4)^2 + (y_3 - y_4)^2]^{\frac{1}{2}}}, \quad \bar{\beta} = \frac{y_3 - y_4}{[(x_3 - x_4)^2 + (y_3 - y_4)^2]^{\frac{1}{2}}} \tag{10.37}$$

And the direction cosines of the outward unit normal vector are

$$l = -\bar{\beta}, \quad m = \bar{\alpha} \tag{10.38}$$

For side $\xi = -1$, the direction cosines of the unit tangent vector are found to be

$$\bar{\alpha} = \frac{x_4 - x_1}{[(x_4 - x_1)^2 + (y_4 - y_1)^2]^{\frac{1}{2}}}, \quad \bar{\beta} = \frac{y_4 - y_1}{[(x_4 - x_1)^2 + (y_4 - y_1)^2]^{\frac{1}{2}}} \tag{10.39}$$

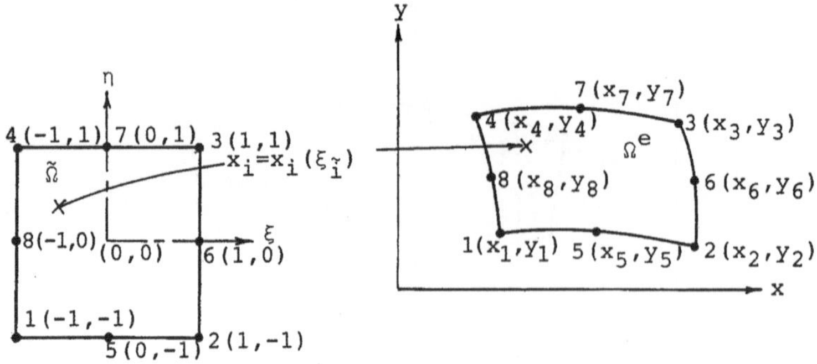

Fig. 10.7. Mapping of the quadratic serendipity element

And the direction cosines of the outward unit normal vector are

$$l = -\bar{\beta}, \quad m = \bar{\alpha} \tag{10.40}$$

Also considering the mapping of the quadratic serendipity element shown in Fig. 10.7. The eight shape functions which define the mapping are expressed as

$$\begin{aligned}
\Psi_1 &= -\frac{1}{4}(1-\xi)(1-\eta)(1+\xi+\eta), \quad \Psi_2 = \frac{1}{4}(1+\xi)(1-\eta)(-1+\xi-\eta), \\
\Psi_3 &= \frac{1}{4}(1+\xi)(1+\eta)(-1+\xi+\eta), \quad \Psi_4 = \frac{1}{4}(1-\xi)(1+\eta)(-1-\xi+\eta), \\
\Psi_5 &= \frac{1}{2}(1-\xi^2)(1-\eta), \quad \Psi_6 = \frac{1}{2}(1+\xi)(1-\eta^2), \\
\Psi_7 &= \frac{1}{2}(1-\xi^2)(1+\eta), \quad \Psi_8 = \frac{1}{2}(1-\xi)(1-\eta^2)
\end{aligned} \tag{10.41}$$

Referring to Fig. 10.8, the position vector of a point on side $\eta = -1$ is

$$\begin{aligned}
\mathbf{R}(\xi) = x\mathbf{i} + y\mathbf{j} &= \left[\frac{1}{2}(\xi^2-\xi)x_1 + \frac{1}{2}(\xi^2+\xi)x_2 + (1-\xi^2)x_5\right]\mathbf{i} \\
&+ \left[\frac{1}{2}(\xi^2-\xi)y_1 + \frac{1}{2}(\xi^2+\xi)y_2 + (1-\xi^2)y_5\right]\mathbf{j}
\end{aligned} \tag{10.42}$$

Then the unit tangent vector $\mathbf{t}$ can be expressed by

$$\mathbf{t} = \frac{\frac{d\mathbf{R}(\xi)}{d\xi}}{\left|\frac{d\mathbf{R}(\xi)}{d\xi}\right|} = \frac{A\mathbf{i} + B\mathbf{j}}{(A^2+B^2)^{\frac{1}{2}}} \tag{10.43}$$

where

$$\begin{aligned}
A &= \frac{1}{2}(2\xi-1)x_1 + \frac{1}{2}(2\xi+1)x_2 - 2\xi x_5, \\
B &= \frac{1}{2}(2\xi-1)y_1 + \frac{1}{2}(2\xi+1)y_2 - 2\xi y_5
\end{aligned} \tag{10.44}$$

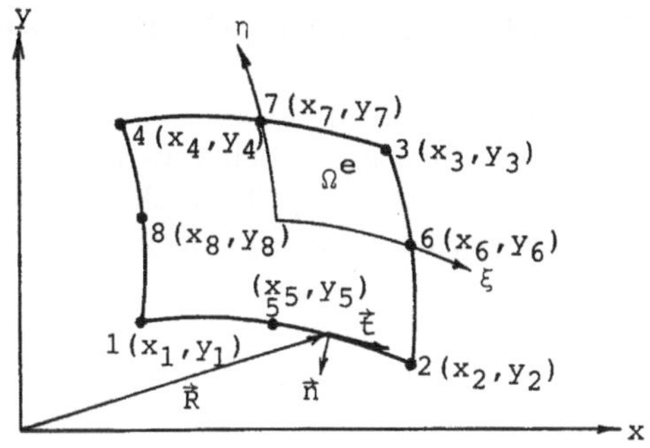

Fig. 10.8. A quadratic serendipity element

Hence, the direction cosines of $\mathbf{t}$ are

$$\bar{\alpha} = \frac{A}{(A^2 + B^2)^{\frac{1}{2}}}, \quad \bar{\beta} = \frac{B}{(A^2 + B^2)^{\frac{1}{2}}} \tag{10.45}$$

Then the direction cosines of the outward unit normal vector can be obtained by using $\bar{\alpha}$ and $\bar{\beta}$ in Eq. (10.25). For side $\xi = 1$, A and B for defining the direction cosines of the unit tangent vector, expressed by Eqs. (10.36), are found to be

$$\begin{aligned} A &= \frac{1}{2}(2\eta - 1)x_2 + \frac{1}{2}(2\eta + 1)x_3 - 2\eta x_6, \\ B &= \frac{1}{2}(2\eta - 1)y_2 + \frac{1}{2}(2\eta + 1)y_3 - 2\eta y_6 \end{aligned} \tag{10.46}$$

Then the direction cosines of the outward unit normal vector can be obtained by using $\bar{\alpha}$ and $\bar{\beta}$ in Eq. (10.25). For side $\eta = 1$, A and B for defining the direction cosines of the unit tangent vector, expressed by Eqs. (10.37), are found to be

$$\begin{aligned} A &= \frac{1}{2}(2\xi - 1)x_4 + \frac{1}{2}(2\xi + 1)x_3 - 2\xi x_7, \\ B &= \frac{1}{2}(2\xi - 1)y_4 + \frac{1}{2}(2\xi + 1)y_3 - 2\xi y_7 \end{aligned} \tag{10.47}$$

Then the direction cosines of the outward unit normal vector can be obtained by using $\bar{\alpha}$ and $\bar{\beta}$ in Eq. (10.25). For side $\xi = -1$, A and B for defining the direction cosines of the unit tangent vector, expressed by Eqs. (10.36), are found to be

$$\begin{aligned} A &= \frac{1}{2}(2\eta - 1)x_1 + \frac{1}{2}(2\eta + 1)x_4 - 2\eta x_8, \\ B &= \frac{1}{2}(2\eta - 1)y_1 + \frac{1}{2}(2\eta + 1)y_4 - 2\eta y_8 \end{aligned} \tag{10.48}$$

And the direction cosines of the outward unit normal vector can be obtained by using $\bar{\alpha}$ and $\bar{\beta}$ in Eq. (10.25).

10.3.2 Secant Relation for Calculating the Direction Cosines

The unit tangent vector at a discrete point on the element boundary is constructed by the secant approximation which uses two or three consecutive discrete points on the element boundary. Then the direction cosines of the corresponding outward unit normal vector can be obtained by using the transformation law for first-order Cartesian tensors. Consider three counterclockwise consecutive nodes i, j, and k, viewed at a point inside the element, on one side of an element. Then the unit tangent vector $\mathbf{t}_j$ can be approximated by

$$\mathbf{t}_j = \frac{(x_k - x_i)\mathbf{i} + (y_k - y_i)\mathbf{j}}{[(x_k - x_i)^2 + (y_k - y_i)^2]^{\frac{1}{2}}} = \bar{\alpha}\mathbf{i} + \bar{\beta}\mathbf{j} \tag{10.49}$$

If i is the starting point of the element side, the unit tangent vector $\mathbf{t}_i$ can be expressed by

$$\mathbf{t}_i = \frac{(x_j - x_i)\mathbf{i} + (y_j - y_i)\mathbf{j}}{[(x_j - x_i)^2 + (y_j - y_i)^2]^{\frac{1}{2}}} = \bar{\alpha}\mathbf{i} + \bar{\beta}\mathbf{j} \tag{10.50}$$

And if k is the end point of the element side, the unit tangent vector $\mathbf{t}_k$ can be expressed by

$$\mathbf{t}_k = \frac{(x_k - x_j)\mathbf{i} + (y_k - y_j)\mathbf{j}}{[(x_k - x_j)^2 + (y_k - y_j)^2]^{\frac{1}{2}}} = \bar{\alpha}\mathbf{i} + \bar{\beta}\mathbf{j} \tag{10.51}$$

Using $\bar{\alpha}$ and $\bar{\beta}$, the direction cosines of the outward unit normal vector $\mathbf{n} = l\mathbf{i} + m\mathbf{j}$ can be obtained

$$l = \bar{\alpha} cos\frac{\pi}{2} + \bar{\beta} sin\frac{\pi}{2} = \bar{\beta}, \quad m = -\bar{\alpha} sin\frac{\pi}{2} + \bar{\beta} cos\frac{\pi}{2} = -\bar{\alpha} \tag{10.52}$$

11

DQEM Analysis of Two-Dimensional Steady-State Field Problems

Potential flows, heat conductions, electrostatic fields, and stress function and warping function formulations of the torsion of solid shafts,..., are field problems. In this chapter, numerical formulation of the DQEM analysis of two-dimensional steady-state field problems is carried out. The quadrilateral and triangular element models are introduced. The assemblage of all discrete equations into the overall algebraic system is discussed. The DQEM is used to the analysis of various field problems. Both Lagrange DQ model and Chebyshev DQ model adopted for sample DQEM analyses.

11.1 Fundamental Relations

The governing mathematical model of steady-state field problems is a boundary value problem of partial differential equation [118–120]. For the two-dimensional nonuniform problems with orthotropic media, the governing quasiharmonic equation is expressed as

$$(\tilde{k}_x \phi_{,x})_{,x} + (\tilde{k}_y \phi_{,y})_{,y} + \tilde{Q} = 0 \tag{11.1}$$

where ϕ is the field variable, $\tilde{k}_x = \delta k_x$, $\tilde{k}_y = \delta k_y$, $\tilde{Q} = \delta Q$ are functions of the coordinate variables x and y with δ the thickness of medium and variable in the 2-D space. For the 2-D steady-state heat transfer problem, ϕ represents the temperature, k_x and k_y represent thermal conductivities and Q represents the heat generation rate per unit volume.

There are three possible boundary conditions: Dirichlet, Neumann and Cauchy boundary conditions for the domain A. Let S_D, S_N and S_C denote the Dirichlet, Neumann and Cauchy boundaries, respectively. The three boundary conditions are expressed as:

$$\phi = \bar{\phi}, \quad \text{on} \quad S_D \tag{11.2}$$

$$k_x l\phi_{,x} + k_y m\phi_{,y} = \bar{q}, \quad \text{on} \quad S_N \tag{11.3}$$

$$k_x l\phi_{,x} + k_y m\phi_{,y} = c_h(\phi_0 - \phi), \quad \text{on} \quad S_C \tag{11.4}$$

In the above three equations, $\bar{\phi}$, $\bar{q}$ and ϕ_0 are prescribed field variable, field flux and ambient field variable, respectively, l and m are direction cosines of the outward unit normal vector on the natural boundary, and c_h is a coefficient. Dirichlet boundary condition is a kinematic boundary condition, while Neumann and Cauchy boundary conditions are natural boundary conditions. For the 2-D steady-state heat transfer problem, $\bar{\phi}$, $\bar{q}$, ϕ_0 and c_h are prescribed temperature, boundary heat flux into the domain, ambient temperature and convective heat transfer coefficient, respectively.

Solution of the boundary value problem of the governing equation provides the following defined two components of the internal field flux

$$q_x = -k_x\phi_{,x}, \quad q_y = -k_y\phi_{,y} \tag{11.5}$$

For the 2-D steady-state heat transfer problem, q_x and q_y are the two components of internal heat flux.

11.2 DQEM Formulation

Consider that the number of natural coordinates used to define the element e is two. By using Eqs. (10.2) and (10.3) to transform $\phi^e_{,x}$, $\phi^e_{,y}$, $\phi^e_{,xx}$, $\phi^e_{,yy}$, $\tilde{k}^e_{x,x}$ and $\tilde{k}^e_{y,y}$ then substituting them into Eq. (11.1) leads to the following governing equation:

$$F^e_1(\xi,\eta)\phi^e_{,\xi\xi} + F^e_2(\xi,\eta)\phi^e_{,\xi\eta} + F^e_3(\xi,\eta)\phi^e_{,\eta\eta} + F^e_4(\xi,\eta)\phi^e_{,\xi} + F^e_5(\xi,\eta)\phi^e_{,\eta} + \tilde{q}^e = 0 \tag{11.6}$$

where

$$\begin{aligned}
&F^e_1(\xi,\eta) = \tilde{k}^e_x\xi^2_{,x} + \tilde{k}^e_y\xi^2_{,y}, \quad F^e_2(\xi,\eta) = 2(\tilde{k}^e_x\xi_{,x}\eta_{,x} + \tilde{k}^e_y\xi_{,y}\eta_{,y}),\\
&F^e_3(\xi,\eta) = \tilde{k}^e_x\eta^2_{,x} + \tilde{k}^e_y\eta^2_{,y},\\
&F^e_4(\xi,\eta) = \tilde{k}^e_x\xi_{,xx} + \tilde{k}^e_y\xi_{,yy} + \tilde{k}^e_{x,\xi}\xi^2_{,x} + \tilde{k}^e_{x,\eta}\xi_{,x}\eta_{,x} + \tilde{k}^e_{y,\xi}\xi^2_{,y} + \tilde{k}^e_{y,\eta}\xi_{,y}\eta_{,y},\\
&F^e_5(\xi,\eta) = \tilde{k}^e_x\eta_{,xx} + \tilde{k}^e_y\eta_{,yy} + \tilde{k}^e_{x,\xi}\xi_{,x}\eta_{,x} + \tilde{k}^e_{x,\eta}\eta^2_{,x} + \tilde{k}^e_{y,\xi}\xi_{,y}\eta_{,y} + \tilde{k}^e_{y,\eta}\eta^2_{,y}
\end{aligned} \tag{11.7}$$

Let n denote the element having at least one edge on the Neumann boundary, and c denote the element having at least one edge on the Cauchy boundary. The use of Eq. (10.2) in Eqs. (11.3) and (11.4) leads to the following two transformed equations

$$(k^n_x l^n\xi_{,x} + k^n_y m^n\xi_{,y})\phi^n_{,\xi} + (k^n_x l^n\eta_{,x} + k^n_y m^n\eta_{,y})\phi^n_{,\eta} = \bar{q}^n \tag{11.8}$$

$$(k_x^c l^c \xi_{,x} + k_y^c m^c \xi_{,y})\phi_{,\xi}^c + (k_x^c l^c \eta_{,x} + k_y^c m^c \eta_{,y})\phi_{,\eta}^c + c_h^c \phi^c = c_h^c \bar{\phi}_0^c \tag{11.9}$$

Equation (11.5) can also be transformed by using Eq. (10.2)

$$q_x^e = -k_x^e(\xi_{,x}\phi_{,\xi}^e + \eta_{,x}\phi_{,\eta}^e), \quad q_y^e = -k_y^e(\xi_{,y}\phi_{,\xi}^e + \eta_{,y}\phi_{,\eta}^e) \tag{11.10}$$

The kinematic transition condition on the inter-element boundary $\partial\Omega_D^{r,s}$ of two adjacent elements r and s is the continuity of the field variable ϕ which is expressed as

$$\phi^r = \phi^s \tag{11.11}$$

or the assumption of field variable ϕ which is expressed as

$$\phi^r = \phi^s = \bar{\phi}^{r,s} \tag{11.12}$$

where $\bar{\phi}^{r,s}$ is the prescribed field variable. By using Eq. (11.8), the natural transition condition involving the field flux into the domain on the inter-element boundary $\partial\Omega_N^{r,s}$ can be written as

$$\begin{aligned}&\left(k_x^r l^r \xi_{,x}^r + k_y^r m^r \xi_{,y}^r\right)\phi_{,\xi}^r + \left(k_x^r l^r \eta_{,x}^r + k_y^r m^r \eta_{,y}^r\right)\phi_{,\eta}^r \\ &+ \left(k_x^s l^s \xi_{,x}^s + k_y^s m^s \xi_{,y}^s\right)\phi_{,\xi}^s + \left(k_x^s l^s \eta_{,x}^s + k_y^s m^s \eta_{,y}^s\right)\phi_{,\eta}^s = \tilde{q}^{r,s}\end{aligned} \tag{11.13}$$

where $\tilde{q}^{r,s}$ is the field flux into the domain. By using Eq. (11.9), another natural transition condition on the inter-element boundary $\partial\Omega_C^{r,s}$ can also be defined

$$\begin{aligned}&\left(k_x^r l^r \xi_{,x}^r + k_y^r m^r \xi_{,y}^r\right)\phi_{,\xi}^r + \left(k_x^r l^r \eta_{,x}^r + k_y^r m^r \eta_{,y}^r\right)\phi_{,\eta}^r \\ &+ \left(k_x^s l^s \xi_{,x}^s + k_y^s m^s \xi_{,y}^s\right)\phi_{,\xi}^s + \left(k_x^s l^s \eta_{,x}^s + k_y^s m^s \eta_{,y}^s\right)\phi_{,\eta}^s \\ &+ c_h^{r,s}\phi^{r,s} = c_h^{r,s}\tilde{\phi}_0^{r,s}\end{aligned} \tag{11.14}$$

where $\tilde{\phi}_0^{r,s}$ is a prescribed field variable.

11.2.1 Quadrilateral Element

Using the DQ which uses the field variables at the element nodes to define the DQ discretization in Eq. (11.6) at node (α, β) leads to the following equation

$$\begin{aligned}&F_1^e(\xi_\alpha, \eta_\beta) D_{\alpha m}^{e\xi^2}\phi_{m\beta}^e + F_2^e(\xi_\alpha, \eta_\beta) D_{\beta n}^{e\eta} D_{\alpha m}^{e\xi}\phi_{mn}^e + F_3^e(\xi_\alpha, \eta_\beta) D_{\beta n}^{e\eta^2}\phi_{\alpha n}^e \\ &+ F_4^e(\xi_\alpha, \eta_\beta) D_{\alpha m}^{e\xi}\phi_{m\beta}^e + F_5^e(\xi_\alpha, \eta_\beta) D_{\beta n}^{e\eta}\phi_{\alpha n}^e + \tilde{Q}_{\alpha\beta}^e = 0\end{aligned} \tag{11.15}$$

where $D_{\alpha m}^{e\xi^2}$ and $D_{\beta n}^{e\eta^2}$ are weighting coefficients for the second order derivatives in ξ and η directions, respectively, and $D_{\alpha m}^{e\xi}$ and $D_{\beta n}^{e\eta}$ are corresponding weighting coefficients for the first order derivatives. The components of the internal field flux at nodes in an element e can also be expressed by using the DQ and Eq. (11.10)

$$q^e_{x,\alpha\beta} = -k^e_{x(\alpha)(\beta)}\xi^e_{(\alpha)(\beta),x}D^{e\xi}_{\alpha m}\phi^e_{m\beta} - k^e_{x(\alpha)(\beta)}\eta^e_{(\alpha)(\beta),x}D^{e\eta}_{\beta n}\phi^e_{\alpha n},$$
$$q^e_{y,\alpha\beta} = -k^e_{y(\alpha)(\beta)}\xi^e_{(\alpha)(\beta),y}D^{e\xi}_{\alpha m}\phi^e_{m\beta} - k^e_{y(\alpha)(\beta)}\eta^e_{(\alpha)(\beta),y}D^{e\eta}_{\beta n}\phi^e_{\alpha n} \quad (11.16)$$

Consider the inter-element boundary which is the $\xi = 1$ side of element r and the $\xi = 0$ side of element s. Using Eq. (11.11), the discrete continuity conditions of field variable are expressed by

$$\phi^r_{N^r_\xi \beta} = \phi^s_{1\beta}, \quad \beta = 1, 2, ..., N^r_\eta \quad (11.17)$$

and the condition of prescribed field variable at a node on the inter-element boundary is

$$\phi^r_{N^r_\xi \beta} = \phi^s_{1\beta} = \bar{\phi}^{r,s}_\beta \quad (11.18)$$

Using the DQ in Eq. (11.13), the related natural transition condition at a node on the inter-element boundary can be obtained

$$\begin{aligned}
&\left(k^r_{xN^r_\xi(\beta)}l^r_{N^r_\xi(\beta)}\xi^r_{N^r_\xi(\beta),x} + k^r_{yN^r_\xi(\beta)}m^r_{N^r_\xi(\beta)}\xi^r_{N^r_\xi(\beta),y}\right)D^{r\xi}_{N^r_\xi m}\phi^r_{m\beta}\\
&+\left(k^r_{xN^r_\xi(\beta)}l^r_{N^r_\xi(\beta)}\eta^r_{N^r_\xi(\beta),x} + k^r_{yN^r_\xi(\beta)}m^r_{N^r_\xi(\beta)}\eta^r_{N^r_\xi(\beta),y}\right)D^{r\eta}_{\beta n}\phi^r_{N^r_\xi n}\\
&+\left(k^s_{x1(\beta)}l^s_{1(\beta)}\xi^s_{1(\beta),x} + k^s_{y1(\beta)}m^s_{1(\beta)}\xi^s_{1(\beta),y}\right)D^{s\xi}_{1m}\phi^s_{m\beta}\\
&+\left(k^s_{x1(\beta)}l^s_{1(\beta)}\eta^s_{1(\beta),x} + k^s_{y1(\beta)}m^s_{1(\beta)}\eta^s_{1(\beta),y}\right)D^{s\eta}_{\beta n}\phi^s_{1n} = \tilde{q}^{r,s}_\beta
\end{aligned} \quad (11.19)$$

Using the DQ in Eq. (11.14), the related natural transition condition at a node on the inter-element boundary can also be obtained

$$\begin{aligned}
&\left(k^r_{xN^r_\xi(\beta)}l^r_{N^r_\xi(\beta)}\xi^r_{N^r_\xi(\beta),x} + k^r_{yN^r_\xi(\beta)}m^r_{N^r_\xi(\beta)}\xi^r_{N^r_\xi(\beta),y}\right)D^{r\xi}_{N^r_\xi m}\phi^r_{m\beta}\\
&+\left(k^r_{xN^r_\xi(\beta)}l^r_{N^r_\xi(\beta)}\eta^r_{N^r_\xi(\beta),x} + k^r_{yN^r_\xi(\beta)}m^r_{N^r_\xi(\beta)}\eta^r_{N^r_\xi(\beta),y}\right)D^{r\eta}_{\beta n}\phi^r_{N^r_\xi n}\\
&+\left(k^s_{x1(\beta)}l^s_{1(\beta)}\xi^s_{1(\beta),x} + k^s_{y1(\beta)}m^s_{1(\beta)}\xi^s_{1(\beta),y}\right)D^{s\xi}_{1m}\phi^s_{m\beta}\\
&+\left(k^s_{x1(\beta)}l^s_{1(\beta)}\eta^s_{1(\beta),x} + k^s_{y1(\beta)}m^s_{1(\beta)}\eta^s_{1(\beta),y}\right)D^{s\eta}_{\beta n}\phi^s_{1n}\\
&+c^{r,s}_{h(\beta)}\phi^r_{N^r_\xi \beta} = c^{r,s}_{h(\beta)}\tilde{\phi}^{r,s}_{0\beta}
\end{aligned} \quad (11.20)$$

Letting element n be an element with the $\xi = 1$ side on the Neumann boundary, the discrete Neumann boundary condition at a node on the boundary can be obtained by using the DQ in Eq. (11.8)

$$\begin{aligned}
&(k^n_{xN^n_\xi(\beta)}l^n_{N^n_\xi(\beta)}\xi^n_{N^n_\xi(\beta),x} + k^n_{yN^n_\xi(\beta)}m^n_{N^n_\xi(\beta)}\xi^n_{N^n_\xi(\beta),y})D^\xi_{N^n_\xi m}\phi^n_{m\beta}\\
&+(k^n_{xN^n_\xi(\beta)}l^n_{N^n_\xi(\beta)}\eta^n_{N^n_\xi(\beta),x} + k^n_{yN^n_\xi(\beta)}m^n_{N^n_\xi(\beta)}\eta^n_{N^n_\xi(\beta),y})D^\eta_{\beta n}\phi^n_{N^n_\xi n}\\
&= \bar{q}_\beta
\end{aligned} \quad (11.21)$$

If the side of the element is on the Cauchy boundary, the discrete Cauchy boundary conditions can be similarly obtained

$$\begin{aligned}
&\left(k^c_{xN^c_\xi(\beta)} l^c_{N^c_\xi(\beta)} \xi^c_{N^c_\xi(\beta),x} + k^c_{yN^c_\xi(\beta)} m^c_{N^c_\xi(\beta)} \xi^c_{N^c_\xi(\beta),y}\right) D^\xi_{N^c_\xi m} \phi^c_{m\beta} \\
&+ \left(k^c_{xN^c_\xi(\beta)} l^c_{N^c_\xi(\beta)} \eta^c_{N^c_\xi(\beta),x} + k^c_{yN^c_\xi(\beta)} m^c_{N^c_\xi(\beta)} \eta^c_{N^c_\xi(\beta),y}\right) D^\eta_{\beta n} \phi^c_{N^c_\xi n} \\
&+ c^c_{h(\beta)} \phi^c_\beta = c^c_{h(\beta)} \bar{\phi}^c_{0\beta}
\end{aligned} \tag{11.22}$$

11.2.2 Triangular Element

In constructing the mapping transformation, the master triangular element in the natural space may be an arbitrary linear triangle. For convenience, the rectangular unit triangle is adopted. Let the natural coordinates be L_1 and L_2. A representative serendipity rectangular unit C^0 triangular element is shown in Fig. 10.1. Using the area coordinates L_1, L_2 and L_3 as coordinate variables and substituting the transformation equations for $\phi^e_{,x}$, $\phi^e_{,y}$, $\tilde{k}^e_{x,x}$, $\tilde{k}^e_{y,y}$, $\phi^e_{,xx}$ and $\phi^e_{,yy}$ into Eq. (11.1) leads to the following equation defined on $\tilde{\Omega}$

$$\begin{aligned}
&F^e_1(L_1,L_2)\phi^e_{,L_1L_1} + F^e_2(L_1,L_2)\phi^e_{,L_2L_2} + F^e_3(L_1,L_2)\phi^e_{,L_3L_3} \\
&+ F^e_4(L_1,L_2)\phi^e_{,L_1L_2} + F^e_5(L_1,L_2)\phi^e_{,L_1L_3} + F^e_6(L_1,L_2)\phi^e_{,L_2L_3} \\
&+ F^e_7(L_1,L_2)\phi^e_{,L_1} + F^e_8(L_1,L_2)\phi^e_{,L_2} + F^e_9(L_1,L_2)\phi^e_{,L_3} + \tilde{Q}^e = 0
\end{aligned} \tag{11.23}$$

where

$$\begin{aligned}
&F^e_1(L_1,L_2) = \tilde{k}^e_x L^2_{1,x} + \tilde{k}^e_y L^2_{1,y}, \quad F^e_2(L_1,L_2) = \tilde{k}^e_x L^2_{2,x} + \tilde{k}^e_y L^2_{2,y}, \\
&F^e_3(L_1,L_2) = \tilde{k}^e_x (L_{1,x} + L_{2,x})^2 + \tilde{k}^e_y (L_{1,y} + L_{2,y})^2, \\
&F^e_4(L_1,L_2) = \tilde{k}^e_x L_{1,x} L_{2,x} + \tilde{k}^e_y L_{1,y} L_{2,y}, \\
&F^e_5(L_1,L_2) = -2[\tilde{k}^e_x L_{1,x}(L_{1,x} + L_{2,x}) + \tilde{k}^e_y L_{1,y}(L_{1,y} + L_{2,y})], \\
&F^e_6(L_1,L_2) = -2[\tilde{k}^e_x L_{2,x}(L_{1,x} + L_{2,x}) + \tilde{k}^e_y L_{2,y}(L_{1,y} + L_{2,y})], \\
&F^e_7(L_1,L_2) = \tilde{k}^e_x L_{1,xx} + \tilde{k}^e_y L_{1,yy} + \tilde{k}^e_{x,x} L_{1,x} + \tilde{k}^e_{y,y} L_{1,y}, \\
&F^e_8(L_1,L_2) = \tilde{k}^e_x L_{2,xx} + \tilde{k}^e_y L_{2,yy} + \tilde{k}^e_{x,x} L_{2,x} + \tilde{k}^e_{y,y} L_{2,y}, \\
&F^e_9(L_1,L_2) = -\tilde{k}^e_x (L_{1,xx} + L_{2,xx}) - \tilde{k}^e_y (L_{1,yy} + L_{2,yy}) \\
&\qquad\qquad -\tilde{k}^e_{x,x}(L_{1,x} + L_{2,x}) - \tilde{k}^e_{y,y}(L_{1,y} + L_{2,y})
\end{aligned} \tag{11.24}$$

The transformed Neumann boundary condition is

$$\begin{aligned}
&(k^n_x l^n L^n_{1,x} + k^n_y m^n L^n_{1,y})\phi^n_{,L_1} + (k^n_x l^n L^n_{2,x} + k^n_y m^n L^n_{2,y})\phi^n_{,L_2} \\
&- \left[k^n_x l^n (L^n_{1,x} + L^n_{2,x}) + k^n_y m^n (L^n_{1,y} + L^n_{2,y})\right] \phi^n_{,L_3} = \bar{q}^n
\end{aligned} \tag{11.25}$$

And the transformed Cauchy boundary condition is

$$\begin{aligned}
&(k^c_x l^c L^c_{1,x} + k^c_y m^c L^c_{1,y})\phi^c_{,L_1} + (k^c_x l^c L^c_{2,x} + k^c_y m^c L^c_{2,y})\phi^c_{,L_2} \\
&- \left[k^c_x l^c (L^c_{1,x} + L^c_{2,x}) + k^c_y m^c (L^c_{1,y} + L^c_{2,y})\right] \phi^c_{,L_3} + c^c_h \phi^c = c^c_h \bar{\phi}^c_0
\end{aligned} \tag{11.26}$$

The component equations of field flux can also be transformed

$$q_x^e = -k_x^e \left[L_{1,x}^e \phi^e_{,L_1} + L_{2,x}^e \phi^e_{,L_2} - (L_{1,x}^e + L_{2,x}^e)\phi^e_{,L_3} \right],$$

$$q_y^e = -k_y^e \left[L_{1,y}^e \phi^e_{,L_1} + L_{2,y}^e \phi^e_{,L_2} - (L_{1,y}^e + L_{2,y}^e)\phi^e_{,L_3} \right] \tag{11.27}$$

The kinematic transition condition is the same as that of the quadrilateral element formulation. The natural transition condition on $\partial\Omega_N^{r,s}$ can be expressed by

$$\begin{aligned}
&(k_x l^r L_{1,x}^r + k_y m^r L_{1,y}^r)\phi^r_{,L_1} + (k_x l^r L_{2,x}^r + k_y m^r L_{2,y}^r)\phi^r_{,L_2} \\
&- \left[k_x l^r (L_{1,x}^r + L_{2,x}^r) + k_y m^r (L_{1,y}^r + L_{2,y}^r) \right] \phi^r_{,L_3} \\
&+ (k_x l^s L_{1,x}^s + k_y m^s L_{1,y}^s)\phi^s_{,L_1} + (k_x l^s L_{2,x}^s + k_y m^s L_{2,y}^s)\phi^s_{,L_2} \\
&- \left[k_x l^s (L_{1,x}^s + L_{2,x}^s) + k_y m^s (L_{1,y}^s + L_{2,y}^s) \right] \phi^s_{,L_3} = \tilde{q}^{r,s}
\end{aligned} \tag{11.28}$$

The natural transition condition on $\partial\Omega_C^{r,s}$ can be expressed by

$$\begin{aligned}
&(k_x l^r L_{1,x}^r + k_y m^r L_{1,y}^r)\phi^r_{,L_1} + (k_x l^r L_{2,x}^r + k_y m^r L_{2,y}^r)\phi^r_{,L_2} \\
&- \left[k_x l^r (L_{1,x}^r + L_{2,x}^r) + k_y m^r (L_{1,y}^r + L_{2,y}^r) \right] \phi^r_{,L_3} \\
&+ (k_x l^s L_{1,x}^s + k_y m^s L_{1,y}^s)\phi^s_{,L_1} + (k_x l^s L_{2,x}^s + k_y m^s L_{2,y}^s)\phi^s_{,L_2} \\
&- \left[k_x l^s (L_{1,x}^s + L_{2,x}^s) + k_y m^s (L_{1,y}^s + L_{2,y}^s) \right] \phi^s_{,L_3} + c_h^{r,s}\phi^{r,s} \\
&= c_h^{r,s}\tilde{\phi}_0^{r,s}
\end{aligned} \tag{11.29}$$

The Pascal triangular grid and the three-dimensional node identification method are used. It is designed by first using $n+1$ parallel lines to subdivide the domain, in each direction of area coordinates, into n subregions with n being the order of approximate polynomials. The parallel lines define the levels of the Pascal triangular grid with the boundary line being the level 1. The level number will increase following the increase of the value of area coordinate. The highest level is level $n+1$. Let L denote the level number of a level below the highest level. Then $n+2-L$ equally spaced nodes are given to each level. Using the DQ discretization, the governing Eq. (11.23) at a node (α, β, γ) in the element e can be discretized

$$\begin{aligned}
&F_1^e(L_{1\alpha}, L_{2\beta}) D_{\alpha\bar{\alpha}}^{eL_1^2} \phi^e_{\bar{\alpha}\beta\gamma} + F_2^e(L_{1\alpha}, L_{2\beta}) D_{\beta\bar{\beta}}^{eL_2^2} \phi^e_{\alpha\bar{\beta}\gamma} \\
&+ F_3^e(L_{1\alpha}, L_{2\beta}) D_{\gamma\bar{\gamma}}^{eL_3^2} \phi^e_{\alpha\beta\bar{\gamma}} + F_4^e(L_{1\alpha}, L_{2\beta}) D_{\alpha\bar{\beta}}^{eL_1} D_{\beta\bar{\beta}}^{eL_2} \phi^e_{\bar{\alpha}\bar{\beta}\gamma} \\
&+ F_5^e(L_{1\alpha}, L_{2\beta}) D_{\alpha\bar{\alpha}}^{eL_1} D_{\gamma\bar{\gamma}}^{eL_3} \phi^e_{\bar{\alpha}\beta\bar{\gamma}} + F_6^e(L_{1\alpha}, L_{2\beta}) D_{\beta\bar{\beta}}^{eL_2} D_{\gamma\bar{\gamma}}^{eL_3} \phi^e_{\alpha\bar{\beta}\bar{\gamma}} \\
&+ F_7^e(L_{1\alpha}, L_{2\beta}) D_{\alpha\bar{\alpha}}^{eL_1} \phi^e_{\bar{\alpha}\beta\gamma} + F_8^e(L_{1\alpha}, L_{2\beta}) D_{\beta\bar{\beta}}^{eL_2} \phi^e_{\alpha\bar{\beta}\gamma} \\
&+ F_9^e(L_{1\alpha}, L_{2\beta}) D_{\gamma\bar{\gamma}}^{eL_3} \phi^e_{\alpha\beta\bar{\gamma}} + \tilde{Q}^e_{\alpha\beta\gamma} = 0
\end{aligned} \tag{11.30}$$

The components of the field flux at a node (α, β, γ) can also be expressed by using the DQ in Eqs. (11.27)

$$\begin{aligned}
q^e_{x,\alpha\beta\gamma} = -k^e_{x,(\alpha)(\beta)} \big[& L^e_{1(\alpha)(\beta),x} D_{\alpha\bar{\alpha}}^{eL_1} \phi^e_{\bar{\alpha}\beta\gamma} + L^e_{2(\alpha)(\beta),x} D_{\beta\bar{\beta}}^{eL_2} \phi^e_{\alpha\bar{\beta}\gamma} \\
& -\big(L^e_{1(\alpha)(\beta),x} + L^e_{2(\alpha)(\beta),x}\big) D_{\gamma\bar{\gamma}}^{eL_3} \phi^e_{\alpha\beta\bar{\gamma}} \big],
\end{aligned}$$

$$q^e_{y,\alpha\beta\gamma} = -k^e_{y,(\alpha)(\beta)}\left[L^e_{1(\alpha)(\beta),y} D^{eL_1}_{\alpha\bar{\alpha}} \phi^e_{\bar{\alpha}\beta\gamma} + L^e_{2(\alpha)(\beta),y} D^{eL_2}_{\beta\bar{\beta}} \phi^e_{\alpha\bar{\beta}\gamma} - \left(L^e_{1(\alpha)(\beta),y} + L^e_{2(\alpha)(\beta),y}\right) D^{eL_3}_{\gamma\bar{\gamma}} \phi^e_{\alpha\beta\bar{\gamma}}\right] \tag{11.31}$$

Using DQ in Eq. (11.28), the related natural transition condition at a node $(\tilde{\alpha}, \tilde{\beta}, \tilde{\gamma})$ on the inter-element boundary $\partial\Omega^{r,s}_N$ can be obtained

$$\begin{aligned}
&\left(k^r_{x(\tilde{\alpha})(\tilde{\beta})} l^r_{(\tilde{\alpha})(\tilde{\beta})} L^r_{1(\tilde{\alpha})(\tilde{\beta}),x} + k^r_{y(\tilde{\alpha})(\tilde{\beta})} m^r_{(\tilde{\alpha})(\tilde{\beta})} L^r_{1(\tilde{\alpha})(\tilde{\beta}),y}\right) D^{rL_1}_{\tilde{\alpha}\bar{\alpha}} \phi^r_{\bar{\alpha}\tilde{\beta}\tilde{\gamma}} \\
&+ \left(k^r_{x(\tilde{\alpha})(\tilde{\beta})} l^r_{(\tilde{\alpha})(\tilde{\beta})} L^r_{2(\tilde{\alpha})(\tilde{\beta}),x} + k^r_{y(\tilde{\alpha})(\tilde{\beta})} m^r_{(\tilde{\alpha})(\tilde{\beta})} L^r_{2(\tilde{\alpha})(\tilde{\beta}),y}\right) D^{rL_2}_{\tilde{\beta}\bar{\beta}} \phi^r_{\tilde{\alpha}\bar{\beta}\tilde{\gamma}} \\
&- \Big[k^r_{x(\tilde{\alpha})(\tilde{\beta})} l^r_{(\tilde{\alpha})(\tilde{\beta})} \left(L^r_{1(\tilde{\alpha})(\tilde{\beta}),x} + L^r_{2(\tilde{\alpha})(\tilde{\beta})x}\right) \\
&+ k^r_{y(\tilde{\alpha})(\tilde{\beta})} m^r_{(\tilde{\alpha})(\tilde{\beta})} \left(L^r_{1(\tilde{\alpha})(\tilde{\beta}),y} + L^r_{2(\tilde{\alpha})(\tilde{\beta}),y}\right)\Big] D^{rL_3}_{\tilde{\gamma}\bar{\gamma}} \phi^r_{\tilde{\alpha}\tilde{\beta}\bar{\gamma}} \\
&+ \left(k^s_{x(\tilde{\alpha})(\tilde{\beta})} l^s_{(\tilde{\alpha})(\tilde{\beta})} L^s_{1(\tilde{\alpha})(\tilde{\beta}),x} + k^s_{y(\tilde{\alpha})(\tilde{\beta})} m^s_{(\tilde{\alpha})(\tilde{\beta})} L^s_{1(\tilde{\alpha})(\tilde{\beta}),y}\right) D^{sL_1}_{\tilde{\alpha}\bar{\alpha}} \phi^s_{\bar{\alpha}\tilde{\beta}\tilde{\gamma}} \\
&+ \left(k^s_{x(\tilde{\alpha})(\tilde{\beta})} l^s_{(\tilde{\alpha})(\tilde{\beta})} L^s_{2(\tilde{\alpha})(\tilde{\beta}),x} + k^s_{y(\tilde{\alpha})(\tilde{\beta})} m^s_{(\tilde{\alpha})(\tilde{\beta})} L^s_{2(\tilde{\alpha})(\tilde{\beta}),y}\right) D^{sL_2}_{\tilde{\beta}\bar{\beta}} \phi^s_{\tilde{\alpha}\bar{\beta}\tilde{\gamma}} \\
&- \Big[k^s_{x(\tilde{\alpha})(\tilde{\beta})} l^s_{(\tilde{\alpha})(\tilde{\beta})} \left(L^s_{1(\tilde{\alpha})(\tilde{\beta}),x} + L^s_{2(\tilde{\alpha})(\tilde{\beta})x}\right) \\
&+ k^s_{y(\tilde{\alpha})(\tilde{\beta})} m^s_{(\tilde{\alpha})(\tilde{\beta})} \left(L^s_{1(\tilde{\alpha})(\tilde{\beta}),y} + L^s_{2(\tilde{\alpha})(\tilde{\beta}),y}\right)\Big] D^{sL_3}_{\tilde{\gamma}\bar{\gamma}} \phi^s_{\tilde{\alpha}\tilde{\beta}\bar{\gamma}} = \tilde{q}^{r,s}_{\tilde{\alpha}\tilde{\beta}\tilde{\gamma}}
\end{aligned} \tag{11.32}$$

Using DQ in Eq. (11.29), the related natural transition condition at a node $(\tilde{\alpha}, \tilde{\beta}, \tilde{\gamma})$ on the inter-element boundary $\partial\Omega^{r,s}_N$ can be obtained

$$\begin{aligned}
&\left(k^r_{x(\tilde{\alpha})(\tilde{\beta})} l^r_{(\tilde{\alpha})(\tilde{\beta})} L^r_{1(\tilde{\alpha})(\tilde{\beta}),x} + k^r_{y(\tilde{\alpha})(\tilde{\beta})} m^r_{(\tilde{\alpha})(\tilde{\beta})} L^r_{1(\tilde{\alpha})(\tilde{\beta}),y}\right) D^{rL_1}_{\tilde{\alpha}\bar{\alpha}} \phi^r_{\bar{\alpha}\tilde{\beta}\tilde{\gamma}} \\
&+ \left(k^r_{x(\tilde{\alpha})(\tilde{\beta})} l^r_{(\tilde{\alpha})(\tilde{\beta})} L^r_{2(\tilde{\alpha})(\tilde{\beta}),x} + k^r_{y(\tilde{\alpha})(\tilde{\beta})} m^r_{(\tilde{\alpha})(\tilde{\beta})} L^r_{2(\tilde{\alpha})(\tilde{\beta}),y}\right) D^{rL_2}_{\tilde{\beta}\bar{\beta}} \phi^r_{\tilde{\alpha}\bar{\beta}\tilde{\gamma}} \\
&- \Big[k^r_{x(\tilde{\alpha})(\tilde{\beta})} l^r_{(\tilde{\alpha})(\tilde{\beta})} \left(L^r_{1(\tilde{\alpha})(\tilde{\beta}),x} + L^r_{2(\tilde{\alpha})(\tilde{\beta})x}\right) \\
&+ k^r_{y(\tilde{\alpha})(\tilde{\beta})} m^r_{(\tilde{\alpha})(\tilde{\beta})} \left(L^r_{1(\tilde{\alpha})(\tilde{\beta}),y} + L^r_{2(\tilde{\alpha})(\tilde{\beta}),y}\right)\Big] D^{rL_3}_{\tilde{\gamma}\bar{\gamma}} \phi^r_{\tilde{\alpha}\tilde{\beta}\bar{\gamma}} \\
&+ \left(k^s_{x(\tilde{\alpha})(\tilde{\beta})} l^s_{(\tilde{\alpha})(\tilde{\beta})} L^s_{1(\tilde{\alpha})(\tilde{\beta}),x} + k^s_{y(\tilde{\alpha})(\tilde{\beta})} m^s_{(\tilde{\alpha})(\tilde{\beta})} L^s_{1(\tilde{\alpha})(\tilde{\beta}),y}\right) D^{sL_1}_{\tilde{\alpha}\bar{\alpha}} \phi^s_{\bar{\alpha}\tilde{\beta}\tilde{\gamma}} \\
&+ \left(k^s_{x(\tilde{\alpha})(\tilde{\beta})} l^s_{(\tilde{\alpha})(\tilde{\beta})} L^s_{2(\tilde{\alpha})(\tilde{\beta}),x} + k^s_{y(\tilde{\alpha})(\tilde{\beta})} m^s_{(\tilde{\alpha})(\tilde{\beta})} L^s_{2(\tilde{\alpha})(\tilde{\beta}),y}\right) D^{sL_2}_{\tilde{\beta}\bar{\beta}} \phi^s_{\tilde{\alpha}\bar{\beta}\tilde{\gamma}} \\
&- \Big[k^s_{x(\tilde{\alpha})(\tilde{\beta})} l^s_{(\tilde{\alpha})(\tilde{\beta})} \left(L^s_{1(\tilde{\alpha})(\tilde{\beta}),x} + L^s_{2(\tilde{\alpha})(\tilde{\beta})x}\right) \\
&+ k^s_{y(\tilde{\alpha})(\tilde{\beta})} m^s_{(\tilde{\alpha})(\tilde{\beta})} \left(L^s_{1(\tilde{\alpha})(\tilde{\beta}),y} + L^s_{2(\tilde{\alpha})(\tilde{\beta}),y}\right)\Big] D^{sL_3}_{\tilde{\gamma}\bar{\gamma}} \phi^s_{\tilde{\alpha}\tilde{\beta}\bar{\gamma}} \\
&+ c^{r,s}_{h(\tilde{\alpha})(\tilde{\beta})(\tilde{\gamma})} \phi^{r,s}_{\tilde{\alpha}\tilde{\beta}\tilde{\gamma}} = c^{r,s}_{h(\tilde{\alpha})(\tilde{\beta})(\tilde{\gamma})} \tilde{\phi}^{r,s}_{0\tilde{\alpha}\tilde{\beta}\tilde{\gamma}}
\end{aligned} \tag{11.33}$$

It should be noted that $(\tilde{\alpha}, \tilde{\beta}, \tilde{\gamma})$ represents a common node of element r and element s on the inter-element boundary. Consequently, the two sets of $(\tilde{\alpha}, \tilde{\beta}, \tilde{\gamma})$ represent the two local element nodes of element r and element s, separately. Letting element n be an element connected to the Neumann boundary, the discrete Neumann boundary condition at a node $(\tilde{\alpha}, \tilde{\beta}, \tilde{\gamma})$ can be obtained by using the DQ in Eq. (11.25)

$$\begin{aligned}
&\left(k^n_{x(\tilde{\alpha})(\tilde{\beta})} l^n_{(\tilde{\alpha})(\tilde{\beta})} L^n_{1(\tilde{\alpha})(\tilde{\beta}),x} + k^n_{y(\tilde{\alpha})(\tilde{\beta})} m^n_{(\tilde{\alpha})(\tilde{\beta})} L^n_{1(\tilde{\alpha})(\tilde{\beta}),y}\right) D^{nL_1}_{\tilde{\alpha}\bar{\alpha}} \phi^n_{\bar{\alpha}\tilde{\beta}\tilde{\gamma}} \\
&+ \left(k^n_{x(\tilde{\alpha})(\tilde{\beta})} l^n_{(\tilde{\alpha})(\tilde{\beta})} L^n_{2(\tilde{\alpha})(\tilde{\beta}),x} + k^n_{y(\tilde{\alpha})(\tilde{\beta})} m^n_{(\tilde{\alpha})(\tilde{\beta})} L^n_{2(\tilde{\alpha})(\tilde{\beta}),y}\right) D^{nL_2}_{\tilde{\beta}\bar{\beta}} \phi^n_{\tilde{\alpha}\bar{\beta}\tilde{\gamma}} \\
&- \left[k^n_{x(\tilde{\alpha})(\tilde{\beta})} l^n_{(\tilde{\alpha})(\tilde{\beta})} \left(L^n_{1(\tilde{\alpha})(\tilde{\beta}),x} + L^n_{2(\tilde{\alpha})(\tilde{\beta})x}\right)\right. \\
&\quad \left. + k^n_{y(\tilde{\alpha})(\tilde{\beta})} m^n_{(\tilde{\alpha})(\tilde{\beta})} \left(L^n_{1(\tilde{\alpha})(\tilde{\beta}),y} + L^n_{2(\tilde{\alpha})(\tilde{\beta}),y}\right)\right] D^{nL_3}_{\tilde{\gamma}\bar{\gamma}} \phi^n_{\tilde{\alpha}\tilde{\beta}\bar{\gamma}} = \bar{q}^n_{\tilde{\alpha}\tilde{\beta}\tilde{\gamma}}
\end{aligned} \tag{11.34}$$

If the element is connected to the Cauchy boundary, the discrete Cauchy boundary condition at a node $(\tilde{\alpha}, \tilde{\beta}, \tilde{\gamma})$ can also be obtained by using the DQ in Eq. (11.26)

$$\begin{aligned}
&\left(k^c_{x(\tilde{\alpha})(\tilde{\beta})} l^c_{(\tilde{\alpha})(\tilde{\beta})} L^c_{1(\tilde{\alpha})(\tilde{\beta}),x} + k^c_{y(\tilde{\alpha})(\tilde{\beta})} m^c_{(\tilde{\alpha})(\tilde{\beta})} L^c_{1(\tilde{\alpha})(\tilde{\beta}),y}\right) D^{nL_1}_{\tilde{\alpha}\bar{\alpha}} \phi^c_{\bar{\alpha}\tilde{\beta}\tilde{\gamma}} \\
&+ \left(k^c_{x(\tilde{\alpha})(\tilde{\beta})} l^c_{(\tilde{\alpha})(\tilde{\beta})} L^c_{2(\tilde{\alpha})(\tilde{\beta}),x} + k^c_{y(\tilde{\alpha})(\tilde{\beta})} m^c_{(\tilde{\alpha})(\tilde{\beta})} L^c_{2(\tilde{\alpha})(\tilde{\beta}),y}\right) D^{nL_2}_{\tilde{\beta}\bar{\beta}} \phi^c_{\tilde{\alpha}\bar{\beta}\tilde{\gamma}} \\
&- \left[k^c_{x(\tilde{\alpha})(\tilde{\beta})} l^c_{(\tilde{\alpha})(\tilde{\beta})} \left(L^c_{1(\tilde{\alpha})(\tilde{\beta}),x} + L^c_{2(\tilde{\alpha})(\tilde{\beta})x}\right)\right. \\
&\quad \left. + k^c_{y(\tilde{\alpha})(\tilde{\beta})} m^c_{(\tilde{\alpha})(\tilde{\beta})} \left(L^c_{1(\tilde{\alpha})(\tilde{\beta}),y} + L^c_{2(\tilde{\alpha})(\tilde{\beta}),y}\right)\right] D^{nL_3}_{\tilde{\gamma}\bar{\gamma}} \phi^c_{\tilde{\alpha}\tilde{\beta}\bar{\gamma}} \\
&+ c^c_{h(\tilde{\alpha})(\tilde{\beta})(\tilde{\gamma})} \phi^c_{\tilde{\alpha}\tilde{\beta}\tilde{\gamma}} = c^c_{h(\tilde{\alpha})(\tilde{\beta})(\tilde{\gamma})} \bar{\phi}^c_{0\tilde{\alpha}\tilde{\beta}\tilde{\gamma}}
\end{aligned} \tag{11.35}$$

11.3 Assemblage

For the assemblage of discrete equations of quadrilateral elements, by keeping the kinematic transition conditions in mind, then assemble the discrete element governing equations (11.15) for all elements, discrete natural transition conditions (11.19) and (11.20), discrete Neumann boundary conditions (11.21), and discrete Cauchy boundary conditions (11.22), an overall discrete governing/transition/boundary equation can be obtained. It is the overall field equation expresses as

$$\left[\hat{K}\right]\{\Phi\} = \{R_Q\} \tag{11.36}$$

where $\left[\hat{K}\right]$ is the overall field stiffness matrix, $\{\Phi\}$ is the overall field variable vector, and $\{R_Q\}$ is the overall field load vector Considering the Dirichlet boundary conditions and solving the overall field equation system, field variables at all nodes can be obtained. Like FEM, the assemblage is based on an element by element procedure. In assembling the discrete equations of element e, the discrete element governing equations (11.15), and the discrete element boundary field fluxes, expressed by field variables, at the nodes on element boundary edges are directly assembled to the overall discrete equation system. An element basis explicit matrix equation, containing the discrete element governing equations and the discrete element boundary field fluxes placed at the rows with the assigned DOF related to the corresponding discrete element boundary field fluxes, is not necessary to be formed in the assemblage process. This element basis explicit matrix equation is an element field equation which can be expressed by

$$[\hat{k}^e]\{\phi^e\} = \{r_Q^e\} \tag{11.37}$$

where $[\hat{k}^e]$ is a $(N_\xi^e \times N_\eta^e) \times (N_\xi^e \times N_\eta^e)$ element field stiffness matrix,

$$\{\phi^e\} = \lfloor\ \phi_1^e \quad \phi_2^e \quad \phi_3^e \quad \phi_4^e \quad . \quad ... \quad .\ \rfloor^T \tag{11.38}$$

is the element field variable vector, and $\{r_Q^e\}$ is the element field load vector containing $\tilde{Q}_\alpha^e$ at the related internal nodes, and discrete internal element boundary field fluxes. As Eq. (11.37) contains discrete internal element boundary field fluxes at nodes on the four element boundary edges, equilibriums of internal field fluxes and external field fluxes into the medium at nodes on the inter-element boundary of two adjacent elements and the natural boundary are exactly satisfied in the assemblage process. Consequently, the DQEM is different from FEM which needs to explicitly form the element field equation, and which neglects the exactness of Eqs. (11.19) to (11.22).

11.4 Overall Algebraic System

In this section, the design and construction of the overall discrete fundamental equation system is introduced [25–27]. For the DQEM field problem analysis model, the total degrees of freedom must equal the number of discrete constraint equations. The governing equation at a node is a governing equation constraint condition. An interior node can define only one discrete governing equation. The discrete Dirichlet and natural boundary conditions are defined on the Dirichlet and natural boundaries, respectively. A node on the analysis domain boundary (ADB) but not an element corner node can define one boundary condition and one governing equation. At a node on the inter-element boundary, if the node is not an element corner node, in addition to the continuity of field variable one discrete constraint equation of prescribed field variable or natural transition condition and up to two discrete constraint governing equations attached to the two adjacent elements can be defined. An element corner node might be able to define even more constraint equations.

Consider that discrete governing equations are only defined at interior nodes. For an element corner node in the analysis domain which is the common node of N_N natural inter-element boundaries (IEB) and N_K kinematic inter-element boundaries, let N_T denote the number of all constraint conditions. Then, for $N_K \neq 0$, $N_T = N_N + 1$; for $N_K = 0$ and Φ not prescribed at the node, $N_T = N_N$; for $N_K = 0$ and Φ prescribed, $N_T = N_N + 1$. The equation of setting Φ as a prescribed value is also a kinematic constraint condition. Figure 11.1 is a typical element corner node in the analysis domain.

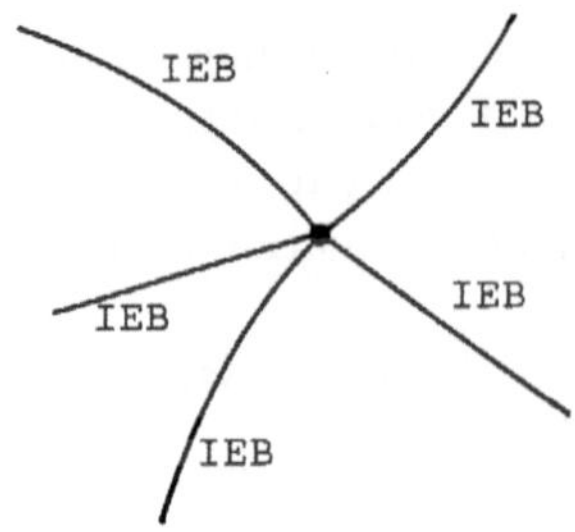

Fig. 11.1. Typical element corner node in the analysis domain

For an element corner node on the analysis domain boundary with two element-based segments of the analysis domain boundary being connected to it, if the two segments are natural boundaries, N_T for various connections are: for $N_K \neq 0$, $N_T = N_N + 3$; for $N_K = 0$ and Φ not prescribed at the node, $N_T = N_N + 2$; for $N_K = 0$ and Φ prescribed, $N_T = N_N + 3$. For this type of element corner node, if only two elements are connected to the node with the inter-element boundary perpendicular to the analysis domain boundary which is a straight line within the two connected elements, then only one natural boundary condition can be applied. If one of the analysis domain boundary segment is natural boundary while the other one is Dirichlet boundary, N_T for various connections are: for $N_K \neq 0$, $N_T = N_N + 2$; for $N_K = 0$, $N_T = N_N + 2$. And if both of the two analysis domain boundary segments are Dirichlet boundaries, N_T for various connections are: for $N_K \neq 0$, $N_T = N_N + 1$; for $N_K = 0$, $N_T = N_N + 1$. A representative of this type of element corner node is shown in Fig. 11.2 which is the type 1 element corner node on the analysis domain boundary.

For an element corner node on the analysis domain boundary with one element-based segment of the analysis domain boundary connected to it, if the segment is natural boundary, N_T for various connections are: for $N_K \neq 0$, $N_T = N_N + 2$; for $N_K = 0$ and Φ not prescribed at the node, $N_T = N_N + 1$; for $N_K = 0$ and Φ prescribed, $N_T = N_N + 2$. If the analysis domain boundary segment is Dirichlet boundary, N_T for various connections are: for $N_K \neq 0$,

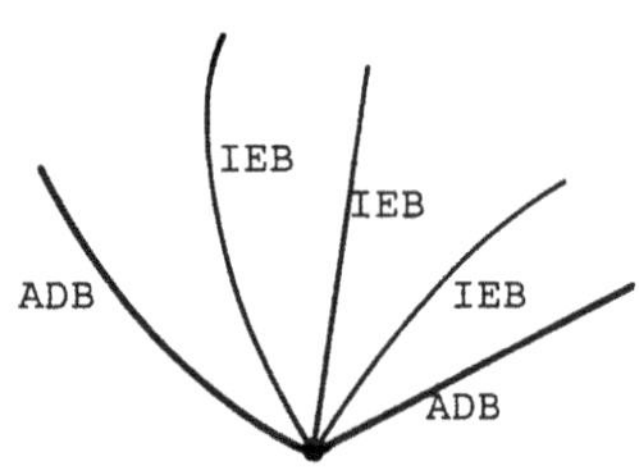

Fig. 11.2. Type 1 element corner node on the analysis domain boundary

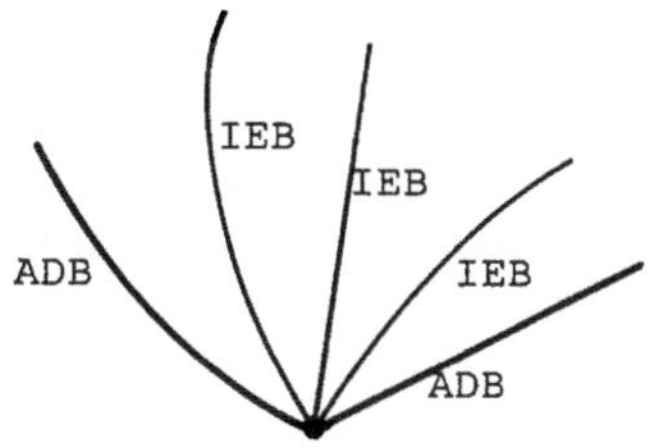

Fig. 11.3. Type 2 element corner node on the analysis domain boundary

$N_T = N_N + 1$; for $N_K = 0$, $N_T = N_N + 1$. A representative of this type of element corner node is shown in Fig. 11.3 which is the type 2 element corner node on the analysis domain boundary.

Consider that only the values of field variable at nodes are used to define the DQ discretization. Then in order to satisfy all constraint conditions at an element corner node, at the assemblage stage, one degree of freedom might not enough. However, we can use more than one constraint condition at that corner node by neglecting certain constraint conditions at interior nodes or at node on the inter-element boundary and other than corner nodes, and giving their degrees of freedom to the inclusion of extra constraint conditions other than the first one.

For the DQEM field problem analysis model, the N_T constraint conditions can partially or fully be satisfied. We can also neglect all of the N_T constraint conditions and give the degree of freedom of that node to the discrete governing equation at that node. The discrete governing equation at the element corner node can be defined as the average of the discrete governing equations of all elements connected to that node.

The various techniques for selecting and implementing the constraint conditions at element corner nodes are flexible. Different approaches lead to different programming efforts. The overall algebraic system obtained by assembling all discrete constraint conditions is the discrete governing/transition/boundary equation system.

The EDQ can adopt the degrees of freedom used to represent the derivatives or partial derivatives of the field variable. In conjunction with the use of the EDQ with which the discretization can be defined at discrete points which are not nodes, the DQEM can also assign the degrees of freedom of the partial derivatives of the field variable to the element boundary nodes.

For analyzing the two-dimensional field problems in order to automatically set the kinematic transition conditions by only using certain degrees of freedom assigned to the element boundary nodes, the degrees of freedom representing the field variable must be assigned to the element boundary nodes. The degrees of freedom representing the partial derivatives of the field variable can also be assigned to the nodes of all neighbor elements on the inter-element boundary and the compatibility conditions of higher order partial derivatives can also be considered. However, if certain external cause such as the fluid flow, conduction heat flux, *etc.* is applied no compatibility condition of partial derivatives can be considered. The discrete governing equations can be defined on the inter-element boundaries as the average discrete governing equations of multiple elements. They can also be defined on the element boundaries without adopting the average treatment. Thus, elements having no interior node can also be used to the DQEM analysis. For analyzing beam or plate problems in order to automatically set the kinematic transition conditions by only using certain degrees of freedom assigned to the element boundary nodes, the degrees of freedom representing the lateral displacement and first order derivative or partial derivatives of the lateral displacement must be assigned to the element boundary nodes. The degrees of freedom representing higher order derivatives or partial derivatives of the displacement can also be assigned to the nodes of all neighbor elements on the inter-element boundary and the compatibility conditions of the higher order derivatives or partial derivatives can also be considered. However, if the moment is applied the highest order of derivative or partial derivative that the compatibility condition can be considered is one. On the other hand, if the lateral force is applied on the inter-element boundary the highest order of derivative or partial derivative that the compatibility condition can be considered is two. The concept can also be used to treat the boundary conditions. It should be noted that if the highest order of derivative or partial derivative assigned to the element boundary nodes is larger than one, the EDQ has to be used.

The philosophy inherent in the outlined techniques for defining discrete connection conditions on the inter-element boundaries, the discrete boundary conditions on the boundary and the discrete constraint conditions at the element corner nodes also holds good for other scientific or engineering problems.

11.5 Problems

11.5.1 Problem 11.1

The steady-state heat conduction of an isotropic medium having a square region $R = \{0 \leq x, y \leq \pi\}$ was solved. For solving the field problem of heat conduction, the field variable ϕ used in the formulation of general field problems represents the temperature T, k_x and k_y represent thermal conductivities, Q represents the heat generation rate per unit volume, $\bar{q}$ and $\tilde{q}$ represent the conduction heat fluxes into the medium on the Neumann boundary and inter-element boundary of two adjacent elements, respectively, the natural boundary condition and inter-element boundary condition involving the transfer coefficient c_h represent the convective heat transfer, q_x and q_y in Eq. (11.5) represent the two components of internal heat flux. The values of Q, k_x, k_y and c_h for this heat conduction problem solved are all equal to 1. Dirichlet conditions are applied on the four sides with $\bar{T}(0, y) = \bar{T}(\pi, y) = \bar{T}(x, \pi) = 0$

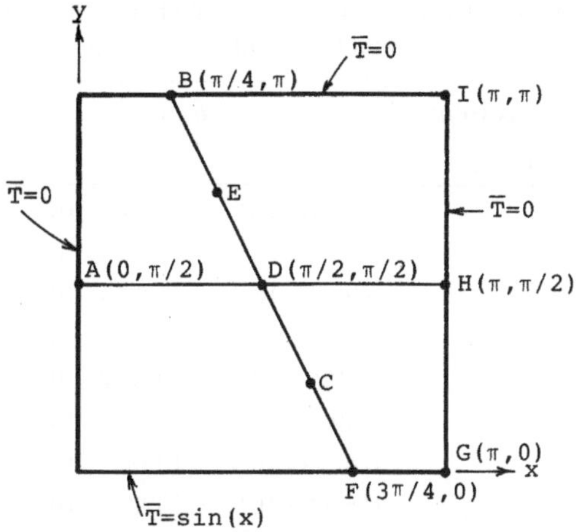

Fig. 11.4. 4-element mesh for the square region

and $\bar{T}(x, 0) = sinx$. In order to demonstrate the generality of the DQEM discrete element analysis technique, four bilinear elements are used to model the problem domain though it can be solved by using only one element to represent the whole domain. The mesh is shown in Fig. 11.4. All four elements have the same type of element grid. The Lagrange DQ model and equally spaced node points in both ξ and η directions are adopted for defining the element grid. One constraint condition at D is applied in the analysis. Three different types of constraint are adopted for the numerical tests to study the convergence. Type 1 constraint is the natural transition condition on DF. Type 2 constraint is the natural transition condition on DA. And type 3 constraint is the discrete governing equation at D. Numerical results of temperatures and heat fluxes at different points are listed in Table 11.1. Analytical solutions are also included for comparison [121–122]. It shows that the results converge fast by gradually increasing the node points in an element for all of the three different types of constraint. The convergence characters of the three different types of constraint make no significant difference. The results converge up to six digit accuracy by increasing the element node points up to 5×5. The convergence test is also carried out by increasing the number of 3×3 grid elements. One constraint condition for each element corner node in the domain is applied. It is the discrete governing equation at the corner node. Figures 11.5 and 11.6 show the 3×3 grid 16-element mesh and 3×3 grid 36-element mesh, respectively. The numerical results are listed in Table 11.2. They also converge well by gradually increasing the number of elements.

Table 11.1. Convergence by increasing the number of nodes per element for the steady-state heat conduction analysis in a square region

Element grid	Constraint type	$T_{,C}$	$q_{x,A}$	$q_{y,B}$
3×3	1	$.389560\times10^{0}$	$-.205404\times10^{0}$	$.433558\times10^{-1}$
	2	$.389375\times10^{0}$	$-.199279\times10^{0}$	$.369948\times10^{-1}$
	3	$.389515\times10^{0}$	$-.198798\times10^{0}$	$.418028\times10^{-1}$
5×5	1	$.419921\times10^{0}$	$-.197209\times10^{0}$	$.624946\times10^{-1}$
	2	$.419921\times10^{0}$	$-.197167\times10^{0}$	$.624467\times10^{-1}$
	3	$.419922\times10^{0}$	$-.196848\times10^{0}$	$.620789\times10^{-1}$
7×7	1	$.418209\times10^{0}$	$-.199277\times10^{0}$	$.612420\times10^{-1}$
	2	$.418209\times10^{0}$	$-.199277\times10^{0}$	$.612422\times10^{-1}$
	3	$.418209\times10^{0}$	$-.199277\times10^{0}$	$.612412\times10^{-1}$
9×9	1	$.418229\times10^{0}$	$-.199268\times10^{0}$	$.612283\times10^{-1}$
	2	$.418229\times10^{0}$	$-.199268\times10^{0}$	$.612283\times10^{-1}$
	3	$.418229\times10^{0}$	$-.199268\times10^{0}$	$.612283\times10^{-1}$
Analytical sol.		$.418229\times10^{0}$	$-.199268\times10^{0}$	$.612281\times10^{-1}$

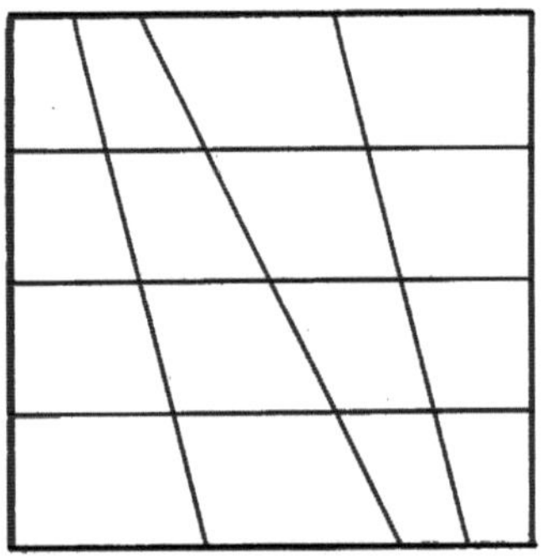

Fig. 11.5. 16-element mesh for the square region

11.5.2 Problem 11.2

A two-dimensional potential flow problem was analyzed. The governing mathematical model of the incompressible inviscid fluid flow is a boundary value problem of partial differential equation. For the velocity potential formulation, the condition of irrotationality and the continuity equation of the two-dimensional incompressible potential flow lead to the Laplace equation. There only two boundaries S_D and S_N exist for this flow problem.

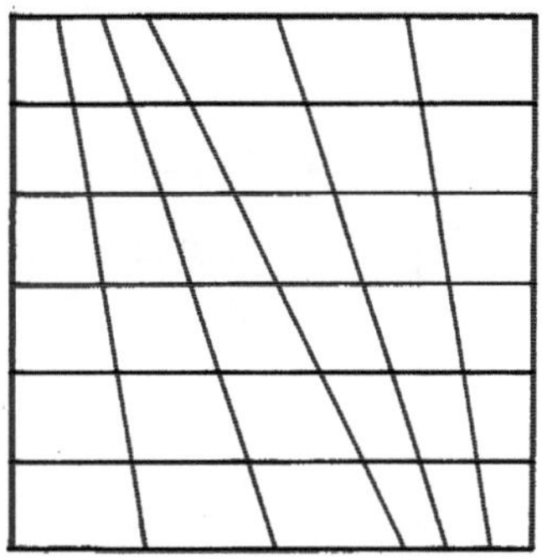

Fig. 11.6. 36-element mesh for the square region

Table 11.2. Convergence by increasing the number of elements for the steady-state heat conduction analysis in a square region

Element grid	Number of elements	$T_{,C}$	$q_{x,A}$	$q_{y,B}$
3×3	3	$.389515\times10^{0}$	$-.198798\times10^{0}$	$.418028\times10^{-1}$
3×3	16	$.421206\times10^{0}$	$-.200779\times10^{0}$	$.574711\times10^{-1}$
3×3	36	$.197862\times10^{0}$	$-.201226\times10^{0}$	$.599995\times10^{-1}$
Analytical sol.		$.418229\times10^{0}$	$-.199268\times10^{0}$	$.612281\times10^{-1}$

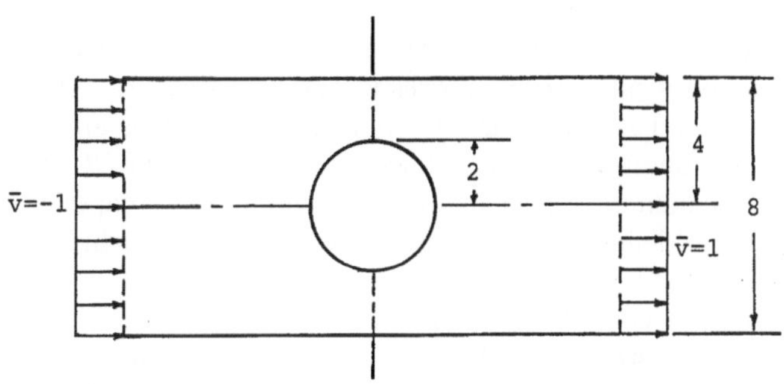

Fig. 11.7. Potential flow past a cylinder in a rectangular channel

For solving the field problem of potential flow, the field variable ϕ used in the formulation of general field problems represents the velocity potential Φ, k_x and k_y equal 1, Q vanishes, $\delta\bar{q}$ and $\delta\tilde{q}$ represent the rates of fluid flow, per unit length along the boundary, out of the Neumann boundary and inter-element boundary of two adjacent elements, respectively, of the analysis

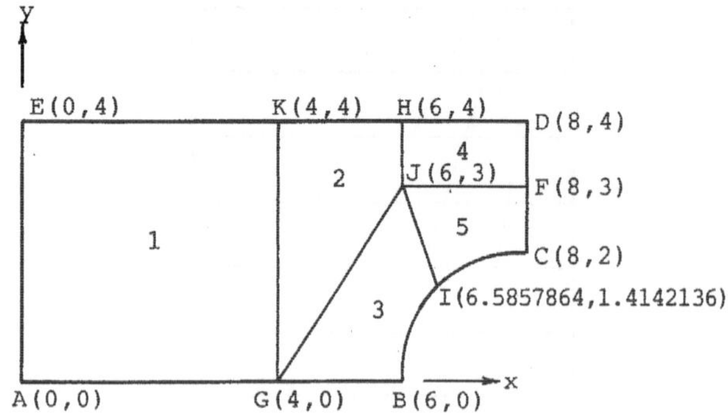

Fig. 11.8. Mesh for analyzing the problem of flow past a cylinder

domain, and the two components of field flux in Eq. (11.5) represent the two components of velocity, $v_x = -q_x$ and $v_y = -q_y$.

The problem solved concerns the flow past a cylinder in a rectangular channel with a uniform inlet flow. The problem is shown in Fig. 11.7. By using centerline symmetry and midstream antisymmetry, one fourth of the domain shown in Fig. 11.8 is used for the analysis. The boundary of the analysis domain consists of four Neumann boundaries and one Dirichlet boundary. The Neumann boundary conditions involve zero normal velocity $\bar{v} = \Phi_{,n} = 0$ along AB, BC and DE, and a uniform inflow $\bar{v} = -1$ along AE. The antisymmetry on CD leads to $v_y = 0$. Thus Φ is constant along CD, and is set to be zero in the analysis. The mesh which is formed by two quadratic serendipity elements, one bilinear element and two regular elements with element sides parallel to the physical coordinate axes is shown in Fig. 11.8. On AB, for all nodes except A and B one Neumann condition per node is applied. The applied Neumann condition at G is defined on element 1. On BC, for all nodes except C one Neumann condition per node is applied. The applied Neumann condition at I is defined on element 3. On CD, one Dirichlet condition is applied to each node. On DE, for all nodes except D and E one Neumann condition per node is applied. On AE, one Neumann condition is applied to each node. One constraint condition at J is considered in the analysis. The DQ and element grid models are the same as those used for the analysis of Problem 11.1. Numerical tests adopting four different types of constraint, separately, are carried out. Type 1 constraint is the natural transition condition on JH. Type 2 constraint is the natural transition condition on JF. Type 3 constraint is the natural transition condition on JG. And type 4 constraint is the averaged discrete governing equation at J. All five elements have the same type of element grid. Equally spaced nodes in both ξ and η directions are adopted for defining the element grid. The p refinement procedure is used to analyze the problem. Numerical results are summarized and listed in Table 11.3. It shows that the velocity potentials and velocities at certain nodes

Table 11.3. Convergence by increasing the number of element nodes for the potential flow analysis

Element grid	Constraint type	$\Phi_{,A}$	$v_{x,C}$	$v_{y,D}$
3×3	1	$-.105347\times10^2$	$.317507\times10^1$	$.211672\times10^1$
	2	$-.105142\times10^2$	$.316755\times10^1$	$.210847\times10^1$
	3	$-.105226\times10^2$	$.317063\times10^1$	$.211185\times10^1$
	4	$-.105431\times10^2$	$.317814\times10^1$	$.212009\times10^1$
5×5	1	$-.986284\times10^1$	$.251048\times10^1$	$.172663\times10^1$
	2	$-.986281\times10^1$	$.251087\times10^1$	$.172633\times10^1$
	3	$-.986282\times10^1$	$.251086\times10^1$	$.172633\times10^1$
	4	$-.986274\times10^1$	$.251093\times10^1$	$.172631\times10^1$
7×7	1	$-.100184\times10^2$	$.261774\times10^1$	$.178964\times10^1$
	2	$-.100183\times10^2$	$.261775\times10^1$	$.178965\times10^1$
	3	$-.100183\times10^2$	$.261774\times10^1$	$.178964\times10^1$
	4	$-.100184\times10^2$	$.261774\times10^1$	$.178964\times10^1$
9×9	1	$-.998552\times10^1$	$.261106\times10^1$	$.177848\times10^1$
	2	$-.998552\times10^1$	$.261106\times10^1$	$.177848\times10^1$
	3	$-.998552\times10^1$	$.261106\times10^1$	$.177848\times10^1$
	4	$-.998552\times10^1$	$.261106\times10^1$	$.177848\times10^1$
11×11	1	$-.999217\times10^1$	$.262216\times10^1$	$.178065\times10^1$
	2	$-.999217\times10^1$	$.262216\times10^1$	$.178065\times10^1$
	3	$-.999217\times10^1$	$.262216\times10^1$	$.178065\times10^1$
	4	$-.999217\times10^1$	$.262216\times10^1$	$.178065\times10^1$
13×13	1	$-.999053\times10^1$	$.262727\times10^1$	$.178007\times10^1$
	2	$-.999053\times10^1$	$.191643\times10^1$	$.178007\times10^1$
	3	$-.999053\times10^1$	$.262727\times10^1$	$.178007\times10^1$
	4	$-.999053\times10^1$	$.191642\times10^1$	$.178007\times10^1$

converge fast by gradually increasing the number of nodes of an element, for all of the four different types of constraint. The convergence characters of the four different types of constraint make no significant difference in the sense that numerical results of all different types of constraint are up to the fifth digit identical by only increasing the element nodes up to 5×5. It should be mentioned that this problem can also be analyzed by simply using two elements to model the analysis domain. It can even be solved by using a single pentagon element to represent the analysis domain and analyze the problem. Since the p version is more efficient than the h version, in the real engineering or scientific application the concept of adaptive discretization can be adopted and the number of elements used to model the analysis domain must be as much as possible small.

11.5.3 Problem 11.3

The problem solved concerns the torsion of a prismatic solid bar having a rectangular cross section. The Prandtl's stress function formulation of torsion theory was used. For solving the field problem of the torsion of a prismatic solid bar by the stress function formulation, the field variable ϕ used in the formulation of general field problems represents the stress function Ψ, k_x and k_y equal 1, Q equals $2G\theta$ with G the shear modulus and θ the angle of twist per unit length, q_x in Eq. (11.5) represents the component of shear stress in the y direction, $q_x \equiv \tau_{zy}$, q_y in Eq. (11.5) represents the negative value of the component of shear stress in the x direction, $q_y \equiv -\tau_{zx}$. Equation (11.3) with $\bar{q} = 0$ represents that the normal shear on the analysis domain boundary vanishes, and Eq. (11.13) with $\tilde{q}^{r,s} = 0$ represents that the normal shears on the inter-element boundary of two adjacent elements r and s are in equilibrium. There only the boundary condition, $\Psi = 0$ on S_D, exists for this problem.

This problem can be solved by using only one rectangular element to represent the problem domain. However, a mesh shown in Fig. 11.9 and having four elements is used for analyzing the problem by adopting the p refinement procedure. Two of the four elements are extremely distorted. All four elements have the same type of element grid. Equally spaced discrete points in both ξ and η directions are adopted for defining the element grid. One constraint condition at the element corner node with the four elements being connected to it is applied. Three different types of constraint are adopted for the numerical tests to study the convergence. The type 1 constraint is the equilibrium of shear stresses on OD. The type 2 constraint is the equilibrium of shear stresses

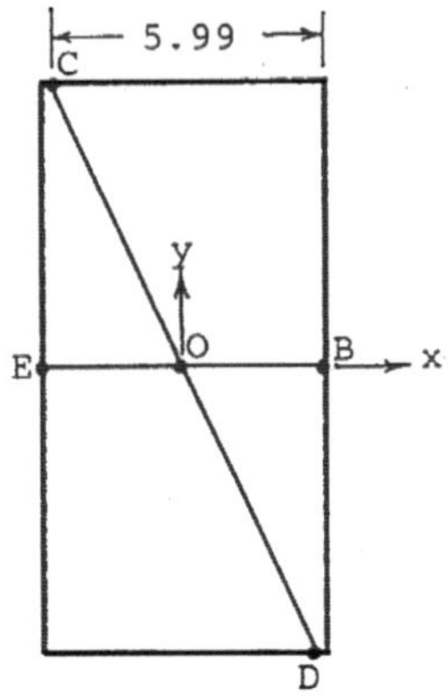

Fig. 11.9. Extremely distorted mesh for the rectangular cross section

on OE. The type 3 constraint is the discrete governing equation at O. Numerical results are summarized and listed in Table 11.4. They are compared with the results of analytical solution [95]. It shows excellent convergence though extremely distorted elements are used.

Table 11.4. Convergence by increasing the number of element nodes for the torsion analysis of a prismatic solid bar using extremely distorted elements

Element grid	Constraint type	$\Psi_{,A}$	$\tau_{zy,B}$
3×3	1	$.846407\times10^1$	$.564271\times10^1$
	2	$.735872\times10^1$	$.600386\times10^1$
	3	$.755576\times10^1$	$.593948\times10^1$
5×5	1	$.803784\times10^1$	$.561706\times10^1$
	2	$.799405\times10^1$	$.563183\times10^1$
	3	$.797751\times10^1$	$.563741\times10^1$
7×7	1	$.816400\times10^1$	$.559335\times10^1$
	2	$.815705\times10^1$	$.559573\times10^1$
	3	$.815529\times10^1$	$.559633\times10^1$
9×9	1	$.818002\times10^1$	$.558581\times10^1$
	2	$.817804\times10^1$	$.558649\times10^1$
	3	$.817730\times10^1$	$.558674\times10^1$
11×11	1	$.818786\times10^1$	$.558345\times10^1$
	2	$.818710\times10^1$	$.558372\times10^1$
	3	$.818677\times10^1$	$.558383\times10^1$
13×13	1	$.819188\times10^1$	$.558253\times10^1$
	2	$.819152\times10^1$	$.558265\times10^1$
	3	$.819136\times10^1$	$.558271\times10^1$
15×15	1	$.819418\times10^1$	$.558212\times10^1$
	2	$.819400\times10^1$	$.558219\times10^1$
	3	$.819391\times10^1$	$.558222\times10^1$
Analytical sol.		$.819877\times10^1$	$.558036\times10^1$

11.5.4 Problem 11.4

The problem solved involves the DQEM analysis of the steady-state heat conduction in a rectangular region with a quarter circular cut. The medium is isotropic with the thermal conductivity equal to 1. The problem is shown in Fig. 11.10. The geometrical parameters are: $a = 6$, $b = 8$, $c = 4$ and $d = 5$.

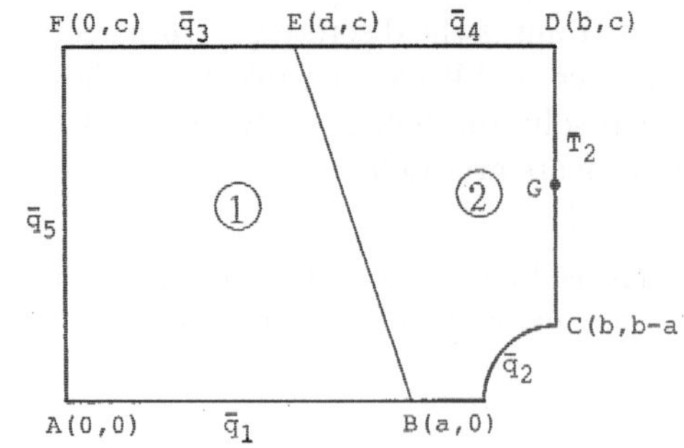

Fig. 11.10. Heat conduction in a rectangular medium with a quarter circular cut

The boundary of the domain consists of four Neumann boundaries and one Dirichlet boundary. The Neumann boundary conditions involve the conduction heat flux into the domain. $\bar{q}$ equals zero along AB, BC, DE and EF, and $\bar{q}$ equals 5 along AF. The temperature is prescribed and zero along CD. And there is no heat generation rate. Two elements are used to model the domain. On AB, for all node points except A and B, one Neumann condition per node is applied. On BC, for all node points except C, one Neumann condition per node is applied. The applied Neumann condition at B is attached to element 2. On CD, each discrete point is set to have one Dirichlet condition. On DE, for all node points except D, one Neumann condition per node is applied. On EF, for all node points except E and F, one Neumann condition per node is applied. On EB, for all node points except E and B, one natural transition condition per node is applied. On AF, each node point is set to have one Neumann condition. The mapping technique is used to generate the grid points in an element and calculate the direction cosines at a node point on the inter-element boundary or natural boundary. Shape functions, expressed by Eq. (10.30), of the C^{1*} element which is a four-node twelve-DOF element, are used for the mapping transformations to approximately describe the shape of curved elements. Lagrange DQ model with equally spaced grid is used for the element basis discretization. Numerical results obtained by the DQEM are summarized and listed in Table 11.5. It shows that the heat fluxes at certain node points converge fast following the increase of node points in an element. It also shows that the value of heat flux in x direction is rather large at C.

11.5.5 Problem 11.5

The problem solved involves the DQEM analysis of the two-dimensional heat conduction of a medium shown in Fig. 11.11 [123]. There are two subdomains having different materials and uniformly distributed heat generation rates. The shape of each subdomain is represented by the quadratic serendipity shape functions. The problem has both the Neumann and Dirichlet boundaries. There is also a constant heat flux into the medium on the

Table 11.5. Convergence of heat conduction in a rectangular medium with a quarter circular cut

Element grid	$q_{x,C}$	$q_{x,G}$	$q_{x,D}$
7×7	$.13092\times10^2$	$.96325\times10^1$	$.89480\times10^1$
9×9	$.13057\times10^2$	$.95730\times10^1$	$.88924\times10^1$
11×11	$.13113\times10^2$	$.95850\times10^1$	$.89034\times10^1$
13×13	$.13134\times10^2$	$.95820\times10^1$	$.89004\times10^1$

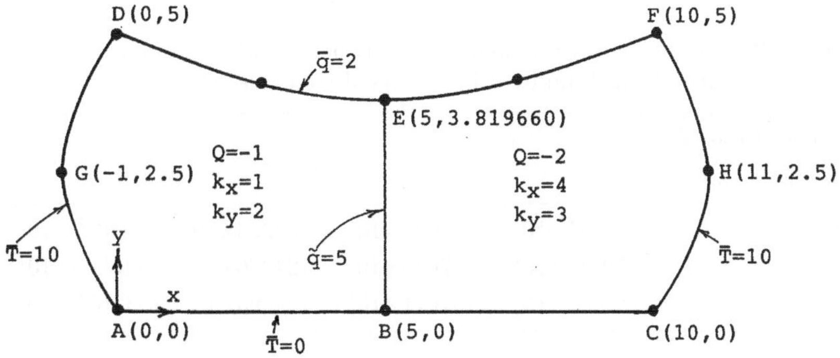

Fig. 11.11. Heat conduction in an irregular orthotropic medium composed of two irregular subdomains

inter-subdomain boundary. At a corner node, one discrete condition equation is considered. At the intersection of the Neumann boundary and the inter-subdomain boundary, the discrete Neumann condition defined on the left subdomain is considered. At D and F, the considered conditions are Dirichlet conditions. At A and C, the considered Dirichlet condition is $\bar{T} = 0$. Chebyshev DQ model with the Chebyshev grid defined by using the roots of Chebyshev polynomials are used to carry out the element basis DQ discretization. Numerical results of temperature $T_{,E}$ at E, and heat fluxes q_{y,B_l} and q_{y,B_r} in the left subdomain and right subdomain , respectively, at B are summarized and listed in Table 11.6. It also shows good convergence.

11.5.6 Problem 11.6

A problem involves the DQEM solution of heat conduction in a half circular region with the radius equal to 3 was solved. The medium is also isotropic with the thermal conductivity equal to 1. The half circle is expressed by $\partial\Omega_N = \{x^2 + y^2 = 9, \quad y \geq 0\}$ which is a Neumann boundary with $\bar{q} = -1/9$, while the boundary segment $\partial\Omega_D = \{-3 \leq x \leq 3, \quad y = 0\}$ is a Dirichlet boundary having a prescribed temperature $\bar{T} = 5/3$. There is no heat

Table 11.6. Convergence of heat conduction in an orthotropic medium

Grid (a subd.)	$T_{,E}$	q_{y,B_l}	q_{y,B_r}
5×5	$.466232\times10^1$	$-.840864\times10^1$	$-.126130\times10^2$
7×7	$.453529\times10^1$	$-.957623\times10^1$	$-.143643\times10^2$
9×9	$.458084\times10^1$	$-.104391\times10^2$	$-.156586\times10^2$
11×11	$.460776\times10^1$	$-.110384\times10^2$	$-.165576\times10^2$
13×13	$.463075\times10^1$	$-.114903\times10^2$	$-.172355\times10^2$
15×15	$.464870\times10^1$	$-.118507\times10^2$	$-.177760\times10^2$

Table 11.7. Convergence of the analysis of 2-D heat conduction in a half circular isotropic medium by using a curved triangular element

Order of Pascal triangular grid	$T_{,A}$	$T_{,B}$	$T_{,C}$	$T_{,D}$
4	$.155564\times10^1$	$.144620\times10^1$	$.133926\times10^1$	$.124040\times10^1$
6	$.151935\times10^1$	$.137616\times10^1$	$.124376\times10^1$	$.113206\times10^1$
8	$.151461\times10^1$	$.137071\times10^1$	$.124666\times10^1$	$.115677\times10^1$

generation rate. The domain is modelled by a twelve-node C^0 triangular element with node 1, node 2 and node 3 located at $(-3,0)$, $(3,0)$ and $(0,3)$, respectively. Numerical results of temperatures at four points $A(0,.6)$, $B(0,1.2)$, $C(0,1.8)$ and $D(0,2.4)$ are listed in Table 11.7. It also shows that the results converge well by gradually increasing the order of the Pascal triangular grid.

12

DQEM Analysis of Two-Dimensional Elasticity Problems

Two-dimensional elasticity problems with orthotropic materials are considered. Numerical formulation DQEM analysis using quadrilateral element is introduced. The assemblage of all discrete equations into the overall algebraic system is discussed. Both Lagrange DQ model and Chebyshev DQ model adopted for sample DQEM analyses.

12.1 Fundamental Relations

Let u and v denote the two displacement components in x and y directions, respectively, and ϵ_{xx}, ϵ_{yy} and γ_{xy} denote the three strain components. Then, the following relations hold good: $\epsilon_{xx} = u_{,x}$, $\epsilon_{yy} = v_{,y}$ and $\gamma_{xy} = u_{,y} + v_{,x}$. Consider orthotropic material and let $\bar{Q}_{11}$, $\bar{Q}_{12}$, $\bar{Q}_{16}$, $\bar{Q}_{22}$, $\bar{Q}_{26}$ and $\bar{Q}_{66}$ denote the reduced stiffnesses, and N_{xx}, N_{yy} and N_{xy} denote the stress resultants. These stress resultants can be related to strains and expressed as [124]

$$\begin{Bmatrix} N_{xx} \\ N_{yy} \\ N_{xy} \end{Bmatrix} = \begin{bmatrix} \bar{B}_{11} & \bar{B}_{12} & \bar{B}_{16} \\ \bar{B}_{12} & \bar{B}_{22} & \bar{B}_{26} \\ \bar{B}_{16} & \bar{B}_{26} & \bar{B}_{66} \end{bmatrix} \begin{Bmatrix} u_{,x} \\ v_{,y} \\ u_{,y} + v_{,x} \end{Bmatrix} \tag{12.1}$$

where $\bar{B}_{ij} = \int_{-h/2}^{h/2} \bar{Q}_{ij} dz$ with h being the thickness of the elastic medium. If the averaging approach is adopted, the effective stiffnesses $\tilde{Q}_{ij}$ are used to calculate $\bar{B}_{ij}$, $\bar{B}_{ij} = t\tilde{Q}_{ij}$. The reduced stiffnesses are obtained by the elastic stiffnesses Q_{ij} through coordinate transformation. Let E_1, E_2, ν_{12}, ν_{21} and G_{13} be the elastic constants, the following relations hold good: $\nu_{12}/E_1 = \nu_{21}/E_2$, $Q_{11} = E_1/(1 - \nu_{12}\nu_{21})$, $Q_{22} = E_2/(1 - \nu_{12}\nu_{21})$, $Q_{12} = \nu_{12}E_2/(1 - \nu_{12}\nu_{21})$, $Q_{66} = G_{13}$.

By introducing the strain-displacement relations and Eq. (12.1) into the equilibrium equations expressed by stress resultants, the equilibrium equations expressed by displacements can be obtained

$$
\begin{aligned}
&(\bar{B}_{11,x}+\bar{B}_{16,y})u_{,x}+(\bar{B}_{16,x}+\bar{B}_{66,y})u_{,y}+\bar{B}_{11}u_{,xx}+\bar{B}_{66}u_{,yy}\\
&+2\bar{B}_{16}u_{,xy}+(\bar{B}_{16,x}+\bar{B}_{66,y})v_{,x}+(\bar{B}_{12,x}\\
&+\bar{B}_{26,y})v_{,y}+\bar{B}_{16}v_{,xx}+\bar{B}_{26}v_{,yy}+(\bar{B}_{12}+\bar{B}_{66})v_{,xy}+h\rho b_x=0,
\end{aligned}
$$

$$
\begin{aligned}
&(\bar{B}_{16,x}+\bar{B}_{12,y})u_{,x}+(\bar{B}_{66,x}+\bar{B}_{26,y})u_{,y}+\bar{B}_{16}u_{,xx}+\bar{B}_{26}u_{,yy}\\
&+(\bar{B}_{12}+\bar{B}_{66})u_{,xy}+(\bar{B}_{66,x}+\bar{B}_{26,y})v_{,x}+(\bar{B}_{26,x}+\bar{B}_{22,y})v_{,y}\\
&+\bar{B}_{66}v_{,xx}+\bar{B}_{22}v_{,yy}+2\bar{B}_{26}v_{,xy}+h\rho b_y=0
\end{aligned}
\tag{12.2}
$$

In the above two equations, ρb_x and ρb_y are the two body forces in the x and the y directions, respectively. The kinematic boundary conditions can be expressed as

$$
u=\bar{u}, \quad v=\bar{v} \tag{12.3}
$$

where $\bar{u}$ and $\bar{v}$ are prescribed values. Let l and m denote the direction cosines of the outward unit normal vector on the natural boundary. Also let $\bar{t}_x$ and $\bar{t}_y$ denote the prescribed traction forces applied on the natural boundary. The natural boundary conditions can be expressed as

$$
\begin{aligned}
&l[\bar{B}_{11}u_{,x}+\bar{B}_{12}v_{,y}+\bar{B}_{16}(u_{,y}+v_{,x})]\\
&+m[\bar{B}_{16}u_{,x}+\bar{B}_{26}v_{,y}+\bar{B}_{66}(u_{,y}+v_{,x})]=\bar{t}_x,
\end{aligned}
$$

$$
\begin{aligned}
&l[\bar{B}_{16}u_{,x}+\bar{B}_{26}v_{,y}+\bar{B}_{66}(u_{,y}+v_{,x})]\\
&+m[\bar{B}_{12}u_{,x}+\bar{B}_{22}v_{,y}+\bar{B}_{26}(u_{,y}+v_{,x})]=\bar{t}_y
\end{aligned}
\tag{12.4}
$$

12.2 DQEM Formulation

12.2.1 Irregular Element

Consider the quadrilateral element. The substitution of mapping transformation relations (10.2) and (10.3) into Eqs. (12.2) leads to the following two equilibrium equations defined on the master element [125]

$$
\begin{aligned}
&E_1^e(\xi,\eta)u^e_{,\xi\xi}+E_2^e(\xi,\eta)u^e_{,\xi\eta}+E_3^e(\xi,\eta)u^e_{,\eta\eta}+E_4^e(\xi,\eta)u^e_{,\xi}\\
&+E_5^e(\xi,\eta)u^e_{,\eta}+F_1^e(\xi,\eta)v^e_{,\xi\xi}+F_2^e(\xi,\eta)v^e_{,\xi\eta}+F_3^e(\xi,\eta)v^e_{,\eta\eta}\\
&+F_4^e(\xi,\eta)v^e_{,\xi}+F_5^e(\xi,\eta)v^e_{,\eta}=-h^e\rho^e b^e_x,
\end{aligned}
$$

$$
\begin{aligned}
&\bar{E}_1^e(\xi,\eta)u^e_{,\xi\xi}+\bar{E}_2^e(\xi,\eta)u^e_{,\xi\eta}+\bar{E}_3^e(\xi,\eta)u^e_{,\eta\eta}+\bar{E}_4^e(\xi,\eta)u^e_{,\xi}\\
&+\bar{E}_5^e(\xi,\eta)u^e_{,\eta}+\bar{F}_1^e(\xi,\eta)v^e_{,\xi\xi}+\bar{F}_2^e(\xi,\eta)v^e_{,\xi\eta}+\bar{F}_3^e(\xi,\eta)v^e_{,\eta\eta}\\
&+\bar{F}_4^e(\xi,\eta)v^e_{,\xi}+\bar{F}_5^e(\xi,\eta)v^e_{,\eta}=-h^e\rho^e b^e_y
\end{aligned}
\tag{12.5}
$$

where

$$E_1^e(\xi,\eta) = \bar{B}_{11}^e \xi_{,x}^2 + \bar{B}_{66}^e \xi_{,y}^2 + 2\bar{B}_{16}^e \xi_{,x}\xi_{,y},$$

$$E_2^e(\xi,\eta) = 2[\bar{B}_{11}^e \xi_{,x}\eta_{,x} + \bar{B}_{66}^e \xi_{,y}\eta_{,y} + \bar{B}_{16}^e(\xi_{,x}\eta_{,y} + \xi_{,y}\eta_{,x})],$$

$$E_3^e(\xi,\eta) = \bar{B}_{11}^e \eta_{,x}^2 + \bar{B}_{66}^e \eta_{,y}^2 + 2\bar{B}_{16}^e \eta_{,x}\eta_{,y},$$

$$E_4^e(\xi,\eta) = \bar{B}_{11}^e \xi_{,xx} + 2\bar{B}_{16}^e \xi_{,xy} + \bar{B}_{66}^e \xi_{,yy} + (\bar{B}_{11,x}^e + \bar{B}_{16,y}^e)\xi_{,x} + (\bar{B}_{16,x}^e + \bar{B}_{66,y}^e)\xi_{,y},$$

$$E_5^e(\xi,\eta) = \bar{B}_{11}^e \eta_{,xx} + 2\bar{B}_{16}^e \eta_{,xy} + \bar{B}_{66}^e \eta_{,yy} + (\bar{B}_{11,x}^e + \bar{B}_{16,y}^e)\eta_{,x} + (\bar{B}_{16,x}^e + \bar{B}_{66,y}^e)\eta_{,y},$$

$$F_1^e(\xi,\eta) = \bar{B}_{16}^e \xi_{,x}^2 + \bar{B}_{26}^e \xi_{,y}^2 + (\bar{B}_{12}^e + \bar{B}_{66}^e)\xi_{,x}\xi_{,y},$$

$$F_2^e(\xi,\eta) = 2\bar{B}_{16}^e \xi_{,x}\eta_{,x} + 2\bar{B}_{26}^e \xi_{,y}\eta_{,y} + (\bar{B}_{12}^e + \bar{B}_{66}^e)(\xi_{,x}\eta_{,y} + \xi_{,y}\eta_{,x}),$$

$$F_3^e(\xi,\eta) = \bar{B}_{16}^e \eta_{,x}^2 + \bar{B}_{26}^e \eta_{,y}^2 + (\bar{B}_{12}^e + \bar{B}_{66}^e)\eta_{,x}\eta_{,y},$$

$$F_4^e(\xi,\eta) = \bar{B}_{16}^e \xi_{,xx} + (\bar{B}_{12}^e + \bar{B}_{66}^e)\xi_{,xy} + \bar{B}_{26}^e \xi_{,yy} + (\bar{B}_{16,x}^e + \bar{B}_{66,y}^e)\xi_{,x} + (\bar{B}_{12,x}^e + \bar{B}_{26,y}^e)\xi_{,y},$$

$$F_5^e(\xi,\eta) = \bar{B}_{16}^e \eta_{,xx} + (\bar{B}_{12}^e + \bar{B}_{66}^e)\eta_{,xy} + \bar{B}_{26}^e \eta_{,yy} + (\bar{B}_{16,x}^e + \bar{B}_{66,y}^e)\eta_{,x} + (\bar{B}_{12,x}^e + \bar{B}_{26,y}^e)\eta_{,y},$$

$$\bar{E}_1^e(\xi,\eta) = \bar{B}_{16}^e \xi_{,x}^2 + \bar{B}_{26}^e \xi_{,y}^2 + (\bar{B}_{12}^e + \bar{B}_{66}^e)\xi_{,x}\xi_{,y},$$

$$\bar{E}_2^e(\xi,\eta) = 2\bar{B}_{16}^e \xi_{,x}\eta_{,x} + 2\bar{B}_{26}^e \xi_{,y}\eta_{,y} + (\bar{B}_{12}^e + \bar{B}_{66}^e)(\xi_{,x}\eta_{,y} + \xi_{,y}\eta_{,x}),$$

$$\bar{E}_3^e(\xi,\eta) = \bar{B}_{16}^e \eta_{,x}^2 + \bar{B}_{26}^e \eta_{,y}^2 + (\bar{B}_{12}^e + \bar{B}_{66}^e)\eta_{,x}\eta_{,y},$$

$$\bar{E}_4^e(\xi,\eta) = \bar{B}_{16}^e \xi_{,xx} + (\bar{B}_{12}^e + \bar{B}_{66}^e)\xi_{,xy} + \bar{B}_{26}^e \xi_{,yy} + (\bar{B}_{16,x}^e + \bar{B}_{12,y}^e)\xi_{,x} + (\bar{B}_{66,x}^e + \bar{B}_{26,y}^e)\xi_{,y},$$

$$\bar{E}_5^e(\xi,\eta) = \bar{B}_{16}^e \eta_{,xx} + (\bar{B}_{12}^e + \bar{B}_{66}^e)\eta_{,xy} + \bar{B}_{26}^e \eta_{,yy} + (\bar{B}_{16,x}^e + \bar{B}_{12,y}^e)\eta_{,x} + (\bar{B}_{66,x}^e + \bar{B}_{26,y}^e)\eta_{,y},$$

$$\bar{F}_1^e(\xi,\eta) = \bar{B}_{66}^e \xi_{,x}^2 + \bar{B}_{22}^e \xi_{,y}^2 + 2\bar{B}_{26}^e \xi_{,x}\xi_{,y},$$

$$\bar{F}_2^e(\xi,\eta) = 2[\bar{B}_{66}^e \xi_{,x}\eta_{,x} + \bar{B}_{22}^e \xi_{,y}\eta_{,y} + \bar{B}_{26}^e(\xi_{,x}\eta_{,y} + \xi_{,y}\eta_{,x})],$$

$$\bar{F}_3^e(\xi,\eta) = \bar{B}_{66}^e \eta_{,x}^2 + \bar{B}_{22}^e \eta_{,y}^2 + 2\bar{B}_{26}^e \eta_{,x}\eta_{,y},$$

$$\bar{F}_4^e(\xi,\eta) = \bar{B}_{66}^e \xi_{,xx} + 2\bar{B}_{26}^e \xi_{,xy} + \bar{B}_{22}^e \xi_{,yy} + (\bar{B}_{66,x}^e + \bar{B}_{26,y}^e)\xi_{,x} + (\bar{B}_{26,x}^e + \bar{B}_{22,y}^e)\xi_{,y},$$

$$\bar{F}_5^e(\xi,\eta) = \bar{B}_{66}^e \eta_{,xx} + 2\bar{B}_{26}^e \eta_{,xy} + \bar{B}_{22}^e \eta_{,yy} + (\bar{B}_{66,x}^e + \bar{B}_{26,y}^e)\eta_{,x} + (\bar{B}_{26,x}^e + \bar{B}_{22,y}^e)\eta_{,y} \tag{12.6}$$

Equation (12.1) can also be transformed by using Eqs. (10.2). The stress resultants in the element defined by the natural coordinates are expressed by

$$N_{xx}^e = I_1^e(\xi,\eta)u_{,\xi}^e + I_2^e(\xi,\eta)u_{,\eta}^e + J_1^e(\xi,\eta)v_{,\xi}^e + J_2^e(\xi,\eta)v_{,\eta}^e,$$

$$N_{yy}^e = \bar{I}_1^e(\xi,\eta)u_{,\xi}^e + \bar{I}_2^e(\xi,\eta)u_{,\eta}^e + \bar{J}_1^e(\xi,\eta)v_{,\xi}^e + \bar{J}_2^e(\xi,\eta)v_{,\eta}^e,$$

$$N_{xy}^e = \tilde{I}_1^e(\xi,\eta)u_{,\xi}^e + \tilde{I}_2^e(\xi,\eta)u_{,\eta}^e + \tilde{J}_1^e(\xi,\eta)v_{,\xi}^e + \tilde{J}_2^e(\xi,\eta)v_{,\eta}^e \qquad (12.7)$$

where

$$I_1^e(\xi,\eta) = \xi_{,x}\bar{B}_{11}^e + \xi_{,y}\bar{B}_{16}^e, \quad I_2^e(\xi,\eta) = \eta_{,x}\bar{B}_{11}^e + \eta_{,y}\bar{B}_{16}^e,$$

$$J_1^e(\xi,\eta) = \xi_{,x}\bar{B}_{16}^e + \xi_{,y}\bar{B}_{12}^e, \quad J_2^e(\xi,\eta) = \eta_{,x}\bar{B}_{16}^e + \eta_{,y}\bar{B}_{12}^e,$$

$$\bar{I}_1^e(\xi,\eta) = \xi_{,x}\bar{B}_{12}^e + \xi_{,y}\bar{B}_{26}^e, \quad \bar{I}_2^e(\xi,\eta) = \eta_{,x}\bar{B}_{12}^e + \eta_{,y}\bar{B}_{26}^e,$$

$$\bar{J}_1^e(\xi,\eta) = \xi_{,x}\bar{B}_{26}^e + \xi_{,y}\bar{B}_{22}^e, \quad \bar{J}_2^e(\xi,\eta) = \eta_{,x}\bar{B}_{26}^e + \eta_{,y}\bar{B}_{22}^e,$$

$$\tilde{I}_1^e(\xi,\eta) = \xi_{,x}\bar{B}_{16}^e + \xi_{,y}\bar{B}_{66}^e, \quad \tilde{I}_2^e(\xi,\eta) = \eta_{,x}\bar{B}_{16}^e + \eta_{,y}\bar{B}_{66}^e,$$

$$\tilde{J}_1^e(\xi,\eta) = \xi_{,x}\bar{B}_{66}^e + \xi_{,y}\bar{B}_{26}^e, \quad \tilde{J}_2^e(\xi,\eta) = \eta_{,x}\bar{B}_{66}^e + \eta_{,y}\bar{B}_{26}^e \qquad (12.8)$$

The substitution of Eqs. (10.2) into Eqs. (12.4) leads to the following transformed natural boundary conditions

$$G_1^e(\xi,\eta)u_{,\xi}^e + G_2^e(\xi,\eta)u_{,\eta}^e + H_1^e(\xi,\eta)v_{,\xi}^e + H_2^e(\xi,\eta)v_{,\eta}^e + \tilde{m}^e\frac{\partial^2 u^e}{\partial t^2} = \bar{t}_x^e,$$

$$\bar{G}_1^e(\xi,\eta)u_{,\xi}^e + \bar{G}_2^e(\xi,\eta)u_{,\eta}^e + \bar{H}_1^e(\xi,\eta)v_{,\xi}^e + \bar{H}_2^e(\xi,\eta)v_{,\eta}^e + \tilde{m}^e\frac{\partial^2 v^e}{\partial t^2} = \bar{t}_y^e \qquad (12.9)$$

where

$$G_1^e(\xi,\eta) = l^e(\xi_{,x}\bar{B}_{11}^e + \xi_{,y}\bar{B}_{16}^e) + m^e(\xi_{,x}\bar{B}_{16}^e + \xi_{,y}\bar{B}_{66}^e),$$

$$G_2^e(\xi,\eta) = l^e(\eta_{,x}\bar{B}_{11}^e + \eta_{,y}\bar{B}_{16}^e) + m^e(\eta_{,x}\bar{B}_{16}^e + \eta_{,y}\bar{B}_{66}^e),$$

$$H_1^e(\xi,\eta) = l^e(\xi_{,x}\bar{B}_{16}^e + \xi_{,y}\bar{B}_{12}^e) + m^e(\xi_{,x}\bar{B}_{66}^e + \xi_{,y}\bar{B}_{26}^e),$$

$$H_2^e(\xi,\eta) = l^e(\eta_{,x}\bar{B}_{16}^e + \eta_{,y}\bar{B}_{12}^e) + m^e(\eta_{,x}\bar{B}_{66}^e + \eta_{,y}\bar{B}_{26}^e),$$

$$\bar{G}_1^e(\xi,\eta) = l^e(\xi_{,x}\bar{B}_{16}^e + \xi_{,y}\bar{B}_{66}^e) + m^e(\xi_{,x}\bar{B}_{12}^e + \xi_{,y}\bar{B}_{26}^e),$$

$$\bar{G}_2^e(\xi,\eta) = l^e(\eta_{,x}\bar{B}_{16}^e + \eta_{,y}\bar{B}_{66}^e) + m^e(\eta_{,x}\bar{B}_{12}^e + \eta_{,y}\bar{B}_{26}^e),$$

$$\bar{H}_1^e(\xi,\eta) = l^e(\xi_{,x}\bar{B}_{66}^e + \xi_{,y}\bar{B}_{26}^e) + m^e(\xi_{,x}\bar{B}_{26}^e + \xi_{,y}\bar{B}_{22}^e),$$

$$\bar{H}_2^e(\xi,\eta) = l^e(\eta_{,x}\bar{B}_{66}^e + \eta_{,y}\bar{B}_{26}^e) + m^e(\eta_{,x}\bar{B}_{26}^e + \eta_{,y}\bar{B}_{22}^e) \qquad (12.10)$$

The kinematic transition conditions on the inter-element boundary $\partial\Omega^{r,s}$ of two adjacent elements r and s are the continuities of displacements which are expressed as

$$u_i^r = u_i^s, \quad \text{on} \quad \partial\Omega^{r,s} \qquad (12.11)$$

If displacements on the inter-element boundary are assumed, the following relation holds

$$u_i^r = u_i^s = \bar{u}_i^{r,s} \quad \text{on} \quad \partial\Omega^{r,s} \qquad (12.12)$$

where $\bar{u}_i^{r,s}$ are prescribed displacements. By using Eqs. (12.9), the natural transition conditions which are equilibrium conditions on the inter-element boundary can also be written as

$$\begin{aligned} &G_1^r(\xi^r,\eta^r)u^r_{,\xi^r} + G_2^r(\xi^r,\eta^r)u^r_{,\eta^r} + H_1^r(\xi^r,\eta^r)v^r_{,\xi^r} + H_2^r(\xi^r,\eta^r)v^r_{,\eta^r} \\ &+ G_1^s(\xi^s,\eta^s)u^s_{,\xi^s} + G_2^s(\xi^s,\eta^s)u^s_{,\eta^s} + H_1^s(\xi^s,\eta^s)v^s_{,\xi^s} + H_2^s(\xi^s,\eta^s)v^s_{,\eta^s} = \tilde{t}_x^{r,s}, \end{aligned}$$

$$\begin{aligned} &\bar{G}_1^r(\xi^r,\eta^r)u^r_{,\xi^r} + \bar{G}_2^r(\xi^r,\eta^r)u^r_{,\eta^r} + \bar{H}_1^r(\xi^r,\eta^r)v^r_{,\xi^r} + \bar{H}_2^r(\xi^r,\eta^r)v^r_{,\eta^r} \\ &+ \bar{G}_1^s(\xi^s,\eta^s)u^s_{,\xi^s} + \bar{G}_2^s(\xi^s,\eta^s)u^s_{,\eta^s} + \bar{H}_1^s(\xi^s,\eta^s)v^s_{,\xi^s} + \bar{H}_2^s(\xi^s,\eta^s)v^s_{,\eta^s} \\ &= \tilde{t}_y^{r,s} \quad \text{on} \quad \partial\Omega^{r,s} \end{aligned} \tag{12.13}$$

where $\tilde{t}_i^{r,s}$ are distributed line forces applied on the inter-element boundary.

12.2.2 Element Basis DQ Discretization

Considering a quadrilateral element and using DQ, the equilibrium conditions (12.5) at a node (α,β) in an element e can be discretized

$$\begin{aligned} &[\,E_1^e(\xi_\alpha,\eta_\beta)D^{e\xi^2}_{\alpha m} + E_4^e(\xi_\alpha,\eta_\beta)D^{e\xi}_{\alpha m}]u^e_{m\beta} + [E_3^e(\xi_\alpha,\eta_\beta)D^{e\eta^2}_{\beta n} \\ &+E_5^e(\xi_\alpha,\eta_\beta)D^{e\eta}_{\beta n}]u^e_{\alpha n} + E_2^e(\xi_\alpha,\eta_\beta)D^{e\eta}_{\beta n}D^{e\xi}_{\alpha m}u^e_{mn} + [F_1^e(\xi_\alpha,\eta_\beta)D^{e\xi^2}_{\alpha m} \\ &+F_4^e(\xi_\alpha,\eta_\beta)D^{e\xi}_{\alpha m}]v^e_{m\beta} + [F_3^e(\xi_\alpha,\eta_\beta)D^{e\eta^2}_{\beta n} + F_5^e(\xi_\alpha,\eta_\beta)D^{e\eta}_{\beta n}]v^e_{\alpha n} \\ &+F_2^e(\xi_\alpha,\eta_\beta)D^{e\eta}_{\beta n}D^{e\xi}_{\alpha m}v^e_{mn} = -(t\rho b_x)^e_{\alpha,\beta}, \end{aligned}$$

$$\begin{aligned} &[\,\bar{E}_1^e(\xi_\alpha,\eta_\beta)D^{e\xi^2}_{\alpha m} + \bar{E}_4^e(\xi_\alpha,\eta_\beta)D^{e\xi}_{\alpha m}]u^e_{m\beta} + [\bar{E}_3^e(\xi_\alpha,\eta_\beta)D^{e\eta^2}_{\beta n} \\ &+\bar{E}_5^e(\xi_\alpha,\eta_\beta)D^{e\eta}_{\beta n}]u^e_{\alpha n} + \bar{E}_2^e(\xi_\alpha,\eta_\beta)D^{e\eta}_{\beta n}D^{e\xi}_{\alpha m}u^e_{mn} + [\bar{F}_1^e(\xi_\alpha,\eta_\beta)D^{e\xi^2}_{\alpha m} \\ &+\bar{F}_4^e(\xi_\alpha,\eta_\beta)D^{e\xi}_{\alpha m}]v^e_{m\beta} + [\bar{F}_3^e(\xi_\alpha,\eta_\beta)D^{e\eta^2}_{\beta n} + \bar{F}_5^e(\xi_\alpha,\eta_\beta)D^{e\eta}_{\beta n}]v^e_{\alpha n} \\ &+\bar{F}_2^e(\xi_\alpha,\eta_\beta)D^{e\eta}_{\beta n}D^{e\xi}_{\alpha m}v^e_{mn} = -(t\rho b_y)^e_{\alpha,\beta}, \end{aligned} \tag{12.14}$$

The stress resultants at a node (α,β) in an element e can also be obtained by using the DQ in Eqs. (12.9)

$$\begin{aligned} N^e_{xx\alpha\beta} = {}& I_1^e(\xi_{\alpha,\beta},\eta_{\alpha,\beta})D^{e\xi}_{\alpha m}u^e_{m\beta} + I_2^e(\xi_{\alpha,\beta},\eta_{\alpha,\beta})D^{e\eta}_{\beta n}u^e_{\alpha n} \\ &+J_1^e(\xi_{\alpha,\beta},\eta_{\alpha,\beta})D^{e\xi}_{\alpha m}v^e_{m\beta} + J_2^e(\xi_{\alpha,\beta},\eta_{\alpha,\beta})D^{e\eta}_{\beta n}v^e_{\alpha n}v^e_{,\eta}, \end{aligned}$$

$$\begin{aligned} N^e_{yy\alpha\beta} = {}& \bar{I}_1^e(\xi_{\alpha,\beta},\eta_{\alpha,\beta})D^{e\xi}_{\alpha m}u^e_{m\beta} + \bar{I}_2^e(\xi_{\alpha,\beta},\eta_{\alpha,\beta})D^{e\eta}_{\beta n}u^e_{\alpha n} \\ &+\bar{J}_1^e(\xi_{\alpha,\beta},\eta_{\alpha,\beta})D^{e\xi}_{\alpha m}v^e_{m\beta} + \bar{J}_2^e(\xi_{\alpha,\beta},\eta_{\alpha,\beta})D^{e\eta}_{\beta n}v^e_{\alpha n}, \end{aligned}$$

$$N^e_{xy\alpha\beta} = \tilde{I}^e_1(\xi_{\alpha,\beta},\eta_{\alpha,\beta})D^{e\xi}_{\alpha m}u^e_{m\beta} + \tilde{I}^e_2(\xi_{\alpha,\beta},\eta_{\alpha,\beta})D^{e\eta}_{\beta n}u^e_{\alpha n}$$
$$+\tilde{J}^e_1(\xi_{\alpha,\beta},\eta_{\alpha,\beta})D^{e\xi}_{\alpha m}v^e_{m\beta} + \tilde{J}^e_2(\xi_{\alpha,\beta},\eta_{\alpha,\beta})D^{e\eta}_{\beta n}v^e_{\alpha n} \quad (12.15)$$

Consider the inter-element boundary of the $\xi = 1$ side of element r and the $\xi = -1$ side of element s. The discrete continuity equations of displacements at a node β on the inter-element boundary can be expressed by

$$u^r_{iN^r_\xi\beta} = u^s_{1\beta}, \quad u^r_{iN^r_\xi\beta} = u^s_{1\beta} = \bar{u}^{r,s}_\beta \quad (12.16)$$

Using the DQ in Eqs. (12.13), the discrete natural transition conditions at the node are expressed as

$$G^r_1(\xi^r_{N^r_\xi\beta},\eta^r_{N^r_\xi\beta})D^{r\xi}_{N^r_\xi m}u^r_{m\beta} + G^r_2(\xi^r_{N^r_\xi\beta},\eta^r_{N^r_\xi\beta})D^{r\eta}_{\beta n}u^r_{N^r_\xi n}$$
$$+ H^r_1(\xi^r_{N^r_\xi\beta},\eta^r_{N^r_\xi\beta})D^{r\xi}_{N^r_\xi m}v^r_{m\beta} + H^r_2(\xi^r_{N^r_\xi\beta},\eta^r_{N^r_\xi\beta})D^{r\eta}_{\beta n}v^r_{N^r_\xi n}$$
$$+ G^s_1(\xi^s_{1\beta},\eta^s_{1\beta})D^{s\xi}_{1m}u^s_{m\beta} + G^s_2(\xi^s_{1\beta},\eta^s_{1\beta})D^{s\eta}_{\beta n}u^s_{1n}$$
$$+ H^s_1(\xi^s_{1\beta},\eta^s_{1\beta})D^{s\xi}_{1m}v^s_{m\beta} + H^s_2(\xi^s_{1\beta},\eta^s_{1\beta})D^{s\eta}_{\beta n}v^s_{1n} = \tilde{t}^{r,s}_{x\beta},$$

$$\bar{G}^r_1(\xi^r_{N^r_\xi\beta},\eta^r_{N^r_\xi\beta})D^{r\xi}_{N^r_\xi m}u^r_{m\beta} + \bar{G}^r_2(\xi^r_{N^r_\xi\beta},\eta^r_{N^r_\xi\beta})D^{r\eta}_{\beta n}u^r_{N^r_\xi n}$$
$$+ \bar{H}^r_1(\xi^r_{N^r_\xi\beta},\eta^r_{N^r_\xi\beta})D^{r\xi}_{N^r_\xi m}v^r_{m\beta} + \bar{H}^r_2(\xi^r_{N^r_\xi\beta},\eta^r_{N^r_\xi\beta})D^{r\eta}_{\beta n}v^r_{N^r_\xi n}$$
$$+ \bar{G}^s_1(\xi^s_{1\beta},\eta^s_{1\beta})D^{s\xi}_{1m}u^s_{m\beta} + \bar{G}^s_2(\xi^s_{1\beta},\eta^s_{1\beta})D^{s\eta}_{\beta n}u^s_{1n}$$
$$+ \bar{H}^s_1(\xi^s_{1\beta},\eta^s_{1\beta})D^{s\xi}_{1m}v^s_{m\beta} + \bar{H}^s_2(\xi^s_{1\beta},\eta^s_{1\beta})D^{s\eta}_{\beta n}v^s_{1n} = \tilde{t}^{r,s}_{y\beta} \quad (12.17)$$

Letting element n be an element with $\xi = 1$ side aligned along the natural boundary, the discrete natural boundary conditions at the node β can be obtained by using DQ in Eqs. (12.7)

$$G^n_1(\xi^n_{N^n_\xi\beta},\eta^n_{N^n_\xi\beta})D^{n\xi}_{N^n_\xi m}u^n_{m\beta} + G^n_2(\xi^n_{N^n_\xi\beta},\eta^n_{N^n_\xi\beta})D^{n\eta}_{\beta n}u^n_{N^n_\xi n}$$
$$+ H^n_1(\xi^n_{N^n_\xi\beta},\eta^n_{N^n_\xi\beta})D^{n\xi}_{N^n_\xi m}v^n_{m\beta} + H^n_2(\xi^n_{N^n_\xi\beta},\eta^n_{N^n_\xi\beta})D^{n\eta}_{\beta n}v^n_{N^n_\xi n} = \bar{t}^n_{x\beta},$$

$$\bar{G}^n_1(\xi^n_{N^n_\xi\beta},\eta^n_{N^n_\xi\beta})D^{n\xi}_{N^n_\xi m}u^n_{m\beta} + \bar{G}^n_2(\xi^n_{N^n_\xi\beta},\eta^n_{N^n_\xi\beta})D^{n\eta}_{\beta n}u^n_{N^n_\xi n}$$
$$+ \bar{H}^n_1(\xi^n_{N^n_\xi\beta},\eta^n_{N^n_\xi\beta})D^{n\xi}_{N^n_\xi m}v^n_{m\beta} + \bar{H}^n_2(\xi^n_{N^n_\xi\beta},\eta^n_{N^n_\xi\beta})D^{n\eta}_{\beta n}v^n_{N^n_\xi n} = \bar{t}^n_{y\beta}$$
$$(12.18)$$

12.3 Assemblage

With the kinematic transition conditions in mind, then assemble the discrete element equilibrium equations (12.14) for all elements, discrete natural transition conditions (12.17), and discrete natural boundary conditions (12.18), an overall discrete equilibrium/transition/boundary equation represented by Eq. (3.18) can be obtained. It is the overall stiffness equation represented by Eq. (3.18). Considering the kinematic boundary conditions and solving the overall discrete equilibrium/transition/boundary equation, displacements at all nodes can be obtained. Like FEM, the assemblage is based on an element by element procedure. In assembling the discrete equations of element e, the discrete element equilibrium equations (12.14), and the discrete element boundary forces, expressed by displacements, at the nodes on the four element boundary edges are directly assembled to the overall discrete equation system. An element basis explicit matrix equation, containing the discrete element equilibrium equations and the discrete element boundary forces placed at the rows with the assigned DOF related to the corresponding discrete element boundary forces, is not necessary to be formed in the assemblage process. This element basis explicit matrix equation is an element stiffness equation represented by Eq. (3.20) in which $[k^e]$ is a $(2N_\xi^e N_\eta^e) \times (2N_\xi^e N_\eta^e)$ element stiffness matrix,

$$\{\delta^e\} = \lfloor\; u_1^e \quad v_1^e \quad u_2^e \quad v_2^e \quad . \quad \ldots \quad . \;\rfloor^T \tag{12.19}$$

is the element displacement vector, and $\{r^e\}$ is the element load vector containing the corresponding loading forces and discrete element boundary forces. As Eq. (3.20) contain discrete element boundary forces at the four element boundary edges, equilibriums of internal forces and external forces at the inter-element boundary of two adjacent elements and the natural boundary are exactly satisfied in the assemblage process. Consequently, the DQEM is different from FEM which needs to form the element stiffness equation, and which neglects the exact equilibriums.

12.4 Overall Algebraic System

Considering the Lagrange family grid, denote N_ξ^e and N_η^e the numbers of levels in ξ and η directions, respectively, in the master element of a physical element. The minimum value of N_ξ^e and N_η^e is 2. The convergence character of this 4-noded element is poor. If the element is a rectangle and the element sides are parallel to x or y axes, mapping results in scaling the lengths of the sides. It can reduce the arithmetic operations in calculating the element stiffness equations. For two adjacent elements having different numbers of nodes on the inter-element boundary, the number of kinematic transition conditions must be larger than the number of natural transition conditions. Let n_d denote the difference between the two node numbers which equals the difference between the two numbers of transition conditions. To set up the kinematic transition conditions, the nodes on the inter-element boundary must be arranged in such a way that only n_d nodes in one element are not coincident with the nodes in the other element. The n_d extra nodes are used to define $2n_d$ extra kinematic transition conditions. In defining the $2n_d$ extra kinematic conditions, the interpolation technique must be used. In addition, the transition conditions can be easily set up by designing the grids of the two adjacent elements in such a way that both elements have the same numbers of nodes on the inter-element boundary no matter what the orders of approximations and grid configurations are.

For the DQEM 2-D elasticity analysis models, the total degrees of freedom must equal the number of discrete constraint equations. The two equilibrium equations defined at a node are equilibrium constraint conditions. An interior node can define only two discrete equilibrium equations. The discrete kinematic and natural boundary conditions are defined on the kinematic and natural boundaries, respectively. A node on the analysis domain boundary (ADB) but not an element corner node can define two boundary conditions and two equilibrium equations. At a node on the inter-element boundary, if the node is not an element corner node, in addition to the continuities of displacements, two discrete constraint equations of prescribed displacements or natural transition conditions and up to four discrete constraint equilibrium equations attached to the two adjacent elements can be defined. An element corner node might be able to define even more constraint equations.

Consider that discrete element equilibrium equations are only defined at interior nodes. For an element corner node in the analysis domain which is the common node of N_N natural inter-element boundaries (IEB) and N_K kinematic inter-element boundaries, let N_T denote the number of all constraint conditions. Then, for $N_K \neq 0$, $N_T = 2N_N + 2$; for $N_K = 0$ and the two displacement components u_i not prescribed at the node, $N_T = 2N_N$; for $N_K = 0$ and one of the two displacement components u_i prescribed at the node, $N_T = 2N_N + 1$; for $N_K = 0$ and the two displacement components u_i prescribed, $N_T = 2N_N + 2$. The two equations of setting u_i as prescribed values are also kinematic constraint conditions. Figure 11.1 is a typical element corner node in the analysis domain.

For an element corner node on the analysis domain boundary (ADB) with two element-based segments of the analysis domain boundary being connected to it, if the two segments are natural boundaries, N_T for various connections are: for $N_K \neq 0$, $N_T = 2N_N + 6$; for $N_K = 0$ and the two displacement components u_i not prescribed at the node, $N_T = 2N_N + 4$; for $N_K = 0$ and one of the two displacement components u_i prescribed at the node, $N_T = 2N_N + 5$; for $N_K = 0$ and the two displacement components u_i prescribed, $N_T = 2N_N + 6$. If one of the analysis domain boundary segment is natural boundary while the other one is kinematic boundary, N_T for various connections are: for $N_K \neq 0$, $N_T = 2N_N + 4$; for $N_K = 0$, $N_T = 2N_N + 4$. And if both of the two analysis domain boundary segments are kinematic boundaries, N_T for various connections are: for $N_K \neq 0$, $N_T = 2N_N + 2$; for $N_K = 0$, $N_T = 2N_N + 2$. A representative of this type of element corner node is shown in Fig. 11.2 which is the type 1 element corner node on the analysis domain boundary.

For an element corner node on the analysis domain boundary with one element-based segment of the analysis domain boundary connected to it, if the segment is natural boundary, N_T for various connections are: for $N_K \neq 0$, $N_T = 2N_N + 4$; for $N_K = 0$ and the two displacement components u_i not prescribed at the node, $N_T = 2N_N + 2$; for $N_K = 0$ and one of the two displacement components u_i prescribed at the node, $N_T = 2N_N + 3$; for $N_K = 0$ and the two displacement components u_i prescribed, $N_T = 2N_N + 4$. If the analysis domain boundary segment is kinematic boundary, N_T for various connections are: for $N_K \neq 0$, $N_T = 2N_N + 2$; for $N_K = 0$, $N_T = 2N_N + 2$. A representative of this type of element corner node is shown in Fig. 11.3 which is the type 2 element corner node on the analysis domain boundary.

Only the values of the two displacement components at nodes are used to define the DQ discretization. Then in order to satisfy all constraint conditions at an element corner node, at the assemblage stage, two degrees of freedom might not be enough. However, we can use more than two constraint conditions at that corner node by neglecting certain constraint conditions at interior nodes or at nodes on the inter-element boundary and other than corner nodes, and giving their degrees of freedom to the inclusion of extra constraint conditions other than the first two.

For the developed DQEM analysis models, the N_T constraint conditions can partially or fully be satisfied. We can also neglect all of the N_T constraint conditions and give the degrees of freedom of that node to the discrete element equilibrium equations at that node. The discrete equilibrium equations at the element corner node can be defined as the averages of the respective discrete equilibrium equations of all elements connected to that node.

The various techniques for selecting and implementing the constraint conditions at element corner nodes are flexible. Different approaches lead to different programming efforts. The overall algebraic system obtained by assembling all discrete constraint conditions is the discrete equilibrium/transition/boundary equation system.

EDQ can also be used to the DQEM analysis. For analyzing the 2-D elasticity problems, in order to automatically set the kinematic transition conditions by only using certain degrees of freedom assigned to the element boundary nodes, the degrees of freedom representing displacements must be assigned to the element boundary nodes. The degrees of freedom representing the partial derivatives of the displacement parameters can also be assigned to the nodes of all neighbor elements on the inter-element boundary and the compatibility conditions of the higher order partial derivatives can also be considered. However, if distributed line loads are applied, no compatibility condition of partial derivatives can be considered. The discrete element equilibrium equations can be defined on the inter-element boundaries as the average discrete equilibrium equations of multiple elements. They can also be defined on the element boundaries without adopting the average treatment. Thus, elements having no interior node can also be used to the DQEM analysis. For analyzing beam or plate problems in order to automatically set the kinematic transition conditions by only using certain degrees of freedom assigned to the element boundary nodes, the degrees of freedom representing the lateral displacement and first order derivative or partial derivatives of the lateral displacement must be assigned to the element boundary nodes. The degrees of freedom representing higher order derivatives or partial derivatives of the displacement can also be assigned to the nodes of all neighbor elements on the inter-element boundary and the compatibility conditions of the higher order derivatives or partial derivatives can also be considered. However, if the moment is applied the highest order of derivative or partial derivative that the compatibility condition can be considered is one. On the other hand, if the lateral force is applied on the inter-element boundary the highest order of derivative or partial derivative that the compatibility condition can be considered is two. The concept can also be used to treat the boundary conditions. It should be noted that if the highest order of derivative or partial derivative assigned to the element boundary nodes is larger than one, the EDQ has to be used.

The philosophy inherent in the outlined techniques for defining discrete connection conditions on the inter-element boundaries, the discrete boundary conditions on the boundary and the discrete constraint conditions at the element corner nodes also holds good for other scientific or engineering problems.

12.5 Problems

12.5.1 Problem 12.1

The problem solved involves the elastic deformation of a square plate having the E glass/epoxy material. For a 0^0 ply plate with a fiber having the volume fraction 0.70, the effective stiffnesses are: $Q_{11} = 9. \times 10^6$ psi, $Q_{12} = .85 \times 10^6$ psi, $Q_{22} = 3.68 \times 10^6$ psi, $Q_{66} = 1.74 \times 10^6$ psi and $Q_{16} = Q_{26} = 0$. They

Table 12.1. Results of the DQEM analysis of a rectangular orthotropic plate

Algorithm	Grid	v at B (in)	v at C (in)
DQEM	3×3	$.1034483\times10^{-2}$	$.5453523\times10^{-3}$
	5×5	$.9396767\times10^{-3}$	$.7424176\times10^{-3}$
	7×7	$.8529769\times10^{-3}$	$.6190472\times10^{-3}$
	9×9	$.8224254\times10^{-3}$	$.5578523\times10^{-3}$
	11×11	$.8120461\times10^{-3}$	$.5358458\times10^{-3}$
FDM	5×5	$.9484796\times10^{-3}$	$.5689211\times10^{-3}$
	7×7	$.8866070\times10^{-3}$	$.5579871\times10^{-3}$
	9×9	$.8562929\times10^{-3}$	$.5479859\times10^{-3}$
	11×11	$.8397994\times10^{-3}$	$.5417193\times10^{-3}$
	13×13	$.8300573\times10^{-3}$	$.5378151\times10^{-3}$
	15×15	$.8239102\times10^{-3}$	$.5353501\times10^{-3}$

are constant through the thickness of the plate. The plate is uniform with the thickness being 1 *in*. The four boundary edges of the plate are: $x = 0$, $x = 10\ ft$, $y = 0$ and $y = 10\ ft$. The two x-edges are fixed. The edge $y = 10\ ft$ is a free edge, while the edge $y = 0$ is a natural boundary edge subjected to a compressive traction force $\bar{t}_y = sin(\pi x/120)\ lb/ft$. The plate was solved by using one DQEM element to model the problem domain. Equally spaced Lagrange DQ is used for the analysis. Numerical results of the displacement v, at two different points $B(5\ ft, 5\ ft)$ and $C(5\ ft, 10\ ft)$, obtained by using various order of DQEM approximation are summarized and listed in Table 12.1. They are compared with the results of the traditional FDM. It also shows that the results of DQEM analysis converge fast. Especially for the vertical displacement of B, the convergence of FDM is very slow.

12.5.2 Problem 12.2

The second problem solved involves the DQEM analysis of an isotropic plate structure subjected to a uniformly distributed load which is shown in Fig. 12.1. The thickness of the plate is $t = 1\ cm$. The material is steel with the material constants: Young's modulus $E = 210\ GPa$, shear modulus $G = 80\ GPa$ and Poisson's ratio $\nu = .30$. Consequently, the stress resultants-strain relation coefficients $\bar{B}_{ij}$ are: $\bar{B}_{11} = \bar{B}_{22} = 2.3076923\ GN/m$, $\bar{B}_{12} = .69230769\ GN/m$, $\bar{B}_{66} = .80769231\ GN/m$ and $\bar{B}_{16} = \bar{B}_{26} = 0$. The plate structure is modelled by one regular square element and one irregular bilinear element. In constructing the overall discrete equilibrium/transition/boundary equation system, all kinematic transition conditions and kinematic boundary conditions at the related nodes are satisfied. All discrete natural transition conditions and natural boundary conditions are satisfied except at B and E. Since at these two nodes

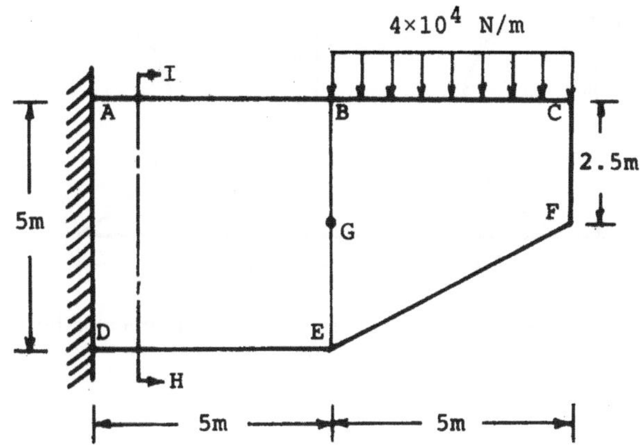

Fig. 12.1. An isotropic plate structure subjected to a uniformly distributed in-plane load

Table 12.2. Results of the DQEM analysis of an isotropic 2-D elasticity problem

DQEM grid lines (an elem.)	u_F (mm)	v_F (mm)	N_{xxG} (N/mm)	N_{xyG} (N/mm)
5×5	$-.604889\times10^0$	$-.131142\times10^1$	$-.169673\times10^3$	$-.498540\times10^2$
7×7	$-.614451\times10^0$	$-.128779\times10^1$	$-.166769\times10^3$	$-.499494\times10^2$
9×9	$-.619910\times10^0$	$-.129438\times10^1$	$-.167796\times10^3$	$-.502051\times10^2$
11×11	$-.622087\times10^0$	$-.129747\times10^1$	$-.168548\times10^3$	$-.504974\times10^2$

only one vertical natural boundary condition can be defined. Each of the two vertical natural boundary conditions at these two nodes is defined on the right element. In addition, two equilibrium equations are defined at all interior nodes. In the analysis, grid lines on the master element are equally spaced. Lagrange polynomials are used to calculate the DQ weighting coefficients. The convergence of numerical solutions can be assured by increasing the order of DQ approximation. Numerical results of displacement components at F and stress resultants at G are summarized and listed in Table 12.2. It shows that the results converge effectively.

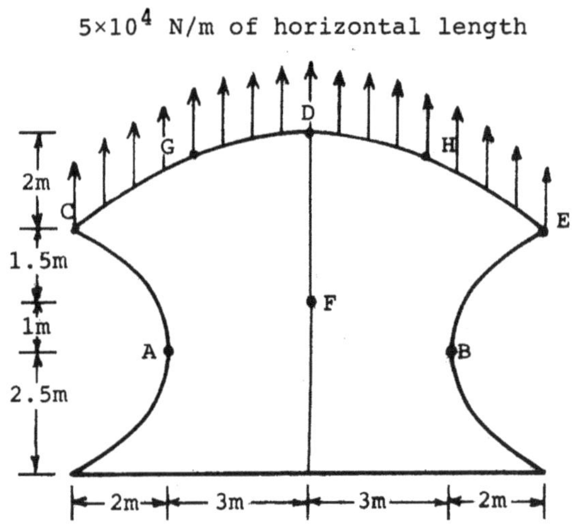

Fig. 12.2. A composite plate structure with two different orthotropic plate elements

12.5.3 Problem 12.3

The problem solved involves the DQEM analysis of a plate structure composed of two orthotropic composite plate elements which is shown in Fig. 12.2. Each of the two plate elements is represented by an eight-node serendipity element. The plate is subjected to a uniformly distributed load, 5×10^4 N/m upward, on the top edge of the plate. The left element is a uniform 90^0 ply composite laminate of T300/5208 graphite-epoxy with the thickness $t = 1$ cm, and $\bar{B}_{11} = 1.54$ GN/m, $\bar{B}_{12} = .03291$ GN/m, $\bar{B}_{22} = .1097$ GN/m and $\bar{B}_{66} = .056$ GN/m, while the right element is a uniform 90^0 ply laminate of T300/934 graphite-epoxy with the thickness t also equal to 1 cm, and $\bar{B}_{11} = 1.645$ GN/m GN/m, $\bar{B}_{12} = .03594$ GN/m, $\bar{B}_{22} = .1198$ and $\bar{B}_{66} = .065$ GN/m. The discrete vertical natural boundary condition at D is defined on the right element. Chebyshev polynomials are used to calculate the DQ weighting coefficients. In an element, the DQ nodes are defined by the roots of Chebyshev polynomials. Numerical results of vertical displacement at D and vertical stress resultants at F are summarized and listed in Table 12.3. The convergence of numerical results can be assured by increasing the order of Chebyshev DQ approximation.

Table 12.3. Results of the DQEM analysis of a curved orthotropic plate

DQEM grid lines (an elem.)	v_F (mm)	N_{yyG} (N/mm)
5×5	$.115484\times10^{0}$	$.519315\times10^{2}$
9×9	$.108449\times10^{0}$	$.510423\times10^{2}$
13×13	$.959177\times10^{-1}$	$.492538\times10^{2}$

13

DQEM Analysis of Kirchhoff-Love Plate Problems

DQEM analysis of Both static deflection and free vibration of Kirchhoff-Love plate structures is carried. In the DQEM discretization, EDQ model with rotational degrees of freedom assigned to nodes on element edges is used to define the discrete element model. The development of this DQEM plate analysis model adopting anisotropic plate theory.

13.1 Static Deflection Analysis

13.1.1 Fundamental Relations

Consider the composite plate problems. Let w denote the lateral displacement in z direction in a right-handed Cartesian rectangular coordinate system xyz with xy plane coincident with the neutral surface of the plate. Figure 13.1 shows the plate with the stress resultants, and external moments and transverse forces, per unit length, applied along the edges $x = 0$ and $y = 0$ and directed toward the related coordinate directions. By neglecting the transverse shear deformation, the relations between displacement w, and internal moments m_x, m_y and m_{xy} per unit length can be expressed by [124]

$$\left\{ \begin{array}{c} m_x \\ m_y \\ m_{xy} \end{array} \right\} = - \begin{bmatrix} \bar{D}_{11} & \bar{D}_{12} & \bar{D}_{16} \\ \bar{D}_{12} & \bar{D}_{22} & \bar{D}_{26} \\ \bar{D}_{16} & \bar{D}_{26} & \bar{D}_{66} \end{bmatrix} \left\{ \begin{array}{c} \frac{\partial^2 w}{\partial x^2} \\ \frac{\partial^2 w}{\partial y^2} \\ 2\frac{\partial^2 w}{\partial x \partial y} \end{array} \right\} \tag{13.1}$$

where $\bar{D}_{ij}$ are plate rigidities which can be expressed by the reduced stiffnesses $\bar{Q}_{ij}$ and the thickness of the plate δ, $\bar{D}_{ij} = \int_{\delta/2}^{\delta/2} \bar{Q}_{ij} z^2 dz$. The internal shear forces q_x and q_y can thus expressed by

$$q_x = \frac{\partial m_x}{\partial x} + \frac{\partial m_{xy}}{\partial y}, \quad q_y = \frac{\partial m_{xy}}{\partial x} + \frac{\partial m_y}{\partial y} \tag{13.2}$$

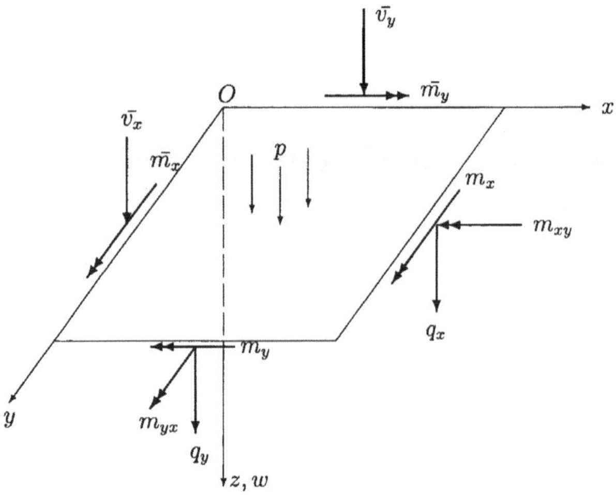

Fig. 13.1. The plate with stress resultants

Let $p(x, y)$ denote the laterally distributed load. The equilibrium equation of the plate can be expressed by

$$\begin{aligned}
&\bar{D}_{11}\frac{\partial^4 w}{\partial x^4} + 4\bar{D}_{16}\frac{\partial^4 w}{\partial x^3 \partial y} + 2(\bar{D}_{12} + 2\bar{D}_{66})\frac{\partial^4 w}{\partial x^2 \partial y^2} + 4\bar{D}_{26}\frac{\partial^4 w}{\partial x \partial y^3} \\
&+ \bar{D}_{22}\frac{\partial^4 w}{\partial y^4} + \left(\frac{\partial \bar{D}_{11}}{\partial x} + \frac{\partial \bar{D}_{16}}{\partial y}\right)\frac{\partial^3 w}{\partial x^3} \\
&+ 2\left[3\frac{\partial \bar{D}_{16}}{\partial x} + \frac{\partial(\bar{D}_{12} + 2\bar{D}_{66})}{\partial y}\right]\frac{\partial^3 w}{\partial x^2 \partial y} \\
&+ 2\left[\frac{\partial(\bar{D}_{12} + 2\bar{D}_{66})}{\partial x} + 3\frac{\partial \bar{D}_{26}}{\partial y}\right]\frac{\partial^3 w}{\partial x \partial y^2} + 2\left(\frac{\partial \bar{D}_{26}}{\partial x} + \frac{\partial \bar{D}_{22}}{\partial y}\right)\frac{\partial^3 w}{\partial y^3} \\
&+ \left(\frac{\partial^2 \bar{D}_{11}}{\partial x^2} + 2\frac{\partial^2 \bar{D}_{16}}{\partial x \partial y} + \frac{\partial^2 \bar{D}_{12}}{\partial y^2}\right)\frac{\partial^2 w}{\partial x^2} \\
&+ 2\left(\frac{\partial^2 \bar{D}_{16}}{\partial x^2} + 2\frac{\partial^2 \bar{D}_{66}}{\partial x \partial y} + \frac{\partial^2 \bar{D}_{26}}{\partial y^2}\right)\frac{\partial^2 w}{\partial x \partial y} \\
&+ \left(\frac{\partial^2 \bar{D}_{12}}{\partial x^2} + 2\frac{\partial^2 \bar{D}_{26}}{\partial x \partial y} + \frac{\partial^2 \bar{D}_{22}}{\partial y^2}\right)\frac{\partial^2 w}{\partial y^2} = p
\end{aligned} \tag{13.3}$$

Consider that the plate is placed in the region with $x \geq 0$ and $x = 0$ a boundary edge. The kinematic boundary conditions are

$$w(y) = \bar{w}(y), \quad \text{and} \quad \frac{\partial w(y)}{\partial x} = -\bar{\theta}_y(y) \tag{13.4}$$

where $\bar{w}(y)$ and $-\bar{\theta}_y(y)$ are prescribed displacement and rotation angle, respectively, applied on the edge and directed toward the related coordinate directions. Assume that the distributed moment $\bar{m}_x(y)$ and transverse force $\bar{v}_x(y)$, per unit length, are applied along the edge shown in Fig. 13.1. If the edge is simply supported, the boundary conditions are the first of Eq. (13.4) and $-m_x(y) = \bar{m}_x(y)$. The second one can be expressed by the lateral displacement

$$\bar{D}_{11}\frac{\partial^2 w}{\partial x^2} + 2\bar{D}_{16}\frac{\partial^2 w}{\partial x \partial y} + \bar{D}_{12}\frac{\partial^2 w}{\partial y^2} = \bar{m}_x(y) \tag{13.5}$$

If the edge is free, the lateral edge force per unit length is $v_x(y) = q_x(y) + \frac{\partial m_{xy}(y)}{\partial y}$. Then, the boundary conditions are represented by Eq. (13.5) and $-v_x(y) = \bar{v}_x(y)$. The second one can be expressed by the lateral displacement

$$\begin{aligned}
&\bar{D}_{11}\frac{\partial^3 w}{\partial x^3} + 4\bar{D}_{16}\frac{\partial^3 w}{\partial x^2 \partial y} + (\bar{D}_{12} + 4\bar{D}_{66})\frac{\partial^3 w}{\partial x \partial y^2} \\
&+ \left(\frac{\partial \bar{D}_{11}}{\partial x} + 2\frac{\partial \bar{D}_{16}}{\partial y}\right)\frac{\partial^2 w}{\partial x^2} + 2\left(\frac{\partial \bar{D}_{16}}{\partial x} + 2\frac{\partial \bar{D}_{66}}{\partial y}\right)\frac{\partial^2 w}{\partial x \partial y} \\
&+ \left(\frac{\partial \bar{D}_{12}}{\partial x} + 2\frac{\partial \bar{D}_{26}}{\partial y}\right)\frac{\partial^2 w}{\partial y^2} = \bar{v}_x(y)
\end{aligned} \tag{13.6}$$

For an edge parallel to x axis, the boundary conditions can be similarly expressed. At a boundary corner, torsional moments add up. Consequently, an extra corner force exists,

$$R = 2m_{xy} = -2\left(\bar{D}_{16}\frac{\partial^2 w}{\partial x^2} + \bar{D}_{26}\frac{\partial^2 w}{\partial y^2} + 2\bar{D}_{66}\frac{\partial^2 w}{\partial x \partial y}\right) \tag{13.7}$$

At the intersection of two fixed edges or two free edges without externally applied torsional moment, the extra corner force vanishes.

13.1.2 DQEM Formulation

In addition to the degrees of freedom for representing the lateral displacement w, the DOF for representing derivatives of the displacement with respect to one coordinate variable at a node on an element boundary edge can also be assigned. The selection of derivative DOF can be flexible. In order to automatically set the compatibility and conformability conditions by only using the DOF assigned to the element boundary nodes, the DOF representing the first order derivative of w with respect to the coordinate variable with the coordinate direction normal to the related element edge must be assigned to the element boundary nodes. In the present plate analysis model, only the DOF which are necessary for automatically setting the kinematic transition conditions and kinematic boundary conditions are assigned to the element boundary nodes.

Since the highest order of partial differentiations with respect to each coordinate variable existing in the fundamental relations is four, the order of approximate displacement with respect to the related coordinate direction must at least be four, and we need to define at least one discrete element equilibrium equation at one discrete point. The discrete points at which discrete element equilibrium equations are defined can be either in the interior of the element or on the element boundary.

In the present analysis model, each element boundary node has the DOF representing w and the first derivative of w with respect to the coordinate variable with the coordinate direction normal to the related element edge. Only interior discrete points are used to define the discrete element equilibrium equations. Consider the EDQ discretization in the natural space with the two natural coordinates ξ and η. Let N_ξ^e and N_η^e denote the number of nodes for defining the discretization with respect to ξ and η, respectively, $\bar{N}_\xi^e$ and $\bar{N}_\eta^e$ denote the number of the corresponding DOF, $\tilde{w}_{j\beta}^e$ and $\tilde{w}_{\alpha k}^e$ denote the related discrete displacement parameters with $j = 1, 2, ..., \bar{N}_\xi^e$, $k = 1, 2, ..., \bar{N}_\eta^e$, and α and β two typical node lines in ξ and η directions, and $D_{jm}^{e\xi^p}$ and $D_{kn}^{e\eta^q}$ denote the corresponding pth order and qth order weighting coefficients.

When defining the discrete fundamental relations, the numbers of grid lines with the discrete element equilibrium equations defined at discrete points on the grid lines in ξ and η directions are $\bar{N}_\xi^e - 4$ and $\bar{N}_\eta^e - 4$, respectively. The DOF assigned to an interior node for defining the EDQ discretization can be flexible. If only DOF representing w and $\frac{dw}{d\xi}$ (or $\frac{dw}{d\eta}$) are assigned to the interior nodes and the Hermite interpolation functions are used to explicitly calculate the weighting coefficients, the Hermite EDQ model is adopted. Let $\hat{N}_\xi^e$ and $\hat{N}_\eta^e$ denote the numbers of the related interior grid lines with the discrete element equilibrium equations defined at the discrete points located on these grid lines, plus the two element boundary nodes in the related coordinate directions. Then the following two relations hold: $\hat{N}_\xi^e = \bar{N}_\xi^e - 2$ and $\hat{N}_\eta^e = \bar{N}_\eta^e - 2$.

Assume that the side lengths of a rectangular element e in the x and y directions are h_x^e and h_y^e, respectively. The EDQ weighting coefficients are defined with the range for both ξ and η being 1. In the analysis, since the physical rotation DOF at nodes on the element edges are used, the elements of the second and last columns of the EDQ weighting coefficient matrices $D_{jm}^{e\xi^p}$ and $D_{kn}^{e\eta^q}$ need to be scaled by using h_x^e and h_y^e, respectively, to multiply the related coefficient elements.

Then, the discrete equilibrium equation at a discrete point (j, k) in the element can be obtained by using the EDQ in Eq. (13.3)

$$\frac{\bar{D}_{11,(j)(k)}^e}{(h_x^e)^4} D_{jm}^{e\xi^4} \tilde{w}_{mk}^e + \frac{4\bar{D}_{16,(j)(k)}^e}{(h_x^e)^3 h_y^e} D_{jm}^{e\xi^3} D_{kn}^{e\eta} \tilde{w}_{mn}^e$$
$$+ \frac{2\left[\bar{D}_{12,(j)(k)}^e + 2\bar{D}_{66,(j)(k)}^e\right]}{(h_x^e)^2 (h_y^e)^2} D_{jm}^{e\xi^2} D_{kn}^{e\eta^2} \tilde{w}_{mn}^e + \frac{4\bar{D}_{26,(j)(k)}^e}{h_x^e (h_y^e)^3} D_{jm}^{e\xi} D_{kn}^{e\eta^3} \tilde{w}_{mn}^e$$

$$
\begin{aligned}
&+\frac{\bar{D}^e_{22,(j)(k)}}{(h^e_y)^4}D^{e\eta^4}_{jn}\tilde{w}^e_{kn}+\frac{2}{(h^e_x)^3}\left[\frac{\partial\bar{D}^e_{11,(j)(k)}}{\partial x}+\frac{\partial\bar{D}^e_{16,(j)(k)}}{\partial y}\right]D^{e\xi^3}_{jm}\tilde{w}^e_{mk}\\
&+\frac{2}{(h^e)^2h^e_y}\left[3\frac{\partial\bar{D}^e_{16,(j)(k)}}{\partial x}+\frac{\partial\left(\bar{D}^e_{12,(j)(k)}+2\bar{D}^e_{66,(j)(k)}\right)}{\partial y}\right]D^{e\xi^2}_{jm}D^{e\eta}_{kn}\tilde{w}^e_{mn}\\
&+\frac{2}{h^e_x(h^e_y)^2}\left[\frac{\partial(\bar{D}^e_{12,(j)(k)}+2\bar{D}^e_{66,(j)(k)})}{\partial x}+3\frac{\partial\bar{D}^e_{26,(j)(k)}}{\partial y}\right]D^{e\xi}_{jm}D^{e\eta^2}_{kn}\tilde{w}^e_{mn}\\
&+\frac{2}{(h^e_y)^3}\left[\frac{\partial\bar{D}^e_{22,(j)(k)}}{\partial y}+\frac{\partial\bar{D}^e_{26,(j)(k)}}{\partial x}\right]D^{e\eta^3}_{kn}\tilde{w}^e_{jn}\\
&+\frac{1}{(h^e_x)^2}\left[\frac{\partial^2\bar{D}^e_{11,(j)(k)}}{\partial x^2}+2\frac{\partial^2\bar{D}^e_{16,(j)(k)}}{\partial x\partial y}+\frac{\partial^2\bar{D}^e_{12,(j)(k)}}{\partial y^2}\right]D^{e\xi^2}_{jm}\tilde{w}^e_{mk}\\
&+\frac{2}{h^e_xh^e_y}\left[\frac{\partial^2\bar{D}^e_{16,(j)(k)}}{\partial x^2}+2\frac{\partial^2\bar{D}^e_{66,(j)(k)}}{\partial x\partial y}+\frac{\partial^2\bar{D}^e_{26,(j)(k)}}{\partial y^2}\right]D^{e\xi}_{jm}D^{e\eta}_{kn}\tilde{w}^e_{mn}\\
&+\frac{1}{(h^e_y)^2}\left[\frac{\partial^2\bar{D}^e_{12,(j)(k)}}{\partial x^2}+2\frac{\partial^2\bar{D}^e_{26,(j)(k)}}{\partial x\partial y}+\frac{\partial^2\bar{D}^e_{22,(j)(k)}}{\partial y^2}\right]D^{e\eta^2}_{kn}\tilde{w}^e_{jn}\\
&=p_{jk}\quad j=2,...,\hat{N}^e_\xi-1;\quad k=2,...,\hat{N}^e_\eta-1
\end{aligned}
\tag{13.8}
$$

The discrete bending moment $m^e_{x,jk}$ at a discrete point (j,k) in the element can be obtained by using the EDQ into the first component equation of the matrix equation (13.1)

$$
\begin{aligned}
m^e_{x,jk}=&-\frac{\bar{D}^e_{11,(j)(k)}}{(h^e_x)^2}D^{e\xi^2}_{jm}\tilde{w}^e_{mk}-2\frac{\bar{D}^e_{12,(j)(k)}}{(h^e_y)^2}D^{e\eta^2}_{kn}\tilde{w}^e_{jn}\\
&-\frac{\bar{D}^e_{16,(j)(k)}}{h^e_xh^e_y}D^{e\xi}_{jm}D^{e\eta}_{kn}\tilde{w}^e_{mn}
\end{aligned}
\tag{13.9}
$$

The other two discrete interior moments at a discrete point in the element can be similarly obtained. The discrete shear force $q^e_{x,jk}$ at a discrete point (j,k) in the element can be obtained by using the EDQ into the first of Eq. (13.2)

$$
\begin{aligned}
q^e_{x,jk}=&-\frac{\bar{D}^e_{11,(j)(k)}}{(h^e)^3}D^{e\xi^3}_{jm}\tilde{w}^e_{mk}-\frac{4\bar{D}^e_{16,(j)(k)}}{(h^e_x)^2h^e_y}D^{e\xi^2}_{jm}D^{e\eta}_{kn}\tilde{w}^e_{mn}\\
&-\frac{\left(\bar{D}^e_{12,(j)(k)}+4\bar{D}^e_{66,(j)(k)}\right)}{h^e_x(h^e_y)^2}D^{e\xi}_{jm}D^{e\eta^2}_{kn}\tilde{w}^e_{mn}\\
&-\frac{1}{(h^e_x)^2}\left[\frac{\partial\bar{D}^e_{11,(j)(k)}}{\partial x}+\frac{\partial\bar{D}^e_{16,(j)(k)}}{\partial y}\right]D^{e\xi^2}_{jm}\tilde{w}^e_{mk}
\end{aligned}
$$

$$
-\frac{2}{h_x^e h_y^e}\left[\frac{\partial \bar{D}^r_{16,(j)(k)}}{\partial x}+2\frac{\partial \bar{D}^e_{66,(j)(k)}}{\partial y}\right] D^{e\xi}_{jm} D^{e\eta}_{kn} \tilde{w}^e_{mn}
$$

$$
-\frac{1}{(h_y^e)^2}\left[\frac{\partial \bar{D}^e_{12,(j)(k)}}{\partial x}+2\frac{\partial \bar{D}^e_{26,(j)(k)}}{\partial y}\right] D^{e\eta^2}_{kn} \tilde{w}^e_{jn} \tag{13.10}
$$

Equation (13.7) at an element corner node can also be discretized by using the EDQ.

In solving a plate problem with multiple subregions, the transition conditions on the inter-element boundary have to be satisfied. These conditions include displacement compatibility, deflection conformability, equilibrium of normal moments and equilibrium of lateral forces. Consider that the inter-element boundary of two adjacent elements r and s shown in Fig. 13.2, with x increasing from r to s is at $x = x^{r,s}$. Assume that the distributed moment $\bar{m}^{r,s}(y)$ and transverse force $\bar{v}^{r,s}(y)$, per unit length, are applied along the inter-element boundary and directed toward the related coordinate directions. The transition conditions can be expressed by

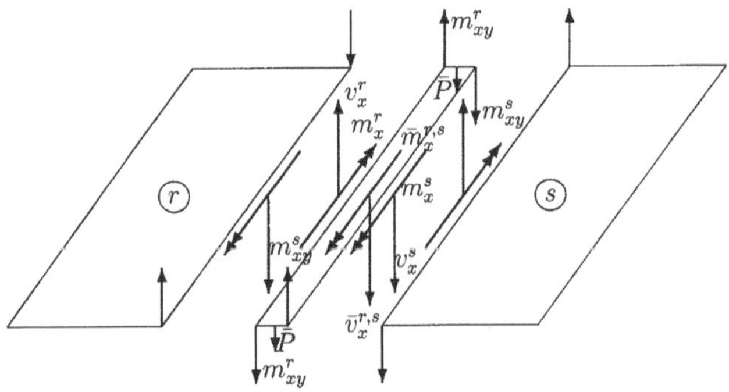

Fig. 13.2. Forces on the inter-element boundary of two adjacent plate element

$$
[w^r(y) - w^s(y)]\,|_{x=x^{r,s}} = 0, \quad \left[\frac{\partial w^r(y)}{\partial x} - \frac{\partial w^s(y)}{\partial x}\right]|_{x=x^{r,s}} = 0,
$$

$$
[v^r_x(y) - v^s_x(y)]\,|_{x=x^{r,s}} = \bar{v}^{r,s}_x(y),
$$

$$
[m^r_x(y) - m^s_x(y)]\,|_{x=x^{r,s}} = \bar{m}^{r,s}_x(y) \tag{13.11}
$$

where $v_x(y) = q_x(y) + \frac{\partial m_{xy}(y)}{\partial y}$. If a forced displacement $\bar{w}^{r,s}$ is applied, the third relation of the above equation is replaced by $w^{r,s} = \bar{w}^{r,s}$, while if a forced rotation $-\bar{\theta}_y(y)$ is applied and directed toward the y coordinate direction the last relation of the above equation is replaced by $\frac{\partial w^{r,s}}{\partial x} = -\bar{\theta}^{r,s}_y(y)$. The first two of Eq. (13.11) at a node on the inter-element boundary can be expressed by

$$\tilde{w}^r_{(\hat{N}^r_\xi - 1)\beta} = \tilde{w}^s_{1\beta}, \quad \tilde{w}^r_{\hat{N}^r_\xi} = \tilde{w}^s_{2\beta} \tag{13.12}$$

The explicit discrete equations of the last two of Eq. (13.11) at a discrete point β on the inter-element boundary can be expressed by

$$\begin{aligned}
&-\frac{\bar{D}^r_{11,\hat{N}^r_\xi(\beta)}}{(h^r_x)^3} D^{r\xi^3}_{\hat{N}^r_\xi m} \tilde{w}^r_{m\beta} - \frac{4\bar{D}^r_{16,\hat{N}^r_\xi(\beta)}}{(h^r_x)^2 h^r_y} D^{r\xi^2}_{\hat{N}^r_\xi m} D^{r\eta}_{\beta n} \tilde{w}^r_{mn} \\
&- \frac{\left(\bar{D}^r_{12,\hat{N}^r_\xi(\beta)} + 4\bar{D}^r_{66,\hat{N}^r_\xi(\beta)}\right)}{h^r_x (h^r_y)^2} D^{r\xi}_{\hat{N}^r_\xi m} D^{r\eta^2}_{\beta n} \tilde{w}^r_{mn} \\
&- \frac{1}{(h^r_x)^2}\left[\frac{\partial \bar{D}^r_{11,\hat{N}^r_\xi(\beta)}}{\partial x} + \frac{\partial \bar{D}^r_{16,\hat{N}^r_\xi(\beta)}}{\partial y}\right] D^{r\xi^2}_{\hat{N}^r_\xi m} \tilde{w}^r_{m\beta} \\
&- \frac{2}{h^r_x h^r_y}\left[\frac{\partial \bar{D}^r_{16,\hat{N}^r_\xi(\beta)}}{\partial x} + 2\frac{\partial \bar{D}^r_{66,\hat{N}^r_\xi(\beta)}}{\partial y}\right] D^{r\xi}_{\hat{N}^r_\xi m} D^{r\eta}_{\beta n} \tilde{w}^r_{mn} \\
&- \frac{1}{(h^r_y)^2}\left[\frac{\partial \bar{D}^r_{12,\hat{N}^r_\xi(\beta)}}{\partial x} + 2\frac{\partial \bar{D}^r_{26,\hat{N}^r_\xi(\beta)}}{\partial y}\right] D^{r\eta^2}_{\beta n} \tilde{w}^r_{\hat{N}^r_\xi n} + \frac{\bar{D}^s_{11,1(\beta)}}{(h^s_x)^3} D^{s\xi^3}_{1m} \tilde{w}^s_{m\beta} \\
&+ \frac{4\bar{D}^s_{16,1(\beta)}}{(h^s_x)^2 h^s_y} D^{s\xi^2}_{1m} D^{s\eta}_{\beta n} \tilde{w}^s_{mn} + \frac{\left(\bar{D}^s_{12,1(\beta)} + 4\bar{D}^s_{66,1(\beta)}\right)}{h^s_x (h^s_y)^2} D^{s\xi}_{1m} D^{s\eta^2}_{\beta n} \tilde{w}^s_{mn} \\
&+ \frac{1}{(h^s_x)^2}\left[\frac{\partial \bar{D}^s_{11,1(\beta)}}{\partial x} + \frac{\partial \bar{D}^s_{16,1(\beta)}}{\partial y}\right] D^{s\xi^2}_{1m} \tilde{w}^s_{m\beta} \\
&+ \frac{2}{h^s_x h^s_y}\left[\frac{\partial \bar{D}^s_{16,1(\beta)}}{\partial x} + 2\frac{\partial \bar{D}^s_{66,1(\beta)}}{\partial y}\right] D^{s\xi}_{1m} D^{s\eta}_{\beta n} \tilde{w}^s_{mn} \\
&+ \frac{1}{(h^s_y)^2}\left[\frac{\partial \bar{D}^s_{12,1(\beta)}}{\partial x} + 2\frac{\partial \bar{D}^s_{26,1(\beta)}}{\partial y}\right] D^{s\eta^2}_{\beta n} \tilde{w}^s_{1n} = \bar{v}^{r,s}_{x,\beta}
\end{aligned} \tag{13.13}$$

and

$$\begin{aligned}
&-\frac{\bar{D}^r_{11,\hat{N}^r_\xi(\beta)}}{(h^r_x)^2} D^{r\xi^2}_{\hat{N}^r_\xi m} \tilde{w}^r_{m\beta} - 2\frac{\bar{D}^r_{12,\hat{N}^r_\xi(\beta)}}{(h^r_y)^2} D^{r\eta^2}_{\beta n} \tilde{w}^r_{\hat{N}^r_\xi n} \\
&- \frac{\bar{D}^r_{16,\hat{N}^r_\xi(\beta)}}{h^r_x h^r_y} D^{r\xi}_{\hat{N}^r_\xi m} D^{r\eta}_{\beta n} \tilde{w}^r_{mn} + \frac{\bar{D}^s_{11,1(\beta)}}{(h^s)^2} D^{s\xi^2}_{1m} \tilde{w}^s_{m\beta} \\
&+ 2\frac{\bar{D}^s_{12,1(\beta)}}{(h^s)^2} D^{s\eta^2}_{\beta n} \tilde{w}^s_{1n} + \frac{\bar{D}^s_{16,1(\beta)}}{h^s_x h^s_y} D^{s\xi}_{1m} D^{s\eta}_{\beta n} \tilde{w}^s_{mn} = \bar{m}^{r,s}_{x,\beta}
\end{aligned} \tag{13.14}$$

Assume that element m has an edge on the kinematic boundary $x = 0$. The kinematic boundary conditions (13.4) at a discrete point $(1, \beta)$ on the edge can be expressed by

$$\tilde{w}_{1\beta}^{m} = \bar{w}_{1\beta}^{m} \quad \text{and} \quad \tilde{w}_{2\beta}^{m} = -\bar{\theta}_{y1\beta}^{m} \tag{13.15}$$

where $\bar{w}_{1\beta}^{m}$ and $-\bar{\theta}_{y1\beta}^{m}$ are prescribed displacement and rotation angle, respectively, at the node $(1,\beta)$ on the kinematic boundary. Also assume that element n has an edge on the natural boundary $x=0$. The natural boundary conditions (13.5) and (13.6) at a discrete point $(1,\beta)$ on the edge can be similarly expressed as

$$\begin{aligned} &\frac{\bar{D}_{11,1(\beta)}^{n}}{(h_x^n)^2} D_{1m}^{n\xi^2} \tilde{w}_{m\beta}^{n} + 2\frac{\bar{D}_{12,1(\beta)}^{n}}{(h_y^n)^2} D_{\beta n}^{n\eta^2} \tilde{w}_{1n}^{n} \\ &+ \frac{\bar{D}_{16,1(\beta)}^{n}}{h_x^n h_y^n} D_{1m}^{n\xi} D_{\beta n}^{n\eta} \tilde{w}_{mn}^{n} = \bar{m}_{1\beta}^{n} \end{aligned} \tag{13.16}$$

and

$$\begin{aligned} &\frac{\bar{D}_{11,1(\beta)}^{n}}{(h_x^n)^3} D_{1m}^{n\xi^3} \tilde{w}_{m\beta}^{n} + \frac{4\bar{D}_{16,1(\beta)}^{n}}{(h_x^n)^2 h_y^n} D_{1m}^{n\xi^2} D_{\beta n}^{n\eta} \tilde{w}_{mn}^{n} \\ &+ \frac{\left(\bar{D}_{12,1(\beta)}^{n} + 4\bar{D}_{66,1(\beta)}^{n}\right)}{h_x^n (h_y^n)^2} D_{1m}^{n\xi} D_{\beta n}^{n\eta^2} \tilde{w}_{mn}^{n} \\ &+ \frac{1}{(h_x^n)^2} \left[\frac{\partial \bar{D}_{11,1(\beta)}^{n}}{\partial x} + \frac{\partial \bar{D}_{16,1(\beta)}^{n}}{\partial y} \right] D_{1m}^{n\xi^2} \tilde{w}_{m\beta}^{n} \\ &+ \frac{2}{h_x^n h_y^n} \left[\frac{\partial \bar{D}_{16,1(\beta)}^{n}}{\partial x} + 2\frac{\partial \bar{D}_{66,1(\beta)}^{n}}{\partial y} \right] D_{1m}^{n\xi} D_{\beta n}^{n\eta} \tilde{w}_{mn}^{n} \\ &+ \frac{1}{(h_y^n)^2} \left[\frac{\partial \bar{D}_{12,1(\beta)}^{n}}{\partial x} + 2\frac{\partial \bar{D}_{26,1(\beta)}^{n}}{\partial y} \right] D_{\beta n}^{n\eta^2} \tilde{w}_{1n}^{n} = \bar{v}_{1\beta}^{n} \end{aligned} \tag{13.17}$$

where $\bar{m}_{1\beta}^{n}$ and $\bar{v}_{1\beta}^{n}$ are prescribed moment and transverse force, respectively, applied at the discrete point $(1,\beta)$ on the natural boundary.

13.1.3 Assemblage

With the discrete kinematic transition conditions in mind, then assemble the discrete element equilibrium equations (13.8) for all elements, discrete natural transition conditions, and discrete natural boundary conditions, an overall discrete equation system can be obtained. It is the overall equilibrium/transition/boundary equation or the overall stiffness equation. This overall stiffness equation can be represented by Eq. (3.18).

Like FEM, the assemblage is based on an element by element procedure. In assembling the discrete equations of element e, the discrete element equilibrium equations (13.8), and the discrete element boundary forces, expressed by displacements, at the nodes on the four element boundary edges are directly assembled to the overall discrete equation system. An element basis explicit element stiffness equation, containing the discrete element equilibrium equations and the discrete element boundary forces placed at the rows with the assigned DOF related to the corresponding discrete element boundary forces, is not necessary to be formed in the assemblage process. This element basis explicit matrix equation is an element stiffness equation which is represented by Eq. (3.20), in which $[k^e]$ is a $\bar{N}_\xi^e \times N_\eta^e + \bar{N}_\eta^e \times N_\xi^e$ element stiffness matrix,

$$\{\delta^e\} = \lfloor \ \tilde{w}_1^e \quad \tilde{w}_2^e \quad . \quad \dots \quad . \ \rfloor^T \tag{13.18}$$

is the element displacement vector, and $\{r^e\}$ is the element load vector containing the corresponding loading forces and discrete element boundary forces.

13.1.4 Problems

A clamped square plate with the length of the edge a subjected to a uniformly distributed load q was analyzed. The plate is isotropic with the flexural rigidity D. The p refinement procedure is used to analyze the problem. One DQEM element is used to model the plate domain. Explicit weighting coefficients for Lagrange DQ model with the node points equally spaced are used to generate the $C^1-C^0-C^1$ EDQ model with the two auxiliary nodes inside the equivalent DQ element. The resulting weighting coefficients are used for the element basis discretization.

Numerical results of the center displacements are listed in Table 13.1. It shows that the DQEM results converge fast to the analytical solutions by gradually increasing the order of EDQ approximation.

Table 13.1. Results of the DQEM analysis of the uniformly loaded clamped plate

Element grid line	DOF	$\pi^6 Dw/(16qa^4)$
5×5	9	$.75517\times10^{-1}$
7×7	25	$.76046\times10^{-1}$
9×9	49	$.76022\times10^{-1}$
11×11	81	$.76028\times10^{-1}$
Analytical solution		$.76028\times10^{-1}$

13.2 Free Vibration Analysis

Let ρ and ω denote the mass density and natural frequency, respectively, of the plate. Then, the differential eigenvalue equation can be obtained by using the modal displacement W and inertia force $\rho\delta\omega^2 W$ to replace the displacement w and externally distributed load p, respectively, in Eq. 13.3. Consider the related dynamic boundary condition of the static boundary condition (13.5). Assume that a distributed mass with $\tilde{M}(y)$ and $\tilde{I}(y)$ the mass and moment of inertia of the mass, respectively, per unit length exists along the edge. Then, the related dynamic boundary condition can be obtained by using the modal displacement W and inertia force $-\tilde{I}(y)\omega^2 \frac{\partial W}{\partial x}$ to replace the displacement w and externally distributed moment $\bar{m}$, respectively, in Eq. 13.5. Also consider the related dynamic boundary condition of the static boundary condition (13.6). Then, the related dynamic boundary condition can be obtained by using the modal displacement W and inertia force $\tilde{M}(y)\omega^2 W$ to replace the displacement w and externally distributed transverse force $\bar{v}$, respectively, in Eq. 13.6. The DQEM formulation for the free vibration analysis of plates can also be carried out using the introduced information [127].

13.2.1 Problems

Free vibration of a paneling shown in Fig. 13.3 with $a = 24\ in$ was also solved. All boundary lines are simply supported. There exists two interior support lines at $y = 12\ in$ and $x = 12\ in$. The two interior support lines and the domain boundary lines separate the plate into three elements. The subdomain with $y \geq 12\ in$ is element 1. The subdomain with $x \leq 12\ in$ and $y \leq 12\ in$ is element 2. The subdomain with $x \geq 12\ in$ is element 3. The material constants are: Young's modulus $E = 1. \times 10^7\ lb/in^2$ and Poisson's ratio $\nu = .33$. The thickness of the plate is $\delta = 0.5\ in$. The values of mass density is $10.52\ slugs/ft^3$. Each panel has the same material and thickness. The DQEM element model used to solve the previous sample problem is adopted for this

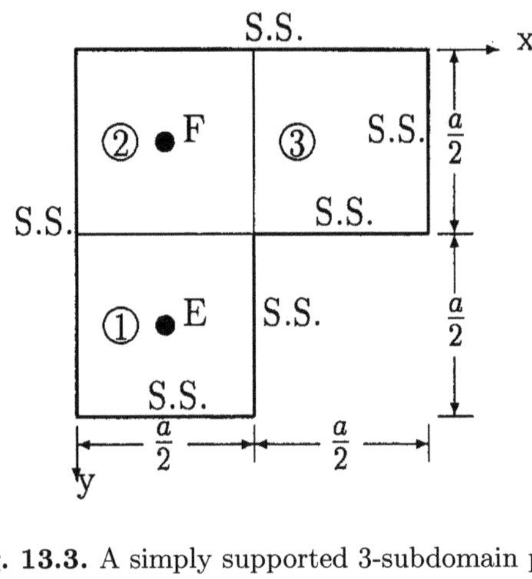

Fig. 13.3. A simply supported 3-subdomain plate

Table 13.2. Natural frequencies of a 3-subdomain paneling (cycles/sec)

Element grid line	ω_1	ω_2	ω_3	ω_4
3×3	$.7513254\times10^7$	$.9322435\times10^7$	$.1321467\times10^8$	
5×5	$.8734809\times10^7$	$.1068452\times10^8$	$.1431324\times10^8$	$.4253762\times10^8$
9×9	$.8658938\times10^7$	$.1061133\times10^8$	$.1427745\times10^8$	$.5412661\times10^8$
13×13	$.8659160\times10^7$	$.1061039\times10^8$	$.1427718\times10^8$	$.5412493\times10^8$

analysis. Three elements are used to model the plate and representing the three subdomains. The first four natural frequencies obtained by the DQEM are listed in Table 13.2. The convergence property is also excellent.

14

DQFDM Analysis

The finite difference operators are derived by the differential quadrature method. They can be obtained by using the weighting coefficients for differential quadrature discretizations. The derivation is straight and easy. By using different order or the same order but different grid differential quadrature discretizations for the same derivative or partial derivative, various finite difference operators for the same differential or partial differential operators can be obtained. Finite difference operators for unequally spaced and irregular grids can also be generated. The derivation of higher order finite difference operators is also easy. Numerical procedures of this DQFDM are summarized. The DQFDM can be used to develop various numerical algorithms for solving various scientific or engineering problems.

14.1 DQ Derivation of Finite Difference Operators

In this section, the derivations of certain difference operators by DQ are illustrated [39]. The Lagrange DQ model is used. For the nondimensional one-dimension grid model having one unit domain length. If the grid has two discrete points, the weighting coefficient matrix is

$$\left[{}^{2}D^{\xi}\right] = \begin{bmatrix} -1 & 1 \\ -1 & 1 \end{bmatrix} \tag{14.1}$$

If the grid has three equally spaced discrete points, the weighting coefficient matrices are

$$\left[{}^{3}D^{\xi}\right] = \begin{bmatrix} -3 & 4 & -1 \\ -1 & 0 & 1 \\ 1 & -4 & 3 \end{bmatrix}, \quad \left[{}^{3}D^{\xi^2}\right] = \begin{bmatrix} 4 & -8 & 4 \\ 4 & -8 & 4 \\ 4 & -8 & 4 \end{bmatrix} \tag{14.2}$$

If the grid has five equally spaced discrete points, the weighting coefficient matrices are

$$\left[{}^{5}D^{\xi^2}\right] = \frac{1}{3} \times \begin{bmatrix} 140 & -416 & 456 & -224 & 44 \\ 44 & -80 & 24 & 16 & -4 \\ -4 & 64 & -120 & 64 & -4 \\ -4 & 16 & 24 & -80 & 44 \\ 44 & -224 & 456 & -416 & 140 \end{bmatrix},$$

$$\left[{}^{5}D^{\xi^3}\right] = \begin{bmatrix} -160 & 576 & -768 & 448 & -96 \\ -96 & 320 & -386 & 192 & -32 \\ -32 & 64 & 0 & -64 & 32 \\ 32 & -192 & 384 & -320 & 96 \\ 96 & -448 & 768 & -576 & 160 \end{bmatrix},$$

$$\left[{}^{5}D^{\xi^4}\right] = \begin{bmatrix} 256 & -1024 & 1536 & -1024 & 256 \\ 256 & -1024 & 1536 & -1024 & 256 \\ 256 & -1024 & 1536 & -1024 & 256 \\ 256 & -1024 & 1536 & -1024 & 256 \\ 256 & -1024 & 1536 & -1024 & 256 \end{bmatrix} \tag{14.3}$$

First consider the problem having only one coordinate variable x. Let h be the distance between two discrete points for three consecutive points $j-1$, j and $j+1$. Then, by scaling and by using the first row of $\left[{}^{2}D^{\xi}\right]$, the forward-difference formula for the first order derivative of the variable function ϕ with respect to the coordinate variable at the discrete point j can be obtained

$$\frac{d\phi}{dx}\mid_j = \frac{1}{h}\frac{d\phi}{d\xi}\mid_j = \frac{1}{h}(\phi_{j+1} - \phi_j) \tag{14.4}$$

The backward-difference formula can also be obtained by using the second row of $\left[{}^{2}D^{\xi}\right]$

$$\frac{d\phi}{dx}\mid_j = \frac{1}{h}\frac{d\phi}{d\xi}\mid_j = \frac{1}{h}(\phi_j - \phi_{j-1}) \tag{14.5}$$

and the central-difference formula can be obtained by using the second row of $\left[{}^{3}D^{\xi}\right]$

$$\frac{d\phi}{dx}\mid_j = \frac{1}{2h}\frac{d\phi}{d\xi}\mid_j = \frac{1}{2h}(\phi_{j+1} - \phi_{j-1}) \tag{14.6}$$

The difference expression for the second order derivative can be obtained by using the second row of $\left[{}^{3}D^{\xi^3}\right]$

$$\frac{d^2\phi}{dx^2}\mid_j = \frac{1}{(2h)^2}\frac{d^2\phi}{d\xi^2}\mid_j = \frac{1}{h^2}(\phi_{j-1} - 2\phi_j + \phi_{j+1}) \tag{14.7}$$

The 5-node grid is used to define the third order derivative. Then, from the third row of $\left[{}^{5}D^{\xi^3}\right]$, the difference expression can be obtained

$$\frac{d^3\phi}{dz^3} = \frac{1}{(4h)^3}\frac{d^3\phi}{d\xi^3}\mid_j = \frac{1}{2h^3}(-\phi_{j-2} + 2\phi_{j-1} - 2\phi_{j+1} + \phi_{j+2}) \tag{14.8}$$

The difference expression for the fourth order derivative can also be obtained by using the third row of $\left[{}^5D^{\xi^4}\right]$

$$\frac{d^4\phi}{dx^4}\Big|_j = \frac{1}{(4h)^4}\frac{d^4\phi}{d\xi^4}\Big|_j = \frac{1}{h^4}(\phi_{j-2} - 4\phi_{j-1} + 6\phi_j - 4\phi_{j+1} + \phi_{j+2}) \qquad (14.9)$$

Next consider the case where ϕ is a function of the two coordinate variables x and y. Assume that the distances between two discrete points in both x and y directions have the same value, h. Let $j-1$, j and $j+1$ be three consecutive discrete points on a grid line in x direction. Also, let $k-1$, k and $k+1$ be three consecutive discrete points on a grid line in y direction. Then, the partial derivative $\frac{\partial^2\phi}{\partial x\partial y}$ at the point (x_j, y_k) can be defined by

$$\frac{\partial^2\phi}{\partial x\partial y}\Big|_{j,k} = \frac{1}{4h^2}\frac{\partial^2\phi}{\partial\xi\partial\eta}\Big|_{j,k} = \frac{1}{4h^2}\,{}^3D^{\eta}_{2n}\,{}^3D^{\xi}_{2m}\,\phi_{(j+m-2)(k+n-2)} \qquad (14.10)$$

The above equation is a difference expression with a range of three on both summed indices m and n. The finite difference operator $\frac{\partial^2}{\partial x\partial y}\big|_{j,k}$ can be represented in the form of the coefficient pattern $4h^2\frac{\partial^2}{\partial x\partial y}\big|_{j,k}$. This coefficient pattern can be expanded by using the second row of $\left[{}^3D^{\xi}\right]$; see Fig. 14.1.

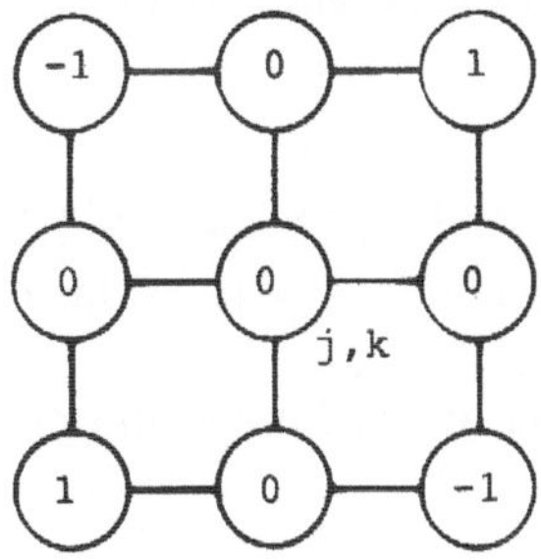

Fig. 14.1. Coefficient pattern $4h^2\frac{\partial^2}{\partial x\partial y}\big|_{j,k}$

The partial derivative $\frac{\partial^3\phi}{\partial x^2\partial y}$ at the point (x_j, y_k) can also be defined by the following difference expression:

$$\frac{\partial^3\phi}{\partial x^2\partial y}\Big|_{j,k} = \frac{1}{8h^3}\frac{\partial^3\phi}{\partial\xi^2\partial\eta}\Big|_{j,k} = \frac{1}{8h^3}\,{}^3D^{\eta}_{2n}\,{}^3D^{\xi^3}_{2m}\,\phi_{(j+m-2)(k+n-2)} \qquad (14.11)$$

where the range on both summed indices m and n is three. Then, the finite difference operator $\frac{\partial^3}{\partial x^2\partial y}\big|_{j,k}$ can be represented in the form of the coefficient pattern $2h^3\frac{\partial^3}{\partial x^2\partial y}\big|_{j,k}$. This coefficient pattern can be expressed by using the second row of $\left[{}^3D^{\xi}\right]$ and $\left[{}^3D^{\xi^2}\right]$; see Fig. 14.2.

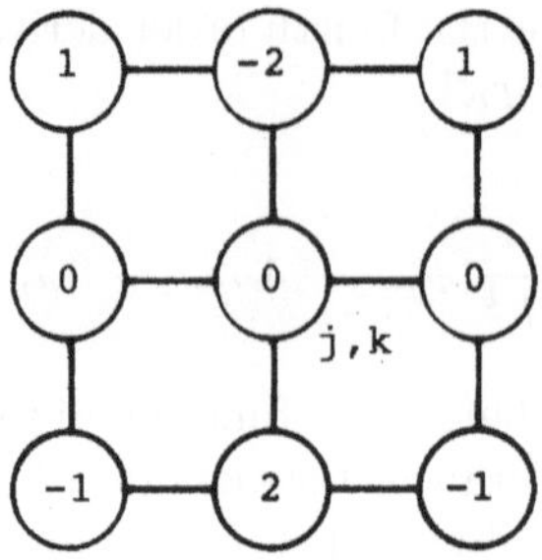

Fig. 14.2. Coefficient pattern $2h^3 \frac{\partial^3}{\partial x^2 \partial y} \mid_{j,k}$

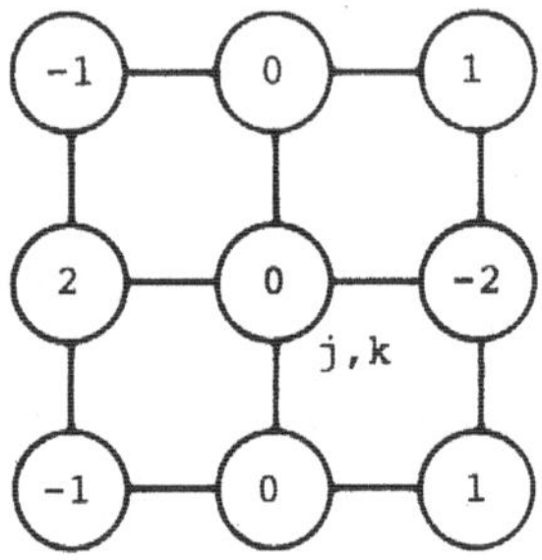

Fig. 14.3. Coefficient pattern $2h^3 \frac{\partial^3}{\partial x \partial y^2} \mid_{j,k}$

The finite difference operator $\frac{\partial^3}{\partial x \partial y^2} \mid_{j,k}$ can be similarly represented in the form of the coefficient pattern $2h^3 \frac{\partial^3}{\partial x \partial y^2} \mid_{j,k}$. This coefficient pattern is shown in Fig. 14.3.

The difference expression of the Laplacian operation $\nabla^2 \phi \mid_{j,k}$ can be defined by

$$\begin{aligned} \nabla^2 \phi \mid_{j,k} &= \frac{1}{4h^2} \left[{}^3D^{\xi^2}_{2m} \phi_{(j+m-2)k} + {}^3D^{\eta^2}_{2n} \phi_{j(k+n-2)} \right] \\ &= \frac{1}{4h^2} \left[{}^3D^{\xi^2}_{2m} \delta_{k(k+n-2)} + {}^3D^{\eta^2}_{2n} \delta_{j(j+m-2)} \right] \phi_{(j+m-2)(k+n-2)} \end{aligned} \tag{14.12}$$

where the range on both summed indices m and n is three. Then, the finite difference operator $\nabla^2 \mid_{j,k}$ can be represented in the form of the coefficient pattern $h^2 \nabla^2 \mid_{j,k}$. This coefficient pattern can be expanded by using the second row of $\left[{}^3D^{\xi^2} \right]$; see Fig. 14.4.

Also let $j-2$, $j-1$, j, $j+1$ and $j+2$ be five consecutive discrete points on a grid line in x direction, and let $k-2$, $k-1$, k, $k+1$ and $k+2$ be five consecutive discrete points on a grid line in y direction. Assume that all line spaces have the same value h. The difference expression of the biharmonic

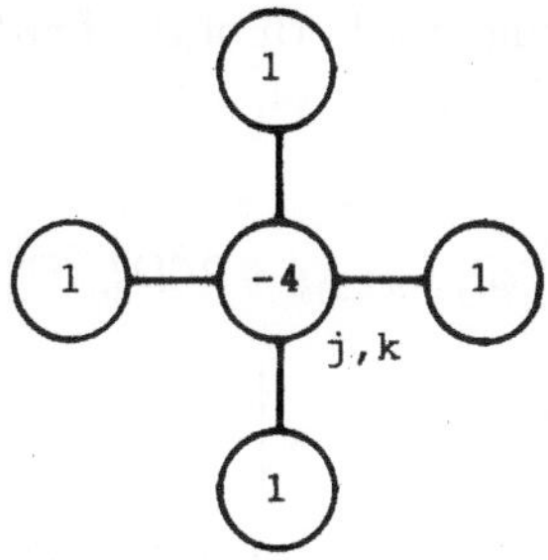

Fig. 14.4. Coefficient pattern $h^2\nabla^2 \mid_{j,k}$

operation $\nabla^4\phi \mid_{j,k}$ can be defined by using five discrete points for the fourth order derivatives and three discrete points for the second order derivatives

$$\nabla^4\phi \mid_{j,k} = \frac{1}{(4h)^4}\left[{}^5D^{\xi^4}_{3m}\phi_{(j+m-3)k} + 32\,{}^3D^{\eta^2}_{2q}\,{}^3D^{\xi^2}_{2p}\phi_{(j+p-2)(k+q-2)} \right.$$
$$\left. + {}^5D^{\eta^4}_{3n}\phi_{j(k+n-3)}\right] \tag{14.13}$$

where the range on the summed indices m and n is five while the range on the summed indices p and q is three. Then, the finite difference operator $\nabla^4 \mid_{j,k}$ can be represented in the form of the coefficient pattern $h^4\nabla^4 \mid_{j,k}$. This coefficient pattern can be expanded by using the second row of $\left[{}^3D^{\xi^2}\right]$ and the third row of $\left[{}^5D^{\xi^4}\right]$; see Fig. 14.5.

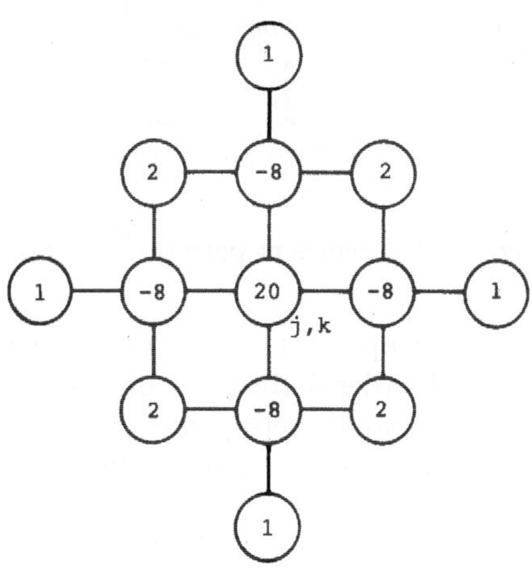

Fig. 14.5. Coefficient pattern $h^4\nabla^4 \mid_{j,k}$

By using five discrete points for both of the fourth order and second order derivatives, another difference expression of the biharmonic operation $\bar{\nabla}^4\phi\mid_{j,k}$ can be defined

$$\begin{aligned}\bar{\nabla}^4\phi\mid_{j,k} &= \frac{1}{(4h)^4}\Big[{}^5D^{\xi^4}_{3m}\phi_{(j+m-3)k} + 2\,{}^5D^{\eta^2}_{3n}\,{}^5D^{\xi^2}_{3m}\phi_{(j+m-3)(k+n-3)} \\ &\qquad + {}^5D^{\eta^4}_{3n}\phi_{j(k+n-3)}\Big] \\ &= \frac{1}{256h^4}\Big[{}^5D^{\xi^4}_{3m}\delta_{k(k+n-3)} + 2\,{}^5D^{\eta^2}_{3n}\,{}^5D^{\xi^2}_{3m} + {}^5D^{\eta^4}_{3n}\delta_{j(j+m-3)}\Big] \\ &\qquad \times\phi_{(j+m-3)(k+n-3)} \end{aligned} \tag{14.14}$$

where the range on both summed indices m and n is five. The corresponding finite difference operator $\bar{\nabla}^4\mid_{j,k}$ can be represented in the form of the coefficient pattern $\frac{h^4}{72}\bar{\nabla}^4\mid_{j,k}$. This coefficient pattern can be expanded by using the third row of $\left[{}^5D^{\xi^2}\right]$ and $\left[{}^5D^{\xi^4}\right]$; see Fig. 14.6.

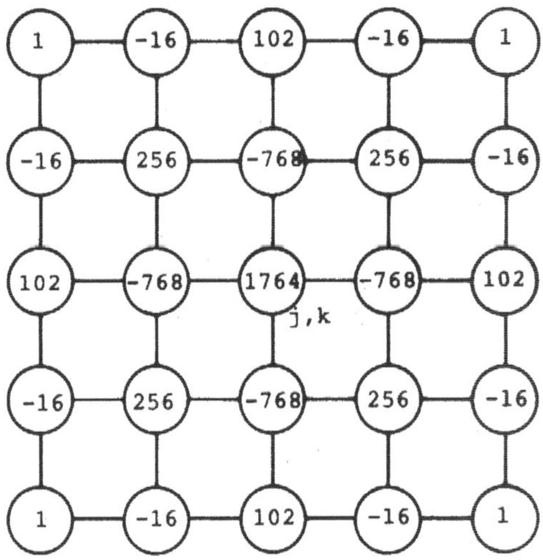

Fig. 14.6. Coefficient pattern $\frac{h^4}{72}\bar{\nabla}^4\mid_{j,k}$

One more finite difference operator $\tilde{\nabla}^4\mid_{j,k}$ can be obtained by using five discrete points for the fourth order derivatives and three discrete points, having the distance $2h$ between two discrete points, for the second order derivatives. This finite difference operator is expressed by

$$\begin{aligned}\tilde{\nabla}^4\phi\mid_{j,k} &= \frac{1}{(4h)^4}\Big[{}^5D^{\xi^4}_{3m}\phi_{(j+m-3)k} + 2\,{}^3D^{\eta^2}_{2q}D^{\xi^2}_{2p}\phi_{(j+2p-4)(k+2q-4)} \\ &\qquad + {}^5D^{\eta^4}_{3n}\phi_{j(k+n-3)}\Big] \end{aligned} \tag{14.15}$$

where the range on the summed indices m and n is five while the range on the summed indices p and q is three. The derivation of higher order finite difference operators is also easy. To illustrate how this concept works, let ..., $j-3$, $j-2$, $j-1$, j, $j+1$, $j+2$, $j+3$, ... be N consecutive discrete points on a grid line in the x direction, and let ..., $j-3$, $j-2$, $j-1$, j, $j+1$, $j+2$, $j+3$, ... be N consecutive discrete points on a grid line in the y direction. Assume that all line spaces have the same value, h. Then, a higher order finite difference expression of the biharmonic operation $\hat{\nabla}^4\phi\mid_{j,k}$ can be defined by using N discrete points for both the fourth and second order derivatives

$$\begin{aligned}\hat{\nabla}^N\phi\mid_{j,k} &= \frac{1}{[(N-1)h]^4}\Big[{}^N D^{\xi^4}_{[(N+1)/2]m}\phi_{(j+m-3)k} \\ &\quad +2\,{}^N D^{\eta^2}_{[(N+1)/2]n}\,{}^N D^{\xi^2}_{[(N+1)/2]m}\phi_{(j+m-3)(k+n-3)} \\ &\quad + {}^N D^{\eta^4}_{[(N+1)/2]n}\phi_{j(k+n-3)}\Big] \\ &= \frac{1}{(N-1)^4h^4}\Big[{}^N D^{\xi^4}_{[(N+1)/2]m}\delta_{k(k+n-3)} \\ &\quad +2\,{}^N D^{\eta^2}_{[(N+1)/2]n}\,{}^N D^{\xi^2}_{[(N+1)/2]m} \\ &\quad + {}^N D^{\eta^4}_{[(N+1)/2]n}\delta_{j(j+m-3)}\Big]\phi_{(j+m-3)(k+n-3)} \qquad (14.16)\end{aligned}$$

14.2 DQFDM Plate Analysis

14.2.1 DQFDM Formulation

Consider the static deflection problem [40]. Let... , $j-3$, $j-2$, $j-1$, j, $j+1$, $j+2$, $j+3$,... be consecutive discrete points on a grid line in the x direction, and let ..., $k-3$, $k-2$, $k-1$, k, $k+1$, $k+2$, $k+3$,... be consecutive discrete points on a grid line in the y direction. Assume that the discrete points in x direction have the same distance h_x between two consecutive points, while the discrete points in y direction have the same distance h_y between two consecutive points. Also assume that $D^{\xi^p}_{mn}$ and $D^{\eta^q}_{mn}$ are the weighting coefficients for the pth and qth order partial derivatives with respect to ξ and η, respectively, with the range for both ξ and η being 1. Let the orders of Lagrange polynomials for defining $D^{\xi^p}_{mn}$ and $D^{\eta^q}_{mn}$ be M and N, respectively. Also let j and k be the rth and sth DQ discrete points, respectively. Then, the finite difference equation of Eq. (13.3) at the discrete point (j,k) can be derived by using the DQ discretization

$$\begin{aligned}&\frac{\bar{D}_{11,jk}}{M^4h_x^4}D^{\xi^4}_{rm}w_{(j+m-r)k} + \frac{4\bar{D}_{16,jk}}{M^3Nh_x^3h_y}D^{\xi^3}_{rm}D^{\eta}_{sn}w_{(j+m-r)(k+n-s)} \\ &+ \frac{2(\bar{D}_{12,jk}+2\bar{D}_{66,jk})}{M^2N^2h_x^2h_y^2}D^{\xi^2}_{rm}D^{\eta^2}_{sn}w_{(j+m-r)(k+n-s)}\end{aligned}$$

$$
\begin{aligned}
&+\frac{4\bar{D}_{26,jk}}{MN^3h_xh_y^3}D^{\xi}_{rm}D^{\eta^3}_{sn}w_{(j+m-r)(k+n-s)}\\
&+\frac{\bar{D}_{22,jk}}{N^4h_y^4}D^{\eta^4}_{sn}w_{j(k+n-s)}+\frac{2}{M^3h_x^3}\left(\frac{\partial\bar{D}_{11,jk}}{\partial x}+\frac{\partial\bar{D}_{16,jk}}{\partial y}\right)D^{\xi^3}_{rm}w_{(j+m-r)k}\\
&+\frac{2}{M^2Nh_x^2h_y}\left[3\frac{\partial\bar{D}_{16,jk}}{\partial x}+\frac{\partial(\bar{D}_{12,jk}+2\bar{D}_{66,jk})}{\partial y}\right]D^{\xi^2}_{rm}D^{\eta}_{sn}w_{(j+m-r)(k+n-s)}\\
&+\frac{2}{MN^2h_xh_y^2}\left[\frac{\partial(\bar{D}_{12,jk}+2\bar{D}_{66,jk})}{\partial x}+3\frac{\partial\bar{D}_{26,jk}}{\partial y}\right]D^{\xi}_{rm}D^{\eta^2}_{sn}w_{(j+m-r)(k+n-s)}\\
&+\frac{2}{N^3h_y^3}\left(\frac{\partial\bar{D}_{22,jk}}{\partial y}+\frac{\partial\bar{D}_{26,jk}}{\partial x}\right)D^{\eta^3}_{sn}w_{j(k+n-s)}\\
&+\frac{1}{M^2h_x^2}\left(\frac{\partial^2\bar{D}_{11,jk}}{\partial x^2}+\frac{\partial^2\bar{D}_{16,jk}}{\partial x\partial y}+\frac{\partial^2\bar{D}_{12,jk}}{\partial y^2}\right)D^{\xi^2}_{rm}w_{(j+m-r)k}\\
&+\frac{2}{MNh_xh_y}\left(\frac{\partial^2\bar{D}_{16,jk}}{\partial x^2}+2\frac{\partial^2\bar{D}_{66,jk}}{\partial x\partial y}+\frac{\partial^2\bar{D}_{26,jk}}{\partial y^2}\right)D^{\xi}_{rm}D^{\eta}_{sn}w_{(j+m-r)(k+n-s)}\\
&+\frac{1}{N^2h_y^2}\left(\frac{\partial^2\bar{D}_{12,jk}}{\partial x^2}+2\frac{\partial^2\bar{D}_{26,jk}}{\partial x\partial y}+\frac{\partial^2\bar{D}_{22,jk}}{\partial y^2}\right)D^{\eta^2}_{sn}w_{j(k+n-s)}=p_{jk}
\end{aligned}
\tag{14.17}
$$

In solving the plate problems, if the distance from the discrete point (j,k) to an x edge is larger than $\frac{(M-3)h_x}{2}$, then r can be equal to $\frac{M-1}{2}$ which is a central difference in the x direction. Otherwise, r can not be equal to $\frac{M-1}{2}$ and the central difference can not be used.

Similar concepts can be adopted for designing the finite difference model in y direction. In addition to the boundary line, one more grid line is necessary for defining the finite difference equations of the boundary conditions. This grid line can be either one auxiliary grid line just outside the plate and parallel to the boundary edge or the interior grid line next to the boundary line. Assume that the auxiliary grid line is added, and that the edge $x=0$ is a boundary line. Then the finite difference equations of the boundary conditions (13.4) for the edge $x=0$ at a discrete point $(1,\beta)$ can be expressed by

$$
w_{2\beta}=\bar{w}_{2\beta}\quad\text{and}\quad\frac{1}{Mh_x}D^{\xi}_{2m}w_{(2+m-r)\beta}=-\bar{\theta}_{y2\beta}
\tag{14.18}
$$

where $-\bar{\theta}_{y2\beta}$ is the prescribed rotation angle along the boundary line toward the y coordinate direction. The finite difference equations of boundary conditions (13.5) and (13.6) at a discrete point $(1,\beta)$ on the edge can be similarly expressed as

$$\frac{\bar{D}_{11,2\beta}}{M^2h_x^2}D_{2m}^{\xi^2}w_{(2+m-r)\beta}+2\frac{\bar{D}_{12,2\beta}}{N^2h_y^2}D_{sn}^{\eta^2}w_{2(\beta+n-s)}$$
$$+\frac{\bar{D}_{16,2\beta}}{MNh_xh_y}D_{2m}^{\xi}D_{sn}^{\eta}w_{(2+m-r)(\beta+n-s)}=\bar{m}_{x,2\beta} \tag{14.19}$$

and

$$\frac{\bar{D}_{11,2\beta}}{M^3h_x^3}D_{2m}^{\xi^3}w_{(2+m-r)\beta}+4\frac{\bar{D}_{16,2\beta}}{M^2Nh_x^2h_y}D_{2m}^{\xi^2}D_{sn}^{\eta}w_{(2+m-r)(\beta+n-s)}$$
$$+\frac{(\bar{D}_{12,2\beta}+4\bar{D}_{66,2\beta})}{MN^2h_xh_y^2}D_{2m}^{\xi}D_{sn}^{\eta^2}w_{(2+m-r)(\beta+n-s)}$$
$$+\frac{1}{M^2h_x^2}\left(\frac{\partial\bar{D}_{11,2\beta}}{\partial x}+\frac{\partial\bar{D}_{16,2\beta}}{\partial y}\right)D_{2m}^{\xi^2}w_{(2+m-r)\beta}$$
$$+\frac{1}{MNh_xh_y}\left(\frac{\partial\bar{D}_{16,2\beta}}{\partial x}+2\frac{\partial\bar{D}_{66,2\beta}}{\partial y}\right)D_{2m}^{\xi}D_{sn}^{\eta}w_{(2+m-r)(\beta+n-s)}$$
$$+\frac{1}{N^2h_y^2}\left(\frac{\partial(\bar{D}_{12,2\beta}}{\partial x}+2\frac{\partial\bar{D}_{26,2\beta}}{\partial y}\right)D_{sn}^{\eta^2}w_{2(\beta+n-s)}=\bar{v}_{x,2\beta} \tag{14.20}$$

In the analysis using higher algorithms, forward or backward discretization is necessary for discretizing a fundamental equation defined at a discrete point on a discrete line near or on the domain boundary. In real application it is not necessary that the values of M and N are fixed in the sense that different terms in the finite difference equations can adopt different values of M and N. It should be mentioned that the partial derivatives of plate rigidities at a discrete point can also be approximated by the finite difference discretization.

It should be noted that the numerical formulation for the DQFDM analysis of plate problems can also be on the element basis. Consequently, finite difference operators can be defined on an element basis and the discrete DQFDM analysis can thus be carried out.

14.2.2 Problems

The problem can be solved with or without adding an auxiliary grid line outside the domain and near the boundary line. If an auxiliary grid line is added , discrete points on this grid line and the boundary line are used to define the two boundary conditions at a discrete point on the boundary line. On the other hand, if an auxiliary grid line is not added, discrete points on the boundary line and the interior grid line next to the boundary line are used to define the two boundary conditions at a discrete point on the boundary line. Various DQFDM algorithms can be constructed by adjusting the values of M and N for different terms in the finite difference equations. In carrying out the numerical tests, four different algorithms are used. In the algorithm A1, both M and N are four for discretizing a fourth order partial derivative with respect to a coordinate variable, while both M and N are two for discretizing a first or second order partial derivative with respect to a coordinate variable. In the algorithm A2, both M and N are four for discretizing a second or fourth order partial derivative, existing in the governing equation, with respect to a coordinate variable. However, in discretizing the boundary conditions, both M and N are two for discretizing a first or second order partial derivative with respect to a coordinate variable. In the algorithm A3, both M and N are four for discretizing a first, second or fourth order partial derivative, existing in the equilibrium equation or boundary conditions, with respect to a coordinate variable. In the algorithm A4, both M and N are four for discretizing a fourth order partial derivative existing in the equilibrium equation and a second order partial derivative existing in the boundary conditions, with respect to a coordinate variable. However, in discretizing the second order partial derivative, existing in the equilibrium equation, with respect to a coordinate variable both M and N are two with the discretizing grid distance twice of other difference discretizations. For all of algorithms A1, A2, A3 and A4, an auxiliary grid line is added outside the domain and near the boundary line.

A simply supported isotropic uniform square plate subjected to a uniformly distributed load p_0 was solved by using Algorithms A1, A2, A3 and A4, separately. The four boundary edges of the square plates are $x = 0$, $x = a$, $y = 0$ and $y = a$. Numerical results of the nondimensional maximum displacement are summarized and listed in Table 14.1. It shows that the results converge to exact solutions and that the algorithm A3 performs the best among the algorithms A1, A2, A3 and A4 which have the same maximum order of approximation due to the consistent discretization of adopting the same M and N for all finite difference discretizations [126].

Table 14.1. The nondimensional maximum displacement $w_{max} \times \frac{\pi^6 D}{16 p_0 a^4}$ of a simply supported isotropic uniform square plate subjected to a uniformly distributed lateral load

DQFDM grid line	Algorithm A1	Algorithm A2	Algorithm A3	Algorithm A4
3×3	$.234714 \times 10^0$	$.199403 \times 10^0$	$.177982 \times 10^0$	$.395308 \times 10^0$
5×5	$.242049 \times 10^0$	$.231221 \times 10^0$	$.229864 \times 10^0$	$.286312 \times 10^0$
7×7	$.243253 \times 10^0$	$.238257 \times 10^0$	$.233587 \times 10^0$	$.263118 \times 10^0$
9×9	$.243638 \times 10^0$	$.240792 \times 10^0$	$.237595 \times 10^0$	$.254837 \times 10^0$
11×11	$.243807 \times 10^0$	$.241976 \times 10^0$	$.239669 \times 10^0$	$.250982 \times 10^0$
13×13	$.243897 \times 10^0$	$.242622 \times 10^0$	$.240886 \times 10^0$	$.248882 \times 10^0$
15×15	$.243950 \times 10^0$	$.243012 \times 10^0$	$.241661 \times 10^0$	$.247613 \times 10^0$
17×17	$.243984 \times 10^0$	$.243265 \times 10^0$	$.242186 \times 10^0$	$.246790 \times 10^0$
19×19	$.244008 \times 10^0$	$.243439 \times 10^0$	$.242557 \times 10^0$	$.246224 \times 10^0$
Exact solution	$.242666 \times 10^0$			

A clamped isotropic uniform square plate subjected to a uniformly distributed load p_0 was also solved. Higher order algorithms are also used to solve this problem. In the algorithm B3, both M and N are six for discretizing a first, second or fourth order partial derivative, existing in the equilibrium equation or boundary conditions, with respect to a coordinate variable. In the algorithm C3, both M and N are eight for discretizing a first, second or fourth order partial derivative existing in the equilibrium equation or boundary conditions, with respect to a coordinate variable. In the algorithm D3, both M and N are ten for discretizing a first, second or fourth order partial derivative, existing in the equilibrium equation or boundary conditions, with respect to a coordinate variable. Algorithms B3, C3 and D3 also adopt the auxiliary grid line. Numerical results of the nondimensional maximum displacement are summarized and listed in Table 14.2. It also shows that the algorithm A3 performs the best among the algorithms A1, A2 and A3, and that the performance of the algorithm can be improved by increasing the order of approximation.

Table 14.2. The nondimensional maximum displacement $w_{max} \times \frac{\pi^6 D}{16 p_0 a^4}$ of a clamped isotropic uniform square plate subjected to a uniformly distributed lateral load

DQFDM grid line	Algorithm A1	Algorithm A2	Algorithm A3	Algorithm B3
5×5	$.108127\times10^{0}$	$.999138\times10^{-1}$	$.100032\times10^{0}$	$.757052\times10^{-1}$
7×7	$.921963\times10^{-1}$	$.887323\times10^{-1}$	$.873039\times10^{-1}$	$.760169\times10^{-1}$
9×9	$.855878\times10^{-1}$	$.836868\times10^{-1}$	$.818896\times10^{-1}$	$.760577\times10^{-1}$
11×11	$.823009\times10^{-1}$	$.810985\times10^{-1}$	$.792915\times10^{-1}$	$.760498\times10^{-1}$
13×13	$.804471\times10^{-1}$	$.796174\times10^{-1}$	$.779108\times10^{-1}$	$.760417\times10^{-1}$
15×15	$.793041\times10^{-1}$	$.786970\times10^{-1}$	$.771176\times10^{-1}$	$.760369\times10^{-1}$
17×17	$.785516\times10^{-1}$	$.780880\times10^{-1}$	$.766343\times10^{-1}$	$.760341\times10^{-1}$
19×19	$.780306\times10^{-1}$	$.776650\times10^{-1}$	$.763264\times10^{-1}$	
Exact solution	$.760279\times10^{-1}$			

A simply supported uniform anisotropic plate subjected to the same uniformly distributed load was also solved by using the algorithm A3. The plate rigidities have the following relations: $\bar{D}_{12} = \bar{D}_{11}$, $\bar{D}_{12} + 2\bar{D}_{66} = 1.061\bar{D}_{11}$ and $\bar{D}_{16} = \bar{D}_{26} = -0.174\bar{D}_{11}$. Numerical results are listed in Table 14.3. It shows that the algorithm A3 converges rather slow for the solution of anisotropic plate problem as compared to the solution of isotropic plate problem. However, it always converge well.

A simply supported isotropic nonuniform square plate subjected to a uniformly distributed load p_0 was also solved. The thickness of the plate is expressed by $\delta(x, y) = \delta_0(1 + x/a)(1 + y/b)$. Denote D_0 as the flexural rigidity

Algorithm C3	Algorithm D3
-	-
$.760585\times10^{-1}$	-
$.760318\times10^{-1}$	$.760254\times10^{-1}$
$.760279\times10^{-1}$	$.760263\times10^{-1}$
$.760278\times10^{-1}$	$.760275\times10^{-1}$
$.760282\times10^{-1}$	$.760283\times10^{-1}$

Table 14.3. The nondimensional displacement $w \times \frac{\bar{D}_{11}}{p_0 a^4}$ of a simply supported anisotropic square plate subjected to a uniformly distributed lateral load

DQFDM grid line	at $(\frac{a}{4}, \frac{a}{4})$	at $(\frac{3a}{8}, \frac{3a}{8})$	at $(\frac{a}{2}, \frac{a}{2})$
5×5	$.194874\times10^{-2}$	-	$.363461\times10^{-2}$
9×9	$.207903\times10^{-2}$	$.337379\times10^{-2}$	$.388439\times10^{-2}$
13×13	$.211296\times10^{-2}$	-	$.394564\times10^{-2}$
21×21	$.213156\times10^{-2}$	-	$.397777\times10^{-2}$
25×25	$.213533\times10^{-2}$	$.346140\times10^{-2}$	$.398422\times10^{-2}$
31×31	-	-	$.399458\times10^{-2}$
Exact solution			$.411\times10^{-2}$

of the plate at the origin $(0, 0)$. Numerical results obtained by the algorithm A3 are summarized and listed in Table 14.4. It also shows that the results converge well.

The free vibration of an isotropic clamped square plate was also solved using the same order of DQFDM approximation, in a specific coordinate direction, to the partial derivatives involved. Let the natural frequency ω_n of

Table 14.4. The nondimensional displacement $w \times \frac{\pi^6 D}{16 p_0 a^4}$ of a simply supported isotropic nonrectangular plate subjected to a uniformly distributed lateral load

DQFDM grid line	at $(\frac{a}{4}, \frac{a}{4})$	at $(\frac{a}{2}, \frac{a}{2})$	at $(\frac{3a}{4}, \frac{3a}{4})$
5×5	$.851025\times10^{-3}$	$.102334\times10^{-2}$	$.380420\times10^{-3}$
7×7	-	$.138524\times10^{-2}$	-
11×11	-	$.178991\times10^{-2}$	-
21×21	$.184495\times10^{-2}$	$.218457\times10^{-2}$	$.791357\times10^{-3}$
25×25	$.189921\times10^{-2}$	$.223957\times10^{-2}$	$.809483\times10^{-3}$
31×31	-	$.233296\times10^{-2}$	-

the nth vibration mode be expressed by $\omega_n = \frac{C_n}{a^2}\sqrt{\frac{D}{\rho\delta}}$ with C_n the frequency factor and δ the depth of the plate. The first four frequency factors obtained by the DQFDM without adding four straight auxiliary grid lines outside the plate domain are listed in Table 14.5. They are compared with those obtained by Leissa [57]. The results of adding four straight auxiliary grid lines outside the plate domain are also obtained and listed in Table 14.6.

The two interior support lines in the simply supported nonrectangular plate shown in Fig. 13.3 were freed and replaced by two line loads. The two line loads are $q_1 = 10sin(\pi x/12)$ applied on the line $y = 12$ in, and

Table 14.5 The first four frequency factors of a clamped square plate (without auxiliary grid lines)

Order of appr.	DQFDM grid line	C_1	C_2	C_3	C_4
4	7×7	.362126×10^2	.701674×10^2	.701674×10^2	.108000×10^3
	9×9	.362358×10^2	.707662×10^2	.707662×10^2	.107348×10^3
	11×11	.361757×10^2	.717800×10^2	.717800×10^2	.107993×10^3
	13×13	.361235×10^2	.723785×10^2	.723785×10^2	.108271×10^3
	15×15	.360868×10^2	.727111×10^2	.727111×10^2	.108352×10^3
	17×17	.360617×10^2	.729048×10^2	.729048×10^2	.108359×10^3
	19×19	.360442×10^2	.730250×10^2	.730250×10^2	.108343×10^3
6	7×7	.368344×10^2	.650341×10^2	.650341×10^2	.901151×10^2
	9×9	.363784×10^2	.734359×10^2	.734359×10^2	.108156×10^3
	11×11	.361481×10^2	.743960×10^2	.743960×10^2	.110417×10^3
	13×13	.360594×10^2	.740078×10^2	.740078×10^2	.109540×10^3
	15×15	.360226×10^2	.737245×10^2	.737245×10^2	.108971×10^3
	17×17	.360056×10^2	.735735×10^2	.735735×10^2	.108656×10^3
	19×19	.359971×10^2	.734949×10^2	.734949×10^2	.108482×10^3
8	11×11	.360123×10^2	.741418×10^2	.741418×10^2	.109729×10^3
	13×13	.359998×10^2	.735130×10^2	.735130×10^2	.108675×10^3
	15×15	.359932×10^2	.734220×10^2	.734220×10^2	.108411×10^3
	17×17	.359898×10^2	.734052×10^2	.734052×10^2	.108316×10^3
	19×19	.359879×10^2	.734005×10^2	.734005×10^2	.108272×10^3
10	11×11	.359929×10^2	.730596×10^2	.730596×10^2	.125345×10^3
	13×13	.359907×10^2	.733625×10^2	.733625×10^2	.108343×10^3
	15×15	.359882×10^2	.734025×10^2	.734025×10^2	.108288×10^3
	17×17	.359868×10^2	.734016×10^2	.734016×10^2	.108254×10^3
	19×19	.359861×10^2	.733985×10^2	.733985×10^2	.108237×10^3
Lessa's sol.		.35992×10^2	.73413×10^2	.73413×10^2	.10827×10^3

Table 14.6 The first four frequency factors of a clamped square plate (with auxiliary grid lines)

Order of appr.	DQFDM subd.	C_1	C_2	C_3	C_4
4	2×2	$.3475629\times10^2$			
	4×4	$.3607781\times10^2$	$.6626038\times10^2$	$.6626038\times10^2$	$.8317439\times10^3$
	6×6	$.3603904\times10^2$	$.7019709\times10^2$	$.7019709\times10^2$	$.1049749\times10^3$
	8×8	$.3604560\times10^2$	$.7145926\times10^2$	$.7145926\times10^2$	$.1065734\times10^3$
	10×10	$.3604122\times10^2$	$.7212046\times10^2$	$.7212046\times10^2$	$.1072856\times10^3$
	12×12	$.3603353\times10^2$	$.7250439\times10^2$	$.7250439\times10^2$	$.1076521\times10^3$
	14×14	$.3602602\times10^2$	$.7274250\times10^2$	$.7274250\times10^2$	$.1078551\times10^3$
	16×16	$.3601958\times10^2$	$.7289844\times10^2$	$.7289844\times10^2$	$.1079743\times10^3$
6	4×4	$.3630833\times10^2$	$.7160144\times10^2$	$.7160144\times10^2$	$.1022950\times10^3$
	6×6	$.3604462\times10^2$	$.7317862\times10^2$	$.7317862\times10^2$	$.1087108\times10^3$
	8×8	$.3598971\times10^2$	$.7334781\times10^2$	$.7334781\times10^2$	$.1084552\times10^3$
	10×10	$.3598075\times10^2$	$.7334579\times10^2$	$.7334579\times10^2$	$.1082798\times10^3$
	12×12	$.3598032\times10^2$	$.7335128\times10^2$	$.7335128\times10^2$	$.1082162\times10^3$
	14×14	$.3598127\times10^2$	$.7336070\times10^2$	$.7336070\times10^2$	$.1082276\times10^3$
	16×16	$.3598223\times10^2$	$.7336915\times10^2$	$.7336915\times10^2$	$.1081947\times10^3$
8	6×6	$.3600669\times10^2$	$.7433538\times10^2$	$.7433538\times10^2$	$.1097962\times10^3$
	8×8	$.3599354\times10^2$	$.7343884\times10^2$	$.7343884\times10^2$	$.1084133\times10^3$
	10×10	$.3598863\times10^2$	$.7339503\times10^2$	$.7339503\times10^2$	$.1082685\times10^3$
	12×12	$.3598676\times10^2$	$.7339449\times10^2$	$.7339449\times10^2$	$.1082385\times10^3$
	14×14	$.3598597\times10^2$	$.7339462\times10^2$	$.7339462\times10^2$	$.1082276\times10^3$
	16×16	$.3598560\times10^2$	$.7339446\times10^2$	$.7339446\times10^2$	$.1082227\times10^3$
10	8×8	$.3599020\times10^2$	$.7335022\times10^2$	$.7335022\times10^2$	$.1082930\times10^3$
	10×10	$.3598723\times10^2$	$.7339637\times10^2$	$.7339637\times10^2$	$.1082685\times10^3$
	12×12	$.3598598\times10^2$	$.7339706\times10^2$	$.7339706\times10^2$	$.1082330\times10^3$
	14×14	$.3598552\times10^2$	$.7339548\times10^2$	$.7339548\times10^2$	$.1082240\times10^3$
	16×16	$.3598553\times10^2$	$.7339464\times10^2$	$.7339464\times10^2$	$.1082202\times10^3$
Lessa's sol.		$.35992\times10^2$	$.73413\times10^2$	$.73413\times10^2$	$.10827\times10^3$

$q_2 = 10sin(\pi y/12)$ applied on the line $x = 12$ in. The material constants are: Young's modulus $E = 1. \times 10^7$ lb/in^2 and Poisson's ratio $\nu = .33$. The thickness of the plate is $\delta = 0.5$ in. The two load lines and domain boundary lines separate the plate into three subregions. The subregion with $y \geq 12$ in is subregion 1. The subregion with $x \leq 12$ in and $y \leq 12$ in is subregion 2. The subregion with $x \geq 12$ in is subregion 3. This nonrectangular plate problem was solved by DQFDM.

In the analysis, the DQFDM uses the same order of approximation to the partial derivatives involved, and no auxiliary grid line added is adopted. On each domain boundary line, the lateral displacement and normal moment must vanish. The active DOF assigned to the two inter-subregion boundary lines are used to define the equilibrium of lateral forces.

The DOF assigned to the four interior grid lines next to the domain boundary lines are used to define the conditions of normal moment free. The DOF assigned to the four interior grid lines next to the two inter-subregion boundary lines are used to define the transition conditions. The DOF assigned to the related interior grid line in subregion 1, except the two DOF assigned to the two end points which are used to define two conditions of normal moment free on the domain boundary, are used to define the condition of slope compatibility. The DOF assigned to the related interior grid line parallel to x axis, in subregion 2, except the two DOF assigned to the two end points, are used to define the equilibrium of moments. The DOF assigned to the end point close to the domain boundary line $x = 0$ is used to define the condition of normal moment free, while the DOF assigned to the other end point is used to define the condition of slope compatibility on the inter-subregion boundary line parallel to y axis. The DOF assigned to the related interior grid line parallel to y axis, in subregion 2, except the two DOF assigned to the two end points, are used to define the condition of slope compatibility. The DOF assigned to the related interior grid line in subregion 3, except the two DOF assigned to the two end points, are used to define the equilibrium of moments.

Numerical results of the displacements at $E(6\ in,\ 12\ in)$ and $F(6\ in,\ 6\ in)$ are summarized and listed in Table 14.6. It shows that a higher-order DQFDM approximation has a better performance.

14.3 DQFDM 2-D Elasticity Problem Analysis

For the 2-D elasticity problems, the partial derivatives involve only the first and second orders with respect to a specific coordinate. Consequently, no auxiliary grid line is necessary for the DQFDM analysis. In addition, the nature of DQFDM formulation is the same as that of DQEM formulation introduced in Chapter 12, except that the span of the DQ grid model for the DQFDM is generally smaller than the element grid, while the DQ grid model for the DQEM is used to represent the element.

14.3.1 Problems

A problem solved involves the plane strain elasticity deformation of a unit square isotropic medium [41]. The four boundary edges are: $x = 0$, $x = 1$, $y = 0$ and $y = 1$. They are fixed edges. The material constants are: Young's modulus $E = 2.5$, Poisson's ratio $\nu = .25$. The body forces are: $\rho b_x = 4\pi^2 sin(\pi x)sin(\pi y) - 2\pi^2 cos(\pi x)cos(\pi y)$, $\rho b_y = 4\pi^2 sin(\pi x)sin(\pi y) -$

$2\pi^2 cos(\pi x)cos(\pi y)$. The exact solutions of displacements are: $u = sin(\pi x)sin(\pi y)$, $v = sin(\pi x)sin(\pi y)$. The displacements u and v are the same at the center of the medium. The DQFDM analysis is carried out by using various DQFDM approximations to solve the problem. Numerical results show that the two displacement components at the center of the medium are the same. They are summarized and listed in Table 14.7. It shows that the DQFDM results converge fast to the exact solution.

Table 14.7. Displacements at the center of a square region having the plain strain deformation behavior

Order of appr.	DQFDM grid line	u_{ctr}
2	3×3	$.1233701\times10^1$
	5×5	$.1048482\times10^1$
	7×7	$.1019951\times10^1$
	9×9	$.1010823\times10^1$
	11×11	$.1006794\times10^1$
	13×13	$.1004665\times10^1$
	15×15	$.1003403\times10^1$
4	5×5	$.9950335\times10^0$
	7×7	$.9987416\times10^0$
	9×9	$.9996459\times10^0$
	11×11	$.9998803\times10^0$
	13×13	$.9999532\times10^0$
6	7×7	$.1000122\times10^1$
	11×11	$.1000011\times10^1$
	13×13	$.1000004\times10^1$
	15×15	$.1000001\times10^1$
Exact solution		$.1000000\times10^1$

A problem involving the in-plane vibration analysis of a square plate having the E glass/epoxy material was also solved. For a 0^0 ply plate with a fiber having the volume fraction 0.70, the effective stiffnesses are: $Q_{11} = 9. \times 10^6$ psi, $Q_{12} = .85 \times 10^6$ psi, $Q_{22} = 3.68 \times 10^6$ psi, $Q_{66} = 1.74 \times 10^6$ psi and $Q_{16} = Q_{26} = 0$. They are constant through the thickness of the plate. The plate is uniform with the thickness being 1 in. The four boundary edges of the plate are: $x = 0$, $x = 10$ ft, $y = 0$ and $y = 10$ ft. The two x-edges are fixed. The two y-edges are free edges. The value of mass density ρ is 4.942259 $slugs/ft^3$. In the analysis, all partial derivatives involved in the DQFDM discretization adopt the same order of approximation in a specific coordinate direction. Numerical computations are carried out using various

Table 14.8. The first two natural frequencies of the free vibration of an anisotropic uniform rectangular plate (cycles/sec)

Order of appr.	DQFDM grid line	ω_1	ω_2
2	7×7	$.2070215\times10^4$	$.3785482\times10^4$
	9×9	$.2105678\times10^4$	$.3740718\times10^4$
	11×11	$.2083700\times10^4$	$.3748755\times10^4$
	13×13	$.2087046\times10^4$	$.3740718\times10^4$
	15×15	$.2089232\times10^4$	$.3742979\times10^4$
4	7×7	$.2109269\times10^4$	$.3721591\times10^4$
	9×9	$.2105678\times10^4$	$.3740718\times10^4$
	11×11	$.2102912\times10^4$	$.3743706\times10^4$
	13×13	$.2100939\times10^4$	$.3743513\times10^4$
	15×15	$.2099510\times10^4$	$.3742758\times10^4$
6	11×11	$.2097632\times10^4$	$.3742400\times10^4$
	13×13	$.2096811\times10^4$	$.3741041\times10^4$
	15×15	$.2096215\times10^4$	$.3740313\times10^4$
8	9×9	$.2080713\times10^4$	$.3754644\times10^4$
	11×11	$.2095710\times10^4$	$.3740861\times10^4$
	13×13	$.2095644\times10^4$	$.3739740\times10^4$
	15×15	$.2096215\times10^4$	$.3740313\times10^4$
11	11×11	$.2089917\times10^4$	$.3744125\times10^4$
	13×13	$.2095041\times10^4$	$.3739494\times10^4$
	15×15	$.2095042\times10^4$	$.3739152\times10^4$

orders of DQFDQ approximation. Numerical results of the first two natural frequencies are summarized and listed in Table 14.8. It also shows that the convergence performance of the DQFDM is excellent.

14.4 DQFDM Analysis of 2-D Heat Conduction Problems

Like the 2-D elasticity problems, the 2-D heat conduction problems also involves only the first and second order partial derivatives with respect to a specific coordinate. Consequently, the nature of the spacial discretization of the DQFDM 2-D heat conduction problems is the same as that of the 2-D elasticity problems. Regards the spacial discretization of the 2-D heat conduction problems, Chapter 11 can be referred.

14.4.1 Problems

A problem involving the 2-D steady state heat conduction in an orthotropic medium, shown in Fig. 14.7 was solved. The values of thermal conductivities k_x and k_y are 1 and 2, respectively. The domain of the medium has three subdomains with the two inter-subdomain boundary lines having a half sine curve distributed heat flux into the medium and one subdomain having a constant heat generation rate. The domain boundary has piecewise constant prescribed temperatures. Numerical results of temperature at three points obtained by DQFDM are summarized and listed in Table 14.9. It shows that the results can converge fast by increasing either the order of approximation or the number of grid lines.

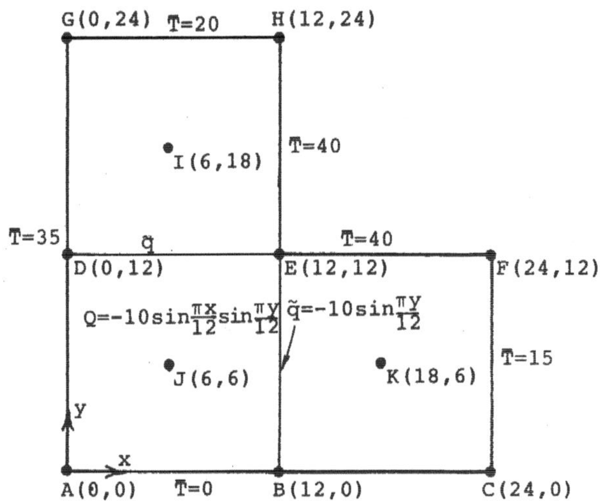

Fig. 14.7. Heat conduction in an orthotropic medium solved by DQFDM

The 2-D heat conduction of a medium having two different isotropic materials shown in Fig. 14.8 was also solved. There are two subdomains having their own materials and uniformly distributed heat generation rates. The shape of each subdomain can be described by the quadratic serendipity shape functions. The problem has both the Neumann and Dirichlet boundaries. There is also a constant heat flux into the medium on the inter-subdomain boundary. At a corner node, one discrete condition equation is considered. At the intersection of the Neumann boundary and the inter-subdomain boundary, the discrete Neumann condition defined on the left subdomain is considered. At D and F, the considered conditions are Dirichlet conditions. At A and C, the considered Dirichlet condition is $\bar{T} = 0$. Various techniques concerning the setup of discrete transition conditions on an inter-subdomain boundary and a corner node are introduced in Section 11.4. Numerical results of temperature

Table 14.9. Convergence of heat conduction in an orthotropic medium solved by DQFDM

Order of appr.	DQFDM grid line (each subregion)	$T_{,I}$	$T_{,J}$	$T_{,K}$
2	3×3	$.5775253\times10^2$	$.1193838\times10^3$	$.3079798\times10^2$
	5×5	$.4462011\times10^2$	$.9022232\times10^2$	$.2365371\times10^2$
	7×7	$.4174511\times10^2$	$.8462334\times10^2$	$.2241048\times10^2$
4	5×5	$.3914354\times10^2$	$.7973436\times10^2$	$.2153298\times10^2$
	7×7	$.3925651\times10^2$	$.7999988\times10^2$	$.2149636\times10^2$
	9×9	$.3933606\times10^2$	$.8012240\times10^2$	$.2148978\times10^2$
6	7×7	$.3944487\times10^2$	$.8027652\times10^2$	$.2151097\times10^2$
	9×9	$.3939666\times10^2$	$.8020003\times10^2$	$.2148732\times10^2$
	11×11	$.3938110\times10^2$	$.8017486\times10^2$	$.2147587\times10^2$
8	9×9	$.3938821\times10^2$	$.8018365\times10^2$	$.2148170\times10^2$
	11×11	$.3937298\times10^2$	$.8103031\times10^2$	$.2147498\times10^2$
	13×13	$.3936902\times10^2$	$.8015594\times10^2$	$.2147306\times10^2$
10	11×11	$.3937603\times10^2$	$.8016577\times10^2$	$.2147587\times10^2$
	13×13	$.3936718\times10^2$	$.8015319\times10^2$	$.2147213\times10^2$
	15×15	$.3936492\times10^2$	$.8014990\times10^2$	$.2147109\times10^2$
12	13×13	$.3936951\times10^2$	$.8015638\times10^2$	$.2147299\times10^2$
	15×15	$.3936390\times10^2$	$.8014839\times10^2$	$.2147061\times10^2$

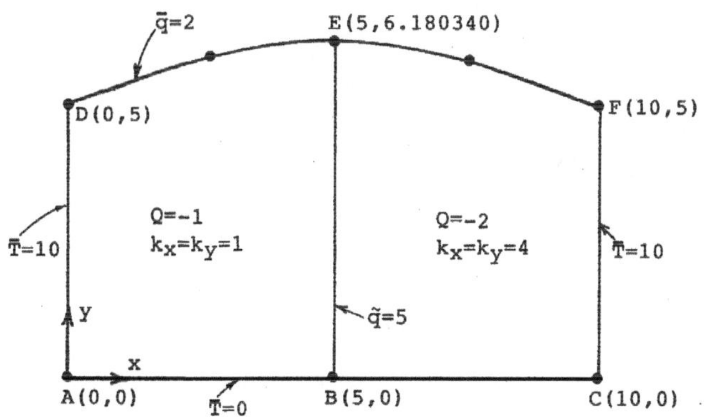

Fig. 14.8. Heat conduction in a medium having two different orthotropic materials solved by DQFDM

$T_{,E}$ at E, and heat fluxes q_{y,B_l} and q_{y,B_r} in the left subdomain and right subdomain , respectively, at B are summarized and listed in Table 14.10. It also shows good convergence.

Table 14.10. Convergence of heat conduction in a medium having two different materials

Order of appr.	DQFDM grid line (each subregion)	$T_{,E}$	q_{y,B_l}	q_{y,B_r}
4	5×5	.884638×10^1	-.183467×10^1	-.733866×10^1
6	7×7	.840760×10^1	-.211950×10^1	-.847801×10^1
8	9×9	.818825×10^1	-.244051×10^1	-.976204×10^1
10	11×11	.805528×10^1	-.270015×10^1	-.108006×10^2
12	13×13	.797701×10^1	-.294005×10^1	-.117602×10^2
	15×15	.786765×10^1	-.318140×10^1	-.127256×10^2
	17×17	.781550×10^1	-.326632×10^1	-.130653×10^2

15

Generalized Coordinate Differential Quadrature Element Method

For the generalized coordinate differential quadrature element method, the element weighting coefficients are generated by a generalized approach with which the element weighting coefficients for elements having arbitrary configurations can straightly be generated. The computation of weighting coefficients is explicit. By using this generalized coordinate differential quadrature technique the governing differential or partial differential equations, the transition conditions of two adjacent elements and the boundary conditions can be discretized. A global algebraic equation system can be obtained by assembling all of the discretized equations.

15.1 Generalized Coordinate Differential Quadrature Discretization

Typical procedures for calculating weighting coefficients for grid models defined by one, two and three coordinates, separately, are summarized [128]. The coordinate variables can be physical or natural.

15.1.1 One-Coordinate Model

The grid configuration of the one-coordinate grid model can be a spacially straight or curved line. The GCDQ discretization for a derivative of order m at discrete point α can be expressed by

$$\frac{d^m \phi_\alpha}{d\xi^m} = G_{\alpha i}^{\xi^m} c_i, \quad i = 1, 2, ..., \tilde{N}_c \tag{15.1}$$

where c_i are generalized coordinates, $\tilde{N}_c$ is the number of generalized coordinates, and $G_{\alpha i}^{\xi^m} c_i$ are weighting coefficients. The variable function can be approximated by

$$\phi(\xi) = \Upsilon_p(\xi) c_p, \quad p = 1, 2, ..., \tilde{N}_c \tag{15.2}$$

where $\Upsilon_p(\xi)$ are appropriate analytical functions. The mth order differentiation of Eq. (15.2) at discrete point α leads to the GCDQ discretization equation (15.1) in which $G^{\xi^m}_{\alpha i}$ can be calculated by the following explicit equation

$$G^{\xi^m}_{\alpha i} = \frac{d^m \Upsilon_i}{d\xi^m} |_\alpha \tag{15.3}$$

The generalized coordinates and appropriate analytical functions can also be expressed by certain other tensors having orders other than one. Using the above equation, the weighting coefficients can be easily obtained by simple algebraic calculations. For $m > 1$, the weighting coefficients $G^{\xi^m}_{i\bar{i}}$ for defining the mth order derivatives at the $\tilde{N}_c$ discrete points can also be obtained by m-1 times of matrix multiplications of the weighting coefficient matrix $G^{\xi}_{j\bar{j}}$ defining the first order derivative GCDQ discretizations at the $\tilde{N}_c$ discrete points

$$G^{\xi^m}_{i\bar{i}} = G^{\xi}_{ij_1} G^{\xi}_{j_1 j_2} ... G^{\xi}_{j_{m-1}\bar{i}} \tag{15.4}$$

15.1.2 Two-Coordinate Grid Model

The grid configuration of a two-coordinate grid model can be a triangle, a quadrilateral or a certain other configuration. By adopting a one-dimensional node identification method, the GCDQ discretization for a partial derivative of order m+n at discrete point α can be expressed by

$$\frac{\partial^{(m+n)} \phi_\alpha}{\partial \xi^m \partial \eta^n} = G^{\xi^m \eta^n}_{\alpha i} c_i, \quad i = 1, 2, ..., \tilde{N}_c \tag{15.5}$$

The variable function can be approximated by

$$\phi(\xi, \eta) = \Upsilon_j(\xi, \eta) c_j, \quad j = 1, 2, ..., \tilde{N}_c \tag{15.6}$$

where c_j are generalized coordinates and $\Upsilon_j(\xi, \eta)$ are appropriate analytical functions. The $(m+n)$th order partial differentiation of Eq. (15.6) at discrete point α leads to the GCDQ discretization equation (15.5) in which $G^{\xi^m \eta^n}_{\alpha i}$ are expressed by

$$G^{\xi^m \eta^n}_{\alpha i} = \frac{\partial^{(m+n)} \Upsilon_i}{\partial \xi^m \partial \eta^n} |_\alpha \tag{15.7}$$

The generalized coordinates and appropriate analytical functions can also be expressed by certain other tensors having orders other than one.

By adopting a two-dimensional node identification method, the GCDQ discretization for a partial derivative of order m+n at discrete point (α, β) can be expressed by

$$\frac{\partial^{(m+n)} \phi_{\alpha\beta}}{\partial \xi^m \partial \eta^n} = G^{\xi^m \eta^n}_{\alpha\beta i} c_i \tag{15.8}$$

The $(m+n)$th order partial differentiation of Eq. (15.6) at discrete point (α, β) leads to the GCDQ discretization equation (15.8) in which the weighting coefficients, $G_{\alpha\beta i}^{\xi^m \eta^n}$, are expressed by

$$G_{\alpha\beta i}^{\xi^m \eta^n} = \frac{\partial^{(m+n)} \Upsilon_i}{\partial \xi^m \partial \eta^n} |_{\alpha\beta} \qquad (15.9)$$

The generalized coordinates and appropriate analytical functions can also be expressed by certain other tensors having orders other than one. If the Lagrange family grid is adopted and the variable function $\phi(\xi, \eta)$ is expressed by

$$\phi(\xi, \eta) = \Upsilon_p(\xi)\Upsilon_q(\eta)c_{pq}, \quad p = 1, 2, ..., \bar{N}_\xi + 1, \quad q = 1, 2, ..., \bar{N}_\eta + 1 \quad (15.10)$$

where $\bar{N}_\xi$ and $\bar{N}_\eta$ are the orders of approximations in ξ and η directions, respectively. The GCDQ discretization equation (15.8) can be rewritten as

$$\frac{\partial^{(m+n)} \phi_{\alpha\beta}}{\partial \xi^m \partial \eta^n} = G_{\alpha\beta ij}^{\xi^m \eta^n} c_{ij} \qquad (15.11)$$

where $G_{\alpha\beta ij}^{\xi^m \eta^n}$ can be obtained by the following equation

$$G_{\alpha\beta ij}^{\xi^m \eta^n} = G_{\alpha i}^{\xi^m} G_{\beta j}^{\eta^n} \qquad (15.12)$$

15.1.3 Three-Coordinate Grid Model

The grid configuration of a three-coordinate grid model can be a triangle with the variable function defined by the area coordinates, a tetrahedron, a triangular prism, a hexahedron or a certain other configuration. By adopting a one-dimensional node identification method, the GCDQ discretization for a partial derivative of order $m+n+o$ at discrete point α can be expressed by

$$\frac{\partial^{(m+n+o)} \phi_\alpha}{\partial \xi^m \eta^n \zeta^o} = G_{\alpha i}^{\xi^m \eta^n \zeta^o} c_i, \quad i = 1, 2, ..., \tilde{N}_c \qquad (15.13)$$

The variable function can be approximated by

$$\phi(\xi, \eta, \zeta) = \Upsilon_j(\xi, \eta, \zeta)c_j, \quad j = 1, 2, ..., \tilde{N}_c \qquad (15.14)$$

where c_j are generalized coordinates and $\Upsilon_j(\xi, \eta, \zeta)$ are appropriate analytical functions. The $(m+n+o)$th order partial differentiation of Eq. (15.14) at discrete point α leads to the generalized GCDQ discretization equation (15.13) in which the weighting coefficients, $G_{\alpha i}^{\xi^m \eta^n \zeta^o}$, are expressed by

$$G_{\alpha i}^{\xi^m \eta^n \zeta^o} = \frac{\partial^{(m+n+o)} \Upsilon_i}{\partial \xi^m \partial \eta^n \partial \zeta^o} |_\alpha \qquad (15.15)$$

The generalized coordinates and appropriate analytical functions can also be expressed by certain other tensors having orders other than one. Consider that the analytical function can be expressed by the inner products of two sets of functions, with one set defined by two coordinate variables while the other set defined by the remaining coordinate variable, and one set of generalized coordinates. A representative of the variable function can be expressed by

$$\phi(\xi,\eta,\zeta) = \Upsilon_j(\xi,\eta)\Upsilon_r(\zeta)c_{jr} \tag{15.16}$$

Then the GCDQ discretization equation (15.13) can be rewritten as

$$\frac{\partial^{(m+n+o)}\phi_\alpha}{\partial\xi^m\eta^n\zeta^o} = G_{\alpha ik}^{\xi^m\eta^n\zeta^o} c_{ik} \tag{15.17}$$

where the weighting coefficients show to have the following form:

$$G_{\alpha ik}^{\xi^m\eta^n\zeta^o} = G_{(\alpha)i}^{\xi^m\eta^n} G_{\alpha k}^{\zeta^o} \tag{15.18}$$

By adopting a two-dimensional node identification method, the GCDQ discretization for a partial derivative of order $m+n+o$ at discrete point (α,γ) can be expressed by

$$\frac{\partial^{(m+n+o)}\phi_{\alpha\gamma}}{\partial\xi^m\partial\eta^n\partial\zeta^o} = G_{\alpha\gamma i}^{\xi^m\eta^n\zeta^o} c_i \tag{15.19}$$

The $(m+n+o)$th order partial differentiation of (15.14) at discrete point (α,γ) leads to the GCDQ discretization equation (15.19) in which the weighting coefficients, $G_{\alpha\gamma i}^{\xi^m\eta^n\zeta^o}$, are expressed by

$$G_{\alpha\gamma i}^{\xi^m\eta^n\zeta^o} = \frac{\partial^{(m+n+o)}\Upsilon_i}{\partial\xi^m\partial\eta^n\partial\zeta^o}|_{\alpha\gamma} \tag{15.20}$$

The generalized coordinates and appropriate analytical functions can also be expressed by certain other tensors having orders other than one. If the variable function can be approximated by Eq. (15.16), the GCDQ discretization equation (15.19) can be rewritten as

$$\frac{\partial^{(m+n+o)}\phi_{\alpha\gamma}}{\partial\xi^m\partial\eta^n\partial\zeta^o} = G_{\alpha\gamma ik}^{\xi^m\eta^n\zeta^o} c_{ik} \tag{15.21}$$

where the weighting coefficients show to have the following form:

$$G_{\alpha\gamma ik}^{\xi^m\eta^n\zeta^o} = G_{\alpha i}^{\xi^m\eta^n} G_{\gamma k}^{\zeta^o} \tag{15.22}$$

By adopting a three-dimensional node identification method, the GCDQ discretization for a partial derivative of order $m+n+o$ at discrete point (α,β,γ) can be expressed by

$$\frac{\partial^{(m+n+o)}\phi_{\alpha\beta\gamma}}{\partial\xi^m\partial\eta^n\zeta^o} = G_{\alpha\beta\gamma i}^{\xi^m\eta^n\zeta^o} c_i \tag{15.23}$$

The $(m+n+o)$th order partial differentiation of (15.14) at discrete point (α, β, γ) leads to the GCDQ discretization equation (15.23) in which the weighting coefficients, $G_{\alpha\beta\gamma i}^{\xi^m \eta^n \zeta^o}$, are expressed by

$$G_{\alpha\beta\gamma i}^{\xi^m \eta^n \zeta^o} = \frac{\partial^{(m+n+o)} \Upsilon_i}{\partial \xi^m \partial \eta^n \partial \zeta^o} |_{\alpha\beta\gamma} \tag{15.24}$$

The generalized coordinates and appropriate analytical functions can also be expressed by certain other tensors having orders other than one. If the variable function can be approximated by Eq. (15.16), the GCDQ discretization equation (15.23) can be rewritten as

$$\frac{\partial^{(m+n+o)} \phi_{\alpha\beta\gamma}}{\partial \xi^m \partial \eta^n \zeta^o} = G_{\alpha\beta\gamma ik}^{\xi^m \eta^n \zeta^o} c_{ik} \tag{15.25}$$

where the weighting coefficients show to have the following form:

$$G_{\alpha\beta\gamma ik}^{\xi^m \eta^n \zeta^o} = G_{\alpha\beta i}^{\xi^m \eta^n} G_{\gamma k}^{\zeta^o} \tag{15.26}$$

If the hexahedral Lagrange family grid is adopted and the variable function $\phi(\xi, \eta, \zeta)$ is expressed by

$$\begin{aligned} \phi(\xi, \eta, \zeta) &= \Upsilon_p(\xi) \Upsilon_q(\eta) \Upsilon_r(\zeta) c_{pqr}, \\ & p = 1, 2, ..., \bar{N}_\xi + 1, \quad q = 1, 2, ..., \bar{N}_\eta + 1, \quad r = 1, 2, ..., \bar{N}_\zeta + 1 \end{aligned} \tag{15.27}$$

where $\bar{N}_\xi$, $\bar{N}_\eta$ and $\bar{N}_\zeta$ are the orders of approximations in ξ, η and ζ directions, respectively. The GCDQ discretization equation can be rewritten as

$$\frac{\partial^{(m+n+o)} \phi_{\alpha\beta\gamma}}{\partial \xi^m \partial \eta^n \zeta^o} = G_{\alpha\beta\gamma ijk}^{\xi^m \eta^n \zeta^o} c_{ijk} \tag{15.28}$$

where $G_{\alpha\beta\gamma ijk}^{\xi^m \eta^n \zeta^o}$ can be obtained by the following equation

$$G_{\alpha\beta\gamma ijk}^{\xi^m \eta^n \zeta^o} = G_{\alpha i}^{\xi^m} G_{\beta j}^{\eta^n} G_{\gamma k}^{\zeta^o} \tag{15.29}$$

15.2 Generalized Coordinate Differential Quadrature Element Analyses

Sample problems solved are introduced to explain the numerical procedures of generalized coordinate differential quadrature based discrete element analysis.

15.2.1 Beam Vibration

GCDQEM Formulation

The GCDQEM beam vibration analysis is illustrated. Considering the prismatic beam, the governing differential eigenvalue equation of beam vibration is expressed as

$$EIW_{,xxxx} - \rho A \omega^2 W(x) = 0 \tag{15.30}$$

Using the GCDQ represented by Eq. (15.1) to discretize Eq. (15.30) at a discrete point α, the following equation can be obtained

$$(EI)^e G^{ex^4}_{\alpha\beta} c^e_\beta - \rho^e \omega^2 \Upsilon^e_\beta(x^e_\alpha) c^e_\beta = 0, \quad \beta = 1, 2, ..., N^e \tag{15.31}$$

where N^e is the number of discrete points or nodes of the element e.

The use of Eq. (15.2) in Eq. (4.9) leads to the following continuity condition of modal displacement at the inter-element boundary of two adjacent elements i and $i+1$

$$\Upsilon^i_\beta(x^i_{N^i}) c^i_\beta = \Upsilon^{i+1}_{\bar{\beta}}(x^{i+1}_1) c^{i+1}_{\bar{\beta}} \tag{15.32}$$

The use of Eq. (15.2) in Eq. (4.10) leads to the following continuity condition of modal deflection slope

$$G^{ix}_{N^i\beta} c^i_\beta - G^{(i+1)x}_{1\bar{\beta}} c^{i+1}_{\bar{\beta}} = 0 \tag{15.33}$$

Assume that a concentrated mass $\tilde{M}^{i,i+1}$ is attached to the inter-element boundary of two adjacent element i and $i+1$. Neglecting the effect of rotary inertia of the concentrated mass, the equilibrium of moments is

$$-(EI)^i W^i_{N^i,xx} + (EI)^{i+1} W^{i+1}_{1,xx} = 0 \tag{15.34}$$

The above equation can be discretized

$$-(EI)^i G^{ix^2}_{N^i\beta} c^i_\beta + (EI)^{i+1} G^{(i+1)x^2}_{1\bar{\beta}} c^{i+1}_{\bar{\beta}} = 0 \tag{15.35}$$

The equilibrium of transverse forces is

$$(EI)^i W^i_{N^i,xxx} - (EI)^{i+1} W^{i+1}_{1,xxx} + \omega^2 \tilde{M}^{i,i+1} W^{i,i+1} = 0 \tag{15.36}$$

The above equation can be discretized

$$(EI)^i G^{ix^3}_{N^i\beta} c^i_\beta - E^{i+1} I^{i+1} G^{(i+1)x^3}_{1\bar{\beta}} c^{i+1}_{\bar{\beta}} - \omega^2 \tilde{M}^{i,i+1} \Upsilon^i_\beta(x^i_{N^i}) c^i_\beta = 0 \tag{15.37}$$

The kinematic boundary conditions can be expressed as $W^m_{I^m} = \bar{W}^m_{I^m}$ and $W^m_{I^m,x} = \bar{W}^m_{I^m,x}$ in which I^m is 1 or N^m. The discrete equation of the first kinematic boundary condition can be obtained using Eq. (15.32)

$$\Upsilon_\beta(x^m_{I^m}) c^m_\beta = \bar{W}^m_{I^m} \tag{15.38}$$

The discrete equation of the second kinematic boundary condition can be obtained using Eq. (15.33)

$$G^{x}_{I^m\beta}c^m_\beta = \bar{w}^m_{I,x}, \quad I = 1 \quad \text{or} \quad N^m \tag{15.39}$$

Let n be an element consisting of the natural boundary. The discrete natural boundary condition of equilibrium of moments can be obtained using Eq. (15.35)

$$(EI)^n G^{nx^2}_{I^n\beta}c^n_\beta = 0, \quad I^n = 1 \quad \text{or} \quad N^n \tag{15.40}$$

Referring to (15.33), the discrete natural boundary condition of the equilibrium of transverse forces is expressed by

$$(EI)^n G^{nx^3}_{I^n\beta}c^n_\beta = \omega^2 M^n \Upsilon^n_\beta(x^n_1)c^n_\beta, \quad \text{for left boundary} \tag{15.41}$$

$$= -\omega^2 \tilde{M}^n \Upsilon^n_\beta(x^n_{N^n})c^n_\beta, \quad \text{for right boundary} \tag{15.42}$$

where $\tilde{M}^n$ is the concentrated mass placed at the natural boundary.

In constructing the numerical model, the local coordinate system defined on each element with the origin located at the first node of the element can be used. Then the nondimensional coordinate defined as the local coordinate variable divided by the length of the element can be used to carry out the GCDQ discretization. By using this approach, all elements have the same one unit nondimensional subdomain. Then only one GCDQ discretization is necessary for elements having the same grid in their nondimensional subdomains. This can reduce the arithmetic operations in carrying out the GCDQ discretizations for all elements.

The problem solved involves the lateral vibration of the fixed-fixed non-prismatic beam shown in Fig. 4.4 in Subsection 4.1.3. The beam is composed of two prismatic segments having the same length 1. The materials of the two segments are also the same with both the Young's modulus and mass density equal to 1. One segment has both the area and moment of inertia of the cross section equal to 1, while the other segment also has the same value of 2 for both the area and the moment of inertia of the cross section. In solving this problem, the effect of rotary inertia is neglected and the elements are equally spaced.

Denote ξ the nondimensional coordinate. The set of polynomials ξ^{k-1}, $k = 1, 2, ..., N^e$ is used to carry out the element basis GCDQ discretization. In a segment, the elements used to model the segment have the same length. Transition or boundary conditions are defined at the element boundary nodes. Though the discrete element eigenvalue equations can also be defined at the element boundary nodes, in this sample analysis the discrete element eigenvalue equations are only defined at interior discrete points. The interior discrete points are equally spaced. Let $h^e = l^e/(N^e + 1)$. The interior discrete points are located at $2h^e, 3h^e, ..., (N^e - 1)h^e$. The natural frequencies obtained are listed in Table 15.1. It shows that the GCDQEM results converge very fast by increasing either the number of elements or discrete points per element.

Table 15.1. The first five natural frequencies of a nonprismatic beam composed of two segments

DOF per element	Number of elements	ω_1	ω_2	ω_3	ω_4	ω_5
5	2	6.4345458	14.078064			
	4	5.5670666	16.898637	38.812658	47.364824	
	6	5.5073212	16.123268	31.841268	56.251621	95.333048
7	2	5.3282936	15.748253	26.369430	39.858780	65.627698
	4	5.4559686	15.490456	28.522681	50.050799	69.157952
	6	5.4613323	15.619719	29.692799	49.341898	69.971431
9	2	5.4661943	15.648038	31.325120	55.544366	85.939373
	4	5.4626645	15.653397	29.969711	50.412941	74.997004
	6	5.4626307	15.650285	29.907237	50.417707	74.364319
11	2	5.4625368	15.649903	29.817807	49.945043	77.346002
	4	5.4626271	15.649944	29.901405	50.379680	74.086114
	6	5.4626274	15.649999	29.903108	50.381025	74.124804

15.2.2 Steady State Field Problems

Numerical simulation involving the steady-state heat conduction in a two-dimensional uniform medium is illustrated. Assume that the medium is isotropic. The governing partial differential equation is expressed as

$$kT_{,xx} + kT_{,yy} + Q = 0 \tag{15.43}$$

The Neumann boundary condition is expressed as

$$\bar{q} = kT_{,n} = klT_{,x} + kmT_{,y} \tag{15.44}$$

Solution of the boundary value problem of Poisson equation provides components of the internal heat flux through the following definition

$$q_x = -kT_{,x}, \quad q_y = -kT_{,y} \tag{15.45}$$

In the GCDQEM analysis, the analysis domain and element configuration can be arbitrary. If the mapping technique is used, the fundamental relations must be transformed into the natural space. In GCDQEM, the GCDQ discretizations can be directly carried out in the physical space for regular or irregular elements. The element must have four or more discrete points. The discrete points at which the discrete element governing equations defined can be either in the interior of the element or on the element boundary. The average discrete governing equation defined at a discrete point on the inter-element boundary can also be a discrete fundamental equation. The one-dimensional node identification method is adopted for carrying out the GCDQ discretization. The discrete equation of (15.42) at node α in the element e is

$$k^e G_{\alpha j}^{x^2} c_j^e + k^e G_{\alpha j}^{y^2} c_j^e + Q_\alpha^e = 0, \quad j = 1, 2, ..., \tilde{N}_c \tag{15.46}$$

The discrete Neumann boundary condition at node $\bar{\alpha}$ can be obtained from (15.43)

$$\bar{q}_{\bar{\alpha}}^n = k^n l_{(\bar{\alpha})}^n G_{\bar{\alpha} j}^x c_j^n + k^n m_{(\bar{\alpha})}^n G_{\bar{\alpha} j}^y c_j^n \tag{15.47}$$

Let the discrete point β of element r and the discrete point $\bar{\beta}$ of element s be a common node On the inter-element boundary, $\partial\Omega^{r,s}$, of two adjacent elements r and s. The discrete kinematic transition condition at this node is the continuity of temperatures which is expressed as

$$T_\beta^r = T_{\bar{\beta}}^s \tag{15.48}$$

The above equation leads to

$$\Upsilon_\gamma^r (x_\beta^r, y_\beta^r) c_\gamma^r = \Upsilon_{\bar{\gamma}}^s (x_{\bar{\beta}}^s, y_{\bar{\beta}}^s) c_{\bar{\gamma}}^s \tag{15.49}$$

The temperature can also be prescribed on the inter-element boundary which is expressed as

$$T_\beta^r = T_{\bar{\beta}}^s = \bar{T}_{\beta,\bar{\beta}}^{r,s} \tag{15.50}$$

where $\bar{T}_{\beta,\bar{\beta}}^{r,s}$ is the prescribed temperature. The above equation leads to

$$\Upsilon_\gamma^r (x_\beta^r, y_\beta^r) c_\gamma^r = \Upsilon_{\bar{\gamma}}^s (x_{\bar{\beta}}^s, y_{\bar{\beta}}^s) c_{\bar{\gamma}}^s = \bar{T}_{\beta,\bar{\beta}}^{r,s} \tag{15.51}$$

By using (15.46), the natural transition condition at the common node can be defined

$$k^r l_{(\beta)}^r G_{\beta p}^x c_p^r + k^r m_{(\beta)}^r G_{\beta p}^y c_p^r + k^s l_{(\bar{\beta})}^s G_{\bar{\beta} q}^x c_q^s + k^s m_{(\bar{\beta})}^s G_{\bar{\beta} q}^y c_q^s = \tilde{q} \tag{15.52}$$

where $\tilde{q}$ is the conduction heat flux into the domain. The components of internal heat flux at the discrete point α in element e can also be obtained from (15.44)

$$q_{x,\alpha}^e = -k^e G_{\alpha j}^x c_j^e, \quad q_{y,\alpha}^e = -k^e G_{\alpha j}^y c_j^e \tag{15.53}$$

Irregular element can be developed with or without using the mapping technique. The generation of mesh and element grid can be referred to Section 4.2. Considering the Lagrange family grid for a quadrilateral element, denote N_ξ and N_η the numbers of level in ξ and η directions, respectively, in the master element of a physical element. The set of polynomials $x^{p-1}y^{q-1}$, $p = 1, 2, ..., N_\xi$, $q = 1, 2, ..., N_\eta$ can be used to construct the first order tensor expressed analytical functions and generate the weighting coefficients. Since this approximation possesses the property of geometric property, the approach of adopting the mapping technique and the approach without adopting the mapping technique result in obtaining the same approximation.

15.2.3 Problems

Steady-State Heat Conduction

The problem solved involves the heat conduction in one quarter of an annular region which is shown in Fig. 15.1. The domain is modelled by four partial annuluses having the same range angle. The thermal conductivity k is 1. The temperatures on the boundaries of two concentric circles of radii 1 and 5 are prescribed which are $\bar{T} = 100$ and $\bar{T} = 40$, respectively. On boundaries FH and GI, conduction heat fluxes of $\bar{q} = -10$ and $\bar{q} = 10$, respectively, are applied. On the inter-element boundary which separates the quarter annulus into two equally ranged partial annuluses, the conduction heat flux of $\tilde{q} = 5$ is also applied. And in the subdomain $FAHKBJ$ the internal heat generation rate $Q = -1$ is applied. In carrying out the numerical tests, equally spaced discrete points in both ξ and η directions are adopted for defining the element grid. All four elements have the same type of element grid. Numerical results of temperatures at five different discrete points are listed in Table 15.2. It shows excellent convergence.

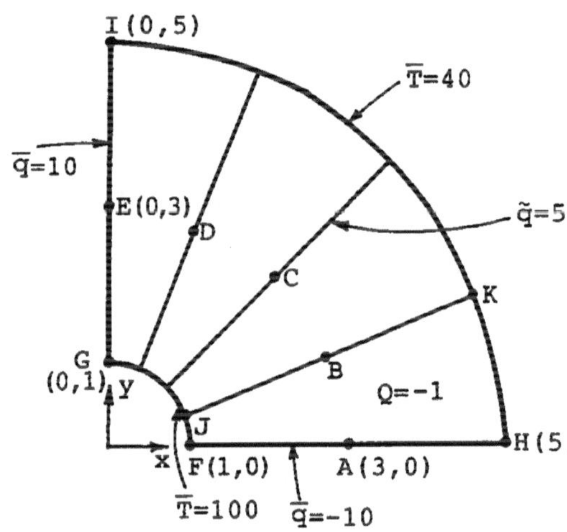

Fig. 15.1. 4-element meshed quarter annular region.

Torsion of a Prismatic Bar Having a Half Elliptic Cross Section

The domain boundary is described by $x^2 + \frac{y^2}{4} = 1$, $y > 0$ and $y = 0$. The half elliptic section is considered to be a triangular grid with the three vertices located at $\bar{A}(-1, 0)$, $A(1, 0)$ and $B(0, 2)$. In calculating the weighting coefficients, the complete polynomials are used. Let n denote the order of

Table 15.2. Convergence by increasing the discrete points in an element for the heat conduction in a quarter annulus

Element grid	T_A	T_B	T_C	T_D	T_E
3×3	.460915×10^2	.558711×10^2	.634581×10^2	.667050×10^2	.751272×10^2
5×5	.457174×10^2	.553609×10^2	.628760×10^2	.661010×10^2	.743623×10^2
9×9	.455149×10^2	.551904×10^2	.627573×10^2	.660273×10^2	.743219×10^2
11×11	.455047×10^2	.551814×10^2	.627523×10^2	.660263×10^2	.743220×10^2

the approximate complete polynomial. The distribution of stress function can be expressed by $\phi(x,y) = \Upsilon_{pk}(x,y)c_{pk}, 0 \le p \le n, 1 \le k \le n+1$ where $\Upsilon_{pk}(x,y) = x^{p+1-k}y^{k-1}$ and c_{pk} are unknown coefficients. Υ_{pk} are complete polynomials. By arranging the complete polynomials into a one-dimensional array and substituting them into (15.3), the weighting coefficients can be calculated. Prandtl's stress function formulation of torsion theory was used. The p refinement procedure adopting the Pascal triangular grid is used to analyze the problem. Numerical results of stress function at $C(0, \frac{2}{3})$ and shear stresses at $O(0,0)$ and $B(0,2)$ are listed in Table 15.3. It also shows that the results converge well by gradually increasing the order of the Pascal triangular grid.

Table 15.3. Numerical results of the torsion of a prismatic bar having a half elliptic cross section

Order of Pascal triangular grid	Φ_C	$\tau_{zx,O}$	$\tau_{zx,O}$
3	.507937×10^0	.285714×10^0	−.571429×10^0
5	.475582×10^0	.407713×10^0	−.489659×10^0
7	.474765×10^0	.423569×10^0	−.546460×10^0

Torsion of a Prismatic Bar Having a Triangular Cross Section

The sides of the triangle are the three straight lines $x = 0$, $y = 0$ and $x+y = 1$. p refinement procedure adopting the Pascal triangular grid is used to analyze problem. The Pascal triangular grid is designed by first using $n+1$ lines parallel to the L_k-axis to subdivide the rectangular unit triangle into n subregions, in each area coordinate direction, with n the order of approximate polynomials. The $n+1$ lines define the levels of the Pascal triangular grid with the side $L_k = 0$ being the level zero. The level number will increase following the increase of the value of L_k. The highest level is level $n+1$. Though the distance between two levels can be different, it is set to be constant in the present numerical analysis. The complete polynomials used for the analysis of last sample problem are also used to define the GCDQ weighting coefficients and carry out the numerical computation. Numerical results of the stress function at $A(\frac{1}{3},\frac{1}{3})$, and the shear stresses at A and $B(0,\frac{1}{2})$ are summarized and listed in Table 15.4. The results converge fast by gradually increasing the order of Pascal triangular grid.

Table 15.4. Results of the torsion of a prismatic bar having a triangular cross section

Order of Pacal triangular grid	Φ_A	$\tau_{zx,A}$	$\tau_{zy,A}$	$\tau_{zy,B}$
3	$.555556\times10^{-1}$	$.000000\times10^{-1}$	$.000000\times10^{0}$	$-.375000\times10^{0}$
5	$.581439\times10^{-1}$	$-.374110\times10^{-1}$	$.374110\times10^{-1}$	$-.421402\times10^{0}$
7	$.580757\times10^{-1}$	$-.389181\times10^{-1}$	$.389181\times10^{-1}$	$-.407827\times10^{0}$
9	$.579874\times10^{-1}$	$-.390231\times10^{-1}$	$.390231\times10^{-1}$	$-.407347\times10^{0}$
11	$.579595\times10^{-1}$	$-.391414\times10^{-1}$	$.391414\times10^{-1}$	$-.408301\times10^{0}$
13	$.579466\times10^{-1}$	$-.391865\times10^{-1}$	$.391865\times10^{-1}$	$-.407915\times10^{0}$
15	$.579397\times10^{-1}$	$-.392097\times10^{-1}$	$.392097\times10^{-1}$	$-.407818\times10^{0}$

15.2.4 Steady Poiseuille Flow in a Pipe with Elliptic Cross Section

The problem concerns the steady uniform incompressible viscous flow in a pipe of elliptic cross section in the direction of z-axis. The governing equation is

$$-\frac{1}{\mu}\frac{\partial p}{\partial z}+\left(\frac{\partial^2 v_z}{\partial x^2}+\frac{\partial^2 v_z}{\partial y^2}\right)=0 \tag{15.54}$$

where μ is the viscosity of the fluid, v_z the velocity component in the z direction and p the distribution of pressure. The cross section of the pipe is defined by $\frac{x^2}{4}+\frac{y^2}{16}=1$. In the analysis, the pressure gradient $\frac{\partial p}{\partial z}$ is set to be -1, and μ is set to be .25. The velocity $v_z(x,y)$ is approximated by

$$v_z(x,y)=c_1+xc_2+yc_3+x^2c_4+y^2c_5+xyc_6+x^3c_7+y^3c_8+x^2yc_9+xy^2c_{10}+\cdots\cdots \tag{15.55}$$

The cross section is represented by one elliptic GCDQ element. Two analyses were carried out by using the four-node element and the five-node element, separately. In the analysis using a four-node element, a discrete governing equation is defined at the node $(0,0)$, and three discrete boundary conditions are defined at the other three points $(2,0)$, $(0,4)$ and $(-2,0)$. In the analysis using a five-node element, one more discrete boundary condition is defined at one additional point $(0,-4)$. Numerical results of velocity at four different

Table 15.5. Results of a steady Poiseuille flow problem

Nodes of nodes	$V_z(,)$	$V_z(,)$	$V_z(,)$	$V_z(,)$
4	$.80000\times10^1$	$.60000\times10^1$	$.40000\times10^1$	$.20000\times10^1$
5	$.64000\times10^1$	$.48000\times10^1$	$.48000\times10^1$	$.32000\times10^1$
Exact solution	$.64000\times10^1$	$.48000\times10^1$	$.48000\times10^1$	$.32000\times10^1$

points are listed in Table 15.5. Exact results were obtained for the five-node analysis.

16

EDQ Based Direct Time Integration Methods

EDQ based time integration algorithms developed for solving the discrete transient equation system of a continuum mechanics problem are introduced. They are direct integration integration methods. Two algorithms are developed. They are time-element by time-element method and stages by stages method. These two time integration methods can be used to solve generic discrete transient equation system of an originally discrete system or a discrete system resulting from the discretization of a transient system of continuum mechanics problems by using a certain discretization technique such as the DQEM, FEM, FDM, ..., etc.

16.1 Second Order Problems

Transient problems with fundamental equations having second order derivatives with respect to the time variable are widely existing in the regime of scientific and engineering. Dynamic response of structures is a representative of these transient problems. The numerical procedures of EDQ based time integration algorithms for structural dynamics are introduced [44].

16.1.1 EDQ Basis Time-Element by Time-Element Integration Algorithm

Consider that an EDQ grid model with the temporal coordinate variable, t. Then, the EDQ grid model can be defined as a time-element. The transient response can be solved by increasing the time, step by step. Each step represents a time-element. Consider that the transient equation system, at a stage of the tth incremental step, of the transient problems of structures is expressed by

$$M_{rs}\ddot{U}_s^t + C_{rs}\dot{U}_s^t + K_{rs}U_s^t = F_r^t \tag{16.1}$$

where C_{rs} is the damping matrix, U_s^t is the overall displacement vector and F_r^t is the overall load vector. Let Δt and τ denote the time increment or the size of time-element and the natural coordinate with respect to the time t. Then, by using the EDQ to discretize $\dot{U}_s^t$ and $\ddot{U}_s^t$ in the above equation, the discrete equation at the time stage p of the tth incremental step can be expressed by the following equation

$$\left(\frac{1}{\Delta t^2} M_{rs} D_{pq}^{\tau^2} + \frac{1}{\Delta t} C_{rs} D_{pq}^{\tau} + K_{rs} \Psi_{pq}\right) \tilde{U}_s^{t,q} = F_r^{t,p} \tag{16.2}$$

where $\tilde{U}_s^{t,q}$ are displacements and/or their derivatives with respect to t, and Ψ_{pq} are the corresponding interpolation functions of the EDQ discretization. In order to solve the above equation, two initial conditions are required. Let $\bar{U}_s^t$ denote the initial displacements of the tth incremental step. The initial condition of displacements is expressed as

$$U_s^{t,1} = \bar{U}_s^t \tag{16.3}$$

Let $\dot{\bar{U}}_s^t$ denote the initial velocities of the tth incremental step. The initial condition of velocities is expressed as

$$\frac{1}{\Delta t} D_{1q}^{\tau} \tilde{U}_s^{t,q} = \dot{\bar{U}}_s^t \tag{16.4}$$

The values of $\bar{U}_s^t$ and $\dot{\bar{U}}_s^t$ can be obtained from the solutions of the $(t-1)$th incremental step. The response histories can be updated by a time-element by time-element procedure.

Various time-elements can be used to develop the direct integration schemes. Consider the Lagrange time-element having L stage nodes. Since no time derivative of displacement is adopted for the time-element, discrete time stages for defining the discrete equations of motion coincide the node stages of the element, and Ψ_{pq} represents the Kronecker delta δ_{pq}. Use the DOF assigned to the first node stage to define the condition of initial displacements, and use the DOF assigned to the second node stages to define the initial condition of velocities. Then, by using Eqs. (16.2), (16.3) and (16.4), the following matrix equation can be obtained

$$\bar{U}_s^t \{\bar{K}_{RI}\} + [\bar{K}_{RR}]\{U^{t,R}\} = \{\bar{F}^{t,R}\} \tag{16.5}$$

where

$$\{U^{t,R}\} = \lfloor\, U_s^{t,2} \;\; U_s^{t,3} \;\; \cdots \;\; U_s^{t,L} \,\rfloor^T, \quad \{\bar{F}^{t,R}\} = \lfloor\, \dot{\bar{U}}_s^t \;\; F_s^{t,3} \;\; \cdots \;\; F_s^{t,L} \,\rfloor^T,$$

$$\begin{aligned}\{\bar{K}_{RI}\} = \lfloor\, & \tfrac{1}{\Delta t}\delta_{rs} D_{11}^{\tau} \quad \tfrac{1}{\Delta t^2} M_{rs} D_{31}^{\tau^2} + \tfrac{1}{\Delta t} C_{rs} D_{31}^{\tau} \\ & \cdots \; \tfrac{1}{\Delta t^2} M_{rs} D_{L1}^{\tau^2} + \tfrac{1}{\Delta t} C_{rs} D_{L1}^{\tau} \,\rfloor^T,\end{aligned}$$

and

$$[\bar{K}_{RR}] = \begin{bmatrix} \frac{1}{\Delta t}\delta_{rs}D^{T}_{12} & \frac{1}{\Delta t}\delta_{rs}D^{T}_{13} & \cdots & \frac{1}{\Delta t}\delta_{rs}D^{T}_{1L} \\ \frac{1}{\Delta t^2}M_{rs}D^{T^2}_{32} + \frac{1}{\Delta t}C_{rs}D^{T}_{32} & \frac{1}{\Delta t^2}M_{rs}D^{T^2}_{33} + \frac{1}{\Delta t}C_{rs}D^{T}_{33} + K_{sr} & \cdots & \frac{1}{\Delta t^2}M_{rs}D^{T^2}_{3L} + \frac{1}{\Delta t}C_{rs}D^{T}_{3L} \\ \vdots & \vdots & \cdots & \vdots \\ \frac{1}{\Delta t^2}M_{rs}D^{T^2}_{L2} + \frac{1}{\Delta t}C_{rs}D^{T}_{L2} & \frac{1}{\Delta t^2}M_{rs}D^{T^2}_{L3} + \frac{1}{\Delta t}C_{rs}D^{T}_{L3} & \cdots & \frac{1}{\Delta t^2}M_{rs}D^{T^2}_{LL} + \frac{1}{\Delta t}C_{rs}D^{T}_{LL} + K_{rs} \end{bmatrix} \tag{16.6}$$

Using Eq. (16.5), displacements $\{U^{t,R}\}$ of the remaining $L-1$ node stages can be found

$$\{U^{t,R}\} = [\bar{K}_{RR}]^{-1}(\{\bar{F}^{t,R}\} - \bar{U}^t_s\{\bar{K}_{RI}\}) \tag{16.7}$$

It should be noted that if the problem is linear and that the size of the time-element is constant, then $[\bar{K}_{RR}]$ is a constant matrix. Consequently, only one decomposition is necessary for updating the response histories if a direct solution scheme is used.

Consider a $C^1 - C^0$ EDQ time-element having E stage nodes and L DOF. Assume that the two DOF assigned to the first stage node represent the displacement and velocity of the stage node, and that each of the remaining $E-1$ stage nodes has one DOF representing the displacement of the stage node. The two DOF assigned to the first stage node are used to define the two initial conditions of an element step, while the DOF assigned to each other stage node is used to define a discrete equation of motion. By using Eqs. (16.2), (16.3) and (16.4), the following matrix equation can be obtained

$$[\tilde{K}_{RI}]\{\tilde{\bar{U}}^{t,I}\} + [\tilde{K}_{RR}]\{\tilde{U}^{t,R}\} = \{\tilde{F}^{t,R}\} \tag{16.8}$$

where

$$\{\tilde{\bar{U}}^{t,I}\} = \lfloor\, \bar{U}^t_s \;\; \dot{\bar{U}}^t_s \,\rfloor^T, \quad \{\tilde{U}^{t,R}\} = \lfloor\, U^{t,2}_s \;\; U^{t,3}_s \;\; \cdots \;\; U^{t,L-1}_s \,\rfloor^T,$$

$$\{\tilde{F}^{t,R}\} = \lfloor\, F^{t,2}_s \;\; F^{t,3}_s \;\; \cdots \;\; F^{t,L-1}_s \,\rfloor^T,$$

$$[\tilde{K}_{RI}] = \begin{bmatrix} \frac{1}{\Delta t^2}M_{rs}D^{T^2}_{21} + \frac{1}{\Delta t}C_{rs}D^{T}_{21} \\ \vdots \\ \frac{1}{\Delta t^2}M_{rs}D^{T^2}_{(L-1)1} + \frac{1}{\Delta t}C_{rs}D^{T}_{(L-1)1} \end{bmatrix}$$

$$\left.\begin{matrix} \frac{1}{\Delta t}M_{rs}D^{\tau^2}_{22} + C_{rs}D^{\tau}_{22} \\ \cdot \\ \cdot \\ \cdot \\ \frac{1}{\Delta t}M_{rs}D^{\tau^2}_{(L-1)2} + C_{rs}D^{\tau}_{(L-1)2} \end{matrix}\right]$$

and

$$[\tilde{K}_{RR}] = \left[\begin{matrix} \frac{1}{\Delta t^2}M_{rs}D^{\tau^2}_{23} + \frac{1}{\Delta t}C_{rs}D^{\tau}_{23} + K_{rs} & \cdots \\ \cdot & \cdots \\ \cdot & \cdots \\ \cdot & \cdots \\ \frac{1}{\Delta t^2}M_{rs}D^{\tau^2}_{(L-1)3} + \frac{1}{\Delta t}C_{rs}D^{\tau}_{(L-1)3} & \cdots \end{matrix}\right.$$
$$\left.\begin{matrix} \frac{1}{\Delta t^2}M_{rs}D^{\tau^2}_{2L} + \frac{1}{\Delta t}C_{rs}D^{\tau}_{2L} \\ \cdot \\ \cdot \\ \cdot \\ \frac{1}{\Delta t^2}M_{rs}D^{\tau^2}_{(L-1)L} + \frac{1}{\Delta t}C_{rs}D^{\tau}_{(L-1)L} + K_{rs} \end{matrix}\right] \tag{16.9}$$

Using Eq. (16.8), displacements $\{U^{t,R}\}$ of the remaining $L-2$ node stages can be found

$$\{\tilde{U}^{t,R}\} = [\tilde{K}_{RR}]^{-1}(\{\tilde{F}^{t,R}\} - [\tilde{K}_{RI}]\{\tilde{\tilde{U}}^{t,I}\}) \tag{16.10}$$

It should be noted that if $L-1$ in the superscripts implies that the $C^1 - C^0$ time-element has $L-1$ stage nodes, while L in the subscripts implies that the time-element has L DOF.

16.1.2 DQ Basis Stages by Stages Integration Algorithm

A different algorithm was developed to solve the discrete transient equilibrium equation system. In this algorithm, the equally spaced Lagrange DQ model is used to discretize $\dot{U}^t_s$ and $\ddot{U}^t_s$. One incremental step solution similar to the time-element by time-element solution procedure of the previous algorithm is first carried out. Since no rotational DOF is assigned to the Lagrange DQ model, the DOF assigned to the second node is used to define the initial condition of velocity. After this solution step, the time variable is increased by adding one time increment. Assume that the Lagrange DQ model has L nodes and that the value of the time increment equals the distance spanning $N+1$ consecutive stages (nodes) of the initial solution step. N further time stages are thus defined. Then, by using this newly increased N stages and the previous L-N stages, the same DQ model can be used to discretize $\dot{U}^t_s$ and $\ddot{U}^t_s$ at the newly increased N stages. The discretized equation is expressed by

$$\left(\frac{1}{\Delta t^2}M_{rs}D^{\tau^2}_{\alpha m} + \frac{1}{\Delta t}C_{rs}D^{\tau}_{\alpha m} + K_{rs}\delta_{\alpha m}\right)\hat{U}^m_s$$
$$= \hat{F}^{\alpha}_r, \quad \alpha = L-N+1, L-N+2, ..., L \tag{16.11}$$

where $\hat{U}_s^m$ is the displacement vector of the temporal discretization using DQ and $\hat{F}_r^\alpha$ is the load vector of the newly increased N stages. Since only displacements at the newly increased N stages are unknowns, matrix partition technique can be used to obtain an equation system with unknowns the displacements at the newly increased N stages. This solution system is smaller than the solution system of the time-element by time-element solution algorithm and the first incremental step solution of the current algorithm which has the displacements of L-2 stages as unknowns. In this direct integration algorithm, the response histories can be updated by a stages by stages solution procedure. For solving transient problems of structures, the maximum value of N will be L-2. If the value of N is larger than 1, the numerical stability is rather poor. The approach of adopting $N = 1$ is a stage by stage method. For the solution of linear problem, both the two solution systems need only one decomposition if a direct solution scheme is used. However, the previous algorithm has better numerical stability.

16.1.3 Problems

Dynamic Response of a Shear-Deformable Axisymmetric Circular Plate

The dynamic response of an axisymmetric orthotropic laminated shear-deformable circular plate subjected to a uniformly distributed dynamic load, $q = 10^6 sin(5t)\ N/m$, was analyzed. The radius of the circular plate is 3 m, while the thickness of the plate is $h = 1\ m$. The laminate is a cross-ply with $[0^0/90^0/0^0]$ which has the orthotropic property in the radial and circumferential directions, with each ply being $\frac{1}{3}$ m thick. The top and bottom plies have the following elastic stiffnesses: $\bar{E}_1 = 14\ GPa$, $\bar{E}_2 = 7\ GPa$, $\bar{G}_{13} = 3.5\ GPa$, $\bar{\nu}_{12} = 0.25$, $\bar{\nu}_{21} = 0.125$, while the middle ply has the following elastic properties: $\bar{E}_1 = 7\ GPa$, $\bar{E}_2 = 14\ GPa$, $\bar{G}_{13} = 7\ GPa$, $\bar{\nu}_{12} = 0.125$, $\bar{\nu}_{21} = 0.25$. Then the reduced stiffnesses of each ply can be calculated by the following relationships: $\bar{Q}_{11} = \frac{\bar{E}_1}{1-\nu_{12}\nu_{21}}$, $\bar{Q}_{22} = \frac{\bar{E}_2}{1-\nu_{12}\nu_{21}}$, $\bar{Q}_{12} = \frac{\bar{\nu}_{12}\bar{E}_2}{1-\nu_{12}\nu_{21}}$. The elastic stiffnesses also have the following relationship: $\frac{\bar{\nu}_{12}}{\bar{E}_1} = \frac{\bar{\nu}_{21}}{\bar{E}_2}$. Consequently, the bending stiffnesses and shear modulus can be obtained by the integrations: $\bar{D}_{ij} = \int_{-h/2}^{h/2} \bar{Q}_{ij} z^2 dz$, $G_{13} = \frac{1}{h}\int_{-h/2}^{h/2} \bar{G}_{13} dz$. The values are: $\bar{D}_{11} = 1.1819992$ $GN{\cdot}m$, $\bar{D}_{22} = .62445241\ GN{\cdot}m$, $\bar{D}_{12} = .15053763\ GN{\cdot}m$, $G_{13} = 4.6666667$ GPa. The mass density is $\rho = 0.1019368\ Gkg/m^3$. One DQEM element adopting the nine-node Lagrange DQ model is used to the spacial discretization of the circular plate. In modelling the structure, the symmetry property at the center is considered for defining the boundary conditions. In carrying out the direct time integration, the Lagrange time-element by time-element method is used. The problem is solved by using various orders of time integration and sizes of time increment. Lateral displacement of four different time stages at the center are summarized and listed in Table 16.1. It shows that the developed algorithms are effective.

Table 16.1. Lateral displacements of the center, at four different time stages, of the orthotropically laminated shear-deformable circular plate subjected to a uniformly distributed dynamic load (mm)

DOF per step	Time increment	$t = .9\ (sec)$	$t = 1.5\ (sec)$	$t = 3.9\ (sec)$
3	.3	−.148676	−.0852588	.135237
	.15	−.180251	−.120042	.128247
	.075	−.183327	−.139845	.118845
4	.3	−.181309	−.166141	.199508
	.15	−.187509	−.162138	.118284
7	.3	−.180211	−.162005	.114624

Dynamic Response of a Deflected Euler-Bernoulli Beam

The problem solved involves the dynamic response of a simply supported beam deflected into an initial position and then released. The length of the beam, area of cross section, moment of inertia of cross section, Young's modulus and mass density are all equal to 1. With $x = 0.$ the origin of the coordinate, the initial deflection of the beam is expressed as $w(x,0) = x - 2x^3 + x^4$. The analytical solution of displacement response is expressed as $w(x,t) = \frac{96}{\pi^5}\sum_{n=0,1,2,\ldots}^{\infty} \frac{1}{(2n+1)^5} \sin\frac{(2n+1)\pi x}{\cos}[(2n+1)\pi]^2 t$. One DQEM element adopting the $13-DOF$ Lagrange DQ model was used to model the beam and carry out the spacial discretization to obtain the discrete equations of motion.

The dynamic response was solved by using the $C^1 - C^0$ time-element by time-element direct time integration procedure. In the analysis, damping effect is neglected. Numerical results of the mid-span displacements at four different time stages are summarized and listed in Table 16.2 with which the convergence and stability can be seen.

Dynamic Response of a Deflected Bar

The problem solved involves the dynamic response of a fixed-free bar deflected into an initial position and then released. The length of the bar, area of cross section, mass density and Young's modulus are all equal to 1. With $x = 0.$ the origin of the coordinate, the initial displacement of the bar is expressed as $u(x,0) = x$. One DQEM element adopting the $11 - DOF$ Chebyshev DQ model was used to model the bar and carry out the spacial discretization to obtain the discrete equations of motion of the bar. In the analysis, the stage by stage solution procedure was used to carry out the time integration. The damping effect is also neglected. Numerical results of using various orders of DQ approximation for the stage by stage procedure are summarized and listed in Table 16.3. The convergence can be seen in the table.

Table 16.2. Mid-span displacements of the beam at four different time stages ($C^1 - C^0$ time-element by time-element method)

DOF per time-element	Size of time-element	$t = .08$	$t = .12$	$t = .28$	$t = .32$
4	.04	.199747	.094101	−.201222	−.190752
	.02	.198464	.084182	−.250999	−.232869
	.01	.198954	.080568	−.281162	−.260233
5	.04	.226180	.128442	−.285116	−.317038
	.02	.229152	.131350	−.281540	−.316651
	.01	.227875	.132752	−.281850	−.314308
6	.04	.197956	.082509	−.265425	−.246877
	.02	.198491	.079470	−.288865	−.267533
	.01	.199701	.077879	−.301616	−.279772
7	.04	.213634	.103578	−.298788	−.303921
	.02	.214513	.103103	−.302044	−.308994
	.01	.214013	.102636	−.303409	−.312018
Anal. sol.		.220073	.118635	−.292681	−.312466

Table 16.3. Displacements of free end at four different time stages

Stages of DQ	Period of DQ	$t = 1.$	$t = 2.$	$t = 3.$	$t = 4.$
3	.25	.578140	.283307	.425257	.614877
	.125	.533093	.200649	.459974	.718572
	.0625	.515750	.141466	.482326	.800278
	.03125	.503459	.097964	.494900	.859948
4	.25	.433064	.083640	.460351	.894515
	.125	.495808	.068693	.513613	.899695
	.0625	.502787	.014689	.514166	1.08634
5	.25	.499217	.029679	.571059	.876318
	.125	.507078	.009580	.490286	.944378

Dynamic Response of a Deformed Membrane

The problem solved involves the dynamic response of a square membrane. Let $u(x, y, t)$ denote the displacement. The dynamic equilibrium equation is expressed by $\frac{\partial^2 u}{\partial x^2} + \frac{\partial^2 u}{\partial y^2} - \frac{\partial^2 u}{\partial t^2} = 0$. The boundary condition is $u = 0$ on the boundary of the membrane for all $t \leq 0$. The membrane is deflected into an initial position $u(x, y, 0) = sin\pi x sin2\pi y$ and then released. One DQEM element adopting the 11×11 Chebyshev DQ model was used to model the membrane and carry out the spacial discretization to obtain the discrete equation of motion of the membrane. In the analysis, Lagrange time-element by time-element procedure was used to carry out the time integration. Let T_1 denote the first natural frequency of the membrane. Numerical results of the displacement at (.5, .25) at four different time stages using various orders of DQ approximation for the Lagrange time-element by time-element procedure are summarized and listed in Table 16.4. It shows that the numerical stability and convergence rate are excellent. The problem was resolved by considering the effect of damping with the damping coefficient equal to 1. Then the dynamic equilibrium equation is expressed by $\frac{\partial^2 u}{\partial x^2} + \frac{\partial^2 u}{\partial y^2} - \frac{\partial^2 u}{\partial t^2} + \frac{\partial u}{\partial t} = 0$. Numerical results obtained are summarized and listed in Table 16.5. It also shows that the numerical stability and convergence rate are excellent.

Van der Pol's Equation

Van del Pol's equation is a self-excited nonlinear vibration equation expressed as

$$m\ddot{u} + c(u^2 - 1)\dot{u} + ku = 0$$

Table 16.4. Displacements of the point (.5, .25) at four different time stages (neglecting the effect of damping)

DOF per step	Time increment	$t/T_1 = .12$	$t/T_1 = .36$	$t/T_1 = .60$	$t/T_1 = .90$
3	.02	.747818	−.435300	−.633678	.413503
	.01	.736821	−.534268	−.708526	.588892
	.005	.732512	−.585537	−.755020	.693230
	.0025	.730648	−.611458	−.780991	.749796
4	.02	.725067	−.646534	−.792620	.823697
	.01	.728096	−.639930	−.805258	.813633
5	.02	.728900	−.638195	−.809501	.810936
	.01	.728962	−.637576	−.809021	.809503
6	.02	.728975	−.637478	−.808969	.809285
7	.02	.728969	−.637492	−.808942	.809307

Table 16.5. Displacements of the point (.5, .25) at four different time stages (considering the effect of damping)

DOF per step	Time increment	$t/T_1 = .12$	$t/T_1 = .36$	$t/T_1 = .60$	$t/T_1 = .90$
3	.02	.747818	−.435300	−.633679	.413504
	.01	.732030	−.592242	−.761318	.707620
	.005	.728301	−.644958	−.815156	.826296
	.0025	.726734	−.671467	−.845271	.890214
4	.02	.721229	−.708923	−.860561	.977129
6	.02	.725361	−.697966	−.877701	.957052

where u is the displacement, m is the mass, c is the damping coefficient and k is the elastic stiffness. A response analysis with $m/c = 10$ was analyzed using both the EDQ based time-element by time-element method and the Newmark β method, separately. In the EDQ based time integration analysis, $C^1 - C^0$ EDQ time-element by time-element procedure, generated by using the third order equivalent Lagrange DQ model with equally spaced nodes, having one auxiliary node outside the physical EDQ model to represent the time-element was used to integrate the discrete equation of motion. The size of time-element for the EDQ base time integration is ten times of the time step of the Newmark β method. In Fig. 16.1, the solid line represents the displacement response obtained by the Newmark β method, while the dot line represents the displacement response obtained by the EDQ based time integration method. It shows that the error of response obtained by the Newmark β method grows faster following the increase of time.

16.2 First Order Problems

DQ can be used to develop integration algorithms for solving discrete transient problems having first order derivatives with respect to the time variable. A representative of transient continuum mechanics problem is the transient response of field problems. The numerical procedures of the DQ based time integration algorithms for field problems are introduced.

16.2.1 DQ Basis Time-Element by Time-Element Integration Algorithm

Consider that the transient equation system, at a stage of the tth incremental step, of the transient problems is expressed by

$$C_{rs}\dot{U}_s^t + K_{rs}U_s^t = F_r^t \tag{16.12}$$

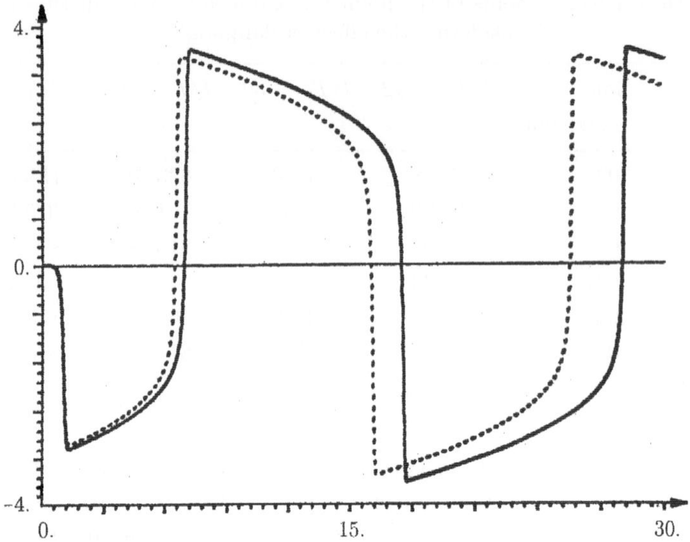

Fig. 16.1. Displacement response of a self-excited Van del Pol's equation. (dot line: EDQ based time integration; solid line: Newmark β method)

where C_{rs} and K_{rs} are coefficient matrices, U_s^t the vector of response variables and F_r^t the vector of external causes. Let Δt and τ denote the time increment or size of time-element and the natural coordinate with respect to the time t. Then, by using the DQ to discretize $\dot{U}_s^t$ in the above equation, the discrete equation at the time stage p of the tth incremental step can be expressed by the following equation

$$\left(\frac{1}{\Delta t}C_{rs}D_{pq}^{\tau} + K_{rs}\delta_{pq}\right)U_s^{t,q} = F_r^{t,p} \tag{16.13}$$

where $U_s^{t,q}$ are response variables. In order to solve the above equation, one initial condition is required. The transient responses can be updated by increasing the time, step by step. Each step represents a time-element. One initial condition is required for solving each incremental step. Let $\bar{U}_s^t$ denote the initial response variables of the tth incremental step. The initial condition is expressed as

$$U_s^{t,1} = \bar{U}_s^t \tag{16.14}$$

The values of $\bar{U}_s^t$ can be obtained from the solutions of the $(t-1)$th incremental step. The response histories can be updated by a time-element by time-element procedure.

The time-element used to develop the direct integration scheme is the Lagrange time-element having L stage nodes. For this time-element, discrete time stages for defining the discrete transient equations coincide the node stages of the element. Use the DOF assigned to the first node stage to define the condition of initial response variables. Then, by using Eqs. (16.12) and (16.14), the following matrix equation can be obtained

$$\bar{U}_s^t\{\bar{K}_{RI}\} + [\bar{K}_{RR}]\{U^{t,R}\} = \{\bar{F}^{t,R}\} \tag{16.15}$$

where

$$\{U^{t,R}\} = \lfloor\, U_s^{t,2} \;\; U_s^{t,3} \;\; \cdots \;\; U_s^{t,L} \,\rfloor^T,$$
$$\{\bar{F}^{t,R}\} = \lfloor\, F_s^{t,2} \;\; F_s^{t,3} \;\; \cdots \;\; F_s^{t,L} \,\rfloor^T,$$
$$\{\bar{K}_{RI}\} = \lfloor\, \tfrac{1}{\Delta t}C_{rs}D_{21}^{\tau} \;\; \tfrac{1}{\Delta t}C_{rs}D_{31}^{\tau} \;\; \cdots \;\; \tfrac{1}{\Delta t}C_{rs}D_{L1}^{\tau} \,\rfloor^T,$$

and

$$[\bar{K}_{RR}] = \begin{bmatrix} \frac{1}{\Delta t}C_{rs}D_{21}^{\tau} & \frac{1}{\Delta t}C_{rs}D_{22}^{\tau} & \cdots & \frac{1}{\Delta t}C_{rs}D_{2L}^{\tau} \\ \frac{1}{\Delta t}C_{rs}D_{31}^{\tau} & \frac{1}{\Delta t}C_{rs}D_{32}^{\tau} + K_{rs} & \cdots & \frac{1}{\Delta t}C_{rs}D_{3L}^{\tau} \\ \cdot & \cdot & \cdots & \cdot \\ \cdot & \cdot & \cdots & \cdot \\ \cdot & \cdot & \cdots & \cdot \\ \frac{1}{\Delta t}C_{rs}D_{L1}^{\tau} & \frac{1}{\Delta t}C_{rs}D_{L2}^{\tau} & \cdots & \frac{1}{\Delta t}C_{rs}D_{LL}^{\tau} + K_{rs} \end{bmatrix} \tag{16.16}$$

Using Eq. (16.15), displacements $\{U^{t,R}\}$ of the remaining $L-1$ node stages can be found

$$\{U^{t,R}\} = [\bar{K}_{RR}]^{-1}(\{\bar{F}^{t,R}\} - \bar{U}_s^t\{\bar{K}_{RI}\}) \tag{16.17}$$

It should be noted that if the problem is linear and that the size of the time-element is constant, then $[\bar{K}_{RR}]$ is a constant matrix. Consequently, only one decomposition is necessary for updating the response histories if a direct solution scheme is used.

16.2.2 DQ Basis Stages by Stages Integration Algorithm

A different algorithm was developed to solve the discrete transient equation system. In this algorithm, the equally spaced grid Lagrange DQ is used to discretize $\dot{U}_s^t$. One incremental step DQ solution similar to the time-element by time-element solution procedure of the previous algorithm is first carried out. The DOF assigned to the first node is used to define the initial condition. After this solution step, a time increment is increased. Assume that the Lagrange DQ model has L nodes and that the value of the time increment equals the distance spanning N+1 consecutive stages (nodes) of the initial solution step. N further time stages are thus defined. Then, by using this newly increased N stages and the previous L-N stages, the same DQ model as the first DQ solution can be used to discretize $\dot{U}_s^t$ at the newly increased N stages. The discretized equation is expressed by

$$\left(\frac{1}{\Delta t}C_{rs}D_{\alpha m}^{\tau} + K_{rs}\delta_{\alpha m}\right)\hat{U}_s^m$$
$$= \hat{F}_r^{\alpha}, \quad \alpha = L-N+1, L-N+2, ..., L \tag{16.18}$$

where $\hat{U}_s^m$ is the response vector of the DQ temporal discretization, and $\hat{F}_r^\alpha$ is the vector of external cause of the newly increased N stages. Since only response variables at the newly increased N stages are unknowns, matrix partition technique can be used to obtain an equation system with unknowns the response variables at the newly increased N stages. This solution system is smaller than the solution system of the time-element by time-element solution algorithm and the first incremental step DQ solution of the current algorithm which has the response variables of L-1 stages as unknowns. In this direct integration algorithm, the response histories can be updated by a stages by stages solution procedure. For solving transient problems, the maximum value of N will be L-1. If the value of N is larger than 1, the numerical stability is rather poor. The approach of adopting $N = 1$ is a stage by stage method.

16.2.3 Problems

Let ρ and σ denote the density and specific heat of the medium, respectively. For the two-dimensional nonuniform problem with orthotropic medium, the governing equation is expressed by

$$(\tilde{k}_x T_{,x})_{,x} + (\tilde{k}_y T_{,y})_{,y} + \tilde{Q} - \rho\sigma \frac{\partial T}{\partial t} = 0 \tag{16.19}$$

The transient heat conduction of the problem shown in Fig. 11.11 is solved. The values of $\sigma\rho$ of the left medium and right medium are 1 and 2, respectively. The same spacial discretization procedures as those used to solve the problem in Section 11.5 with the element grid, 11 × 11, are used. The Lagrange time-element by time-element scheme is used to directly integrate the discrete transient equation system. Numerical results of the temperature $T_{,E}$ at E of four different time stages for the solutions using various orders of time integration and sizes of time increment are summarized and listed in Table 16.6. It shows fast convergence.

Table 16.6. Temperatures of point E at four different time stages

DOF per step	Time increment	$t = .02$	$t = .04$	$t = .06$	$t = .08$
2	.02	$.296029\times10^0$	$.418709\times10^0$	$.501445\times10^2$	$.577310\times10^0$
	.01	$.329958\times10^0$	$.420525\times10^0$	$.506143\times10^2$	$.587442\times10^0$
	.005	$.344314\times10^0$	$.416052\times10^0$	$.510106\times10^2$	$.594914\times10^0$
3	.02	$.369079\times10^0$	$.411212\times10^0$	$.516184\times10^2$	$.602999\times10^0$
	.01	$.348732\times10^0$	$.409533\times10^0$	$.517565\times10^2$	$.604012\times10^0$

References

1. R. Bellman and J. Casti, 1971, "Differential Quadrature and Long-term Integration", *J. Math. Anal. Appl.*, **34**, 234-238.
2. R. Bellman and B.G. Kashef, 1972, "Application of Splines and Differential Quadrature to Partial Differential Equation of Hodgkin-Huxley Type", Proc. 5th Hawaii Intl. Conf. Sys. Sci. Suppl., Hawaii, USA.
3. R. Bellman, B.G. Kashef and J. Casti, 1972, "Differential Quadrature: A technique for the rapid solution of nonlinear partial defferential equations", *J. Comput. Phys.*, **10**, 40-52.
4. R. Bellman, *Methods of Nonlinear Analysis, Volume 2*, Academic Press, New York, 1973.
5. R. Bellman, *Partial Differential Equations*, D Reidel Publishing, Dirdrecht, Netherlands, 1985.
6. R. Bellman, *Methods in Approximation*, D Reidel Publishing, Dirdrecht, Netherlands, 1986.
7. B. Kashef and R. Bellman, 1974, "Solution of the Partial Differential Equation of the Hodgkin-Huxley Model Using Differential Quadrature", *Math. Biosci.*, **19**, 1-8.
8. R. Bellman, B.G. Kashef and R. Vasudevan, 1974, "The Inverse Problem of Estimating Heart Parameters from Cardiograms", *Math. Biosci.*, **19**, 221-230.
9. R. Bellman and R.S. Roth, 1979, "System Identification with Partial Information", *J. Math. Anal. Appls.*, **68**, 321-333.
10. R. Bellman and R.S. Roth, 1979, "A Scanning Technique for System Identification", *J. Math. Anal. Appls.*, **71**, 403-411.
11. G. Naadimuth, R. Bellman, K.M. Wang and E.S. Lee, 1984, "Differential Quadrature and Partial Differential Equations: Some Numerical Results", *J. Math. Anal. Appls.*, **98**, 220-235.
12. J.O. Mingle, 1973, "Computational Considerations in Nonlinear Diffusion", *Intl. J. Numer. Methods Engr.*, **7**, 103-116.
13. J.O. Mingle, 1977, "The Method of Differential Quadrature for Transient Nonlinear Diffusion", *J. Math. Anal. Appl.*, **60**, 559-569.
14. F. Civan and C.M. Sliepcevich, 1983, "Solution of the Poisson Equation by Differential Quadrature", *Intl. J. Numer. Methods Engr.*, **19**, 711-724.
15. F. Civan and C.M. Sliepcevich, 1983, "Application of Differential Quadrature to Transport Processes", *J. Math. Anal. Appls.*, **93**, 206-221.

16. F. Civan and C.M. Sliepcevich, 1984, "Differential Quadrature for Multidimensional Problems", *J. Math. Anal. Appls.*, **101**, 423-443.
17. F. Civan and C.M. Sliepcevich, 1984, "On the Solution of the Thomas-Fermi Equation by Differential Quadrature", *J. Comp. Phys.*, **56**, 343-348.
18. S.K. Jang, C.W. Bert and A.G. Striz, 1989, "Application of Differential Quadrature to Static Analysis of Structural Components", *Intl. J. Numer. Methods Engr.*, **28**, 561-577.
19. C.W. Bert and M. Malik, 1996, "Differential Quadrature Method in Computational Mechanics: a review", *ASME, Appl. Mech. Rev.*, **49**, 1-28.
20. C.N. Chen, 1995, "A Differential Quadrature Element Method", Proc. 1st Intl. Conf. Engr. Comput. Computer Simul., **1**, 25-34, Changsha, CHINA
21. C.N. Chen, 1996, "The Two-dimensional Differential Quadrature Element Method Frame Model", ASME, Proc. 15th Intl. Conf. Offshore Mechs. Arctic Engr., **1**, 283-290, Florence, ITALY
22. C.N. Chen, 1997, "The Two-dimensional Frame Model of the Differential Quadrature Element Method', *Comput. Struct.*, **62**, 555-571.
23. C.N. Chen, 1998, "The Warping Torsion Bar Model of The Differential Quadrature Element Method", *Comput. Struct.*, **66**, 249-257.
24. C.N. Chen, 1998, December, "Solution of Beam on Elastic Foundation by DQEM", *ASCE J. Engr. Mechs.*, **124**, 1381-1384.
25. C.N. Chen, 1998, "Potential Flow Analysis by Using the Irregular Elements of the Differential Quadrature Element Method", ASME, Proc. 17th Intl. Conf. Offshore Mechs. Arctic Engr., Paper No. OMAE98-361, in CD ROM, Lisbon, PORTUGAL.
26. C.N. Chen, 1999, "The Development of Irregular Elements for Differential Quadrature Element Method Steady-state Heat Conduction Analysis", *Comput. Methods Appl. Mechs. Engrg.*, **170**, 1-14.
27. C.N. Chen, 1999, "The Differential Quadrature Element Method Irregular Element Torsion Analysis Model", *Appl. Math. Model.*, **23**, 309-328.
28. A.G. Striz, W.L. Chen and C.W. Bert, 1994, "Static Analysis of Structures by the Quadrature Element Method", *Intl. J. Solid Structures*, **31**, 2807-18.
29. A.G. Striz, W.L. Chen and C.W. Bert, 1995, "High Accuracy Plane Stress and Plate Elements in the Quadrature Element Method", Proc. 36th AIAA/ASME/ASCE/AHS/ASC Struct Struct. Dyn. Mat. Conf. AIAA/ASME Adap. Struct. For., 957-965, New Orleans, USA.
30. C. Shu and B.E. Richards, 1992, "Application of Generalized Differential Quadrature to Solve Two-dimensional Incompressible Navier-Stoke Equation", *Intl. J. Numer. Method Fluids*, **15**, 791-798.
31. K.M. Liew and F.L. Liu, 2000, "Differential Quadrature Method for Vibration Analysis of Shear Deformable Annular Sector Plates", *J. Sound Vib.*, **230**, 335-356.
32. C. Shu and and H. Du, 2000, "Free Vibration Analysis of Curvilinear Quadrilateral Plates by the Differential Quadrature Method", *J. Comput. Phy.*, **163**, 452-466.
33. M. Malik and C.W. Bert, "Vibration Analysis of Plates with Curvilinear Quadrilateral Planforms by DQM Using Blending Functions", *J. Sound Vib.*, **230**, 949-954.
34. C.N. Chen, 1998, "Extended Differential Quadrature", Applied Mechanics in the Americas (Proc. 6th Pan Amer. Congr. Appl. Mechs., Rio de Janeiro, BRAZIL), **6**, 389-392, American Academy of Mechanics, USA.

35. C.N. Chen, 1999, "Differential Quadrature Element Analysis Using Extended Differential Quadrature", *Comput. Math. Appls.*, **39**, 65-79.
36. C.N. Chen, 1999, "Generalization of Differential Quadrature Discretization", *Numerical Algorithms*, **22**, 167-182.
37. C.N. Chen, 2000, "A Generalized Differential Quadrature Element Method", *Comput. Methods Appl. Mechs. Engr.*, **188**, 553-566.
38. C.N. Chen, 1998, "A Generalized Coordinate Differential Quadrature Element Method", Proc. 4th Intl. Conf. Numer. Methods Appls., 775-783, Sofia, BULGARIA.
39. C.N. Chen, 1999, "Finite Difference Discretizations by Differential Quadrature Techniques", *J. Commu. Numer. Methods Engr.*, **15**, 823-833.
40. C.N. Chen, 2000, "Solution of Composite Nonuniform Plate Problems by the Differential Quadrature Finite Difference Method", *Comput. Mechs.*, **26**, 223-228.
41. C.N. Chen, 2001, "Differential Quadrature Finite Difference Method for Structural Mechanics Problems", *J. Commu. Numer. Methods Engr.*, **17**, 423-441.
42. C.N. Chen, 1998, "A Differential Quadrature Finite Element Method", Applied Mechanics in the Americas (Proc. 6th Pan Amer. Congr. Appl. Mechs., Rio de Janeiro, BRAZIL), *6*, 309-312, American Academy of Mechanics, USA.
43. C.N. Chen, 2001, "DQFEM Analyses of Static and Dynamic Nonlinear Elastic-plastic Problems Using a GSR-based Accelerated Constant Stiffness Equilibrium Iteration Technique", *ASME Trans., J. Pressure Vessel Technology*, **123**, 310-317.
44. C.N. Chen, 2002, "Extended GDQ and Related Discrete Element Analysis Methods for Transient Analyses of Continuum Mechanics Problems", ASME, Proc. Pressure Vessels & Piping Conference, PVP-Vol. 441, 37-44, Vancouver, CANADA.
45. C.N. Chen, 2003, "Dynamic Response of Timoshenko Beam Structures Solved by DQEM Using EDQ", *Intl. J. Nonl. Sci. Numer, Simul.*, **4**, 251-263.
46. C.N. Chen, 2004, "Dynamic Response of Shear-deformable Axisymmetric Anisotropic Circular Plate Structures Solved by the DQEM and EDQ Based Time Integration Schemes", *Compos. Struc.*, **64**, 339-348
47. C. Shu and B.E. Richards, 1992, "Parallel Simulation of Incompressible Viscous Flows by Generalized Differential Quadrature", *Comput. Syst. Engr.*, **3**, 271-281.
48. C. Shu, B.C. Khoo, K.S. Yeo and Y.T. Chew, 1994, "Application of GDQ scheme to Simulate Natural Convection in a Square Cavity", *Intl. Commun. Heat Mass Transfer*, **21**, 809-817.
49. M. Abramowitz and I.A. Stegun, 1965, *Handbook of Mathematical Functions*, Dover, New York.
50. F. Stenger, 2000, "Summary of Sinc Numerical Methods", *J. .Comput. Appl. Math.*, **121**, 379-420.
51. C.N. Chen, 1995, "A Solution of Structural Vibration by Using Bernoulli and Euler Functions", Proc. 19th Natl. Conf. Theor. Appl. Mechs., **2**, 69-76, ROC.
52. C.N. Chen, 1996, "A Solution of Structural Buckling by Using Bernoulli and Euler Functions", Proc. 20th Natl. Conf. Theor. Appl. Mechs., 3, 229-236, ROC.
53. C.N. Chen and C.H. Wang, 1997, "Structural Analysis by an Inverse Procedure Using Bernoulli and Euler Functions", *Engr. Struct.*, **19**, 724-732.

54. C.N. Chen, 2004, "Dynamic responses of frame structures solved using DQEM and EDQ based time integration method", ASME, Proc. Pressure Vessels & Piping Conference, PVP-Vol. 482, San Diego, California, USA.
55. C.N. Chen, 2004, "Differential Quadrature, Generalized Methods, Related Discrete Element Analysis Methods and EDQ Based Time Integration Method for Composite Structural Problems", Proc. 7th Intl. Conf. Comput. Struc. Tech. & 4th Intl. Conf. Engr. Comput. Tech., in CD-ROM, Lisbon, PORTUGAL.
56. C.N. Chen, 2005, "DQEM Analysis of In-plane Vibration of Curved Beam Structures", *Adv. Engr. Software*, **36**, 412-424.
57. A.W. Leissa, 1973, "The Free Vibration of Rectangular Plates", J. Sound Vib., **31**, 257-293.
58. A. Jennings, 1966, "A Compact Storage Scheme for the Solution of Linear Simultaneous Equations", *Computer J.*, **9**, 281-285.
59. R.M. Jones, 1972, "Effective Development, Documentation, and Distribution of Computer Programs", *Comput. Struct.*, **2**, 1089-1095.
60. E. Cuthill, 1972, "Several Strategies for Reducing the Bandwidth of Matrices", in *Sparse Matrices and Their Applications*, D.J. Rose and R.A. Willoughby (eds.), Plenum Press, 1972.
61. R.J. Collins, 1973, "Bandwidth Reduction by Automatic Renumbering", *J. Commu. Numer. Methods Engr.*, **17**, 423-441.
62. A.R. Curtis and J.K. Reid, 1971, "The Solution of Large Sparse Unsymmetric Systems of Linear Equations", Proc. IFIP Conf., Ljubljana, YUGOSLAVIA.
63. D.P. Mondkar and G.H. Powell, 1974, "Large Capacity Equation Solver for Structural Analysis", *Comput. Struct.*, **4**, 699-728.
64. D.M. Brandon, 1972, "The Implementation and Use of Sparse Matrix Techniques in General Simulation Programs", *Computer J.*, **17**, 165-171.
65. J.K. Reid, *Large Sparse Sets of Linear Equations*, Academic Press, 1971.
66. J.E. Key, 1973, "Computer Program for Solution of Large, Sparse, Unsymmetric Systems of Linear Equations", *Intl. J. Numer. Methods Engr.*, **6**, 497-509.
67. B.M. Irons, 1970, "A Frontal Solution Program for Finite Element Analysis", *Intl. J. Numer. Methods Engr.*, **2**, 5-32.
68. P. Hood, 1976, "Frontal Solution Program for Unsymmetric Matrices", *Intl. J. Numer. Methods Engr.*, **10**, 379-399.
69. A. Razzaque, 1980, "Automatic Reduction of Frontwidth for Finite Element Analysis", *Intl. J. Numer. Methods Engr.*, **15**, 1315-1324.
70. G. Beer and W. Haas, 1982, "A Partitioned Frontal Solver for Finite Element Analysis", *Intl. J. Numer. Methods Engr.*, **18**, 1623-1654.
71. S.W. Sloan and M.F. Randolph, 1983, "Automatic Element Reordering for Finite Element Analysis with Frontal Solution Schemes", *Intl. J. Numer. Methods Engr.*, **19**, 1153-1181.
72. S.W. Sloan, 1986, "An Algorithm for Profile and Wavefront Reduction of Sparse Matrices", *Intl. J. Numer. Methods Engr.*, **23**, 239-251.
73. D.A.H. Jacob, "The Exploitation of Sparsity by Iterative Methods", in *Sparse Matrices and Their Uses* (I.S. Duff, Ed.), Academic Press, 1981.
74. D.N.G. Allen, *Relaxation Methods in Engineering and Science*, McGraw-Hill, 1954.
75. A. Jennings, 1971, "Accelerating the Convergence of Matrix Iterative Processes", *J. Inst. Math. Appls.*, **8**, 99-110.

76. A. Jennings and G.M. Malik, 1978, " The Solution of Sparse Linear Equations bt the Conjugate Gradient Method", *Intl. J. Numer. Methods Engr.*, **12**, 141-158.
77. J.K. Reid, "On the Method of Conjugate Gradients for the Solution of Large Sparse Systems of Linear Equations", in *Large Sparse Sets of Linear Equation* (Ed. Reid), Academic Press, 1971.
78. J.P. Jackson and P.C. Robinson, 1985, "A Numerical Study of Various Algorithms Related to the PCG Method", *Intl. J. Numer. Methods Engr.*, **21**, 1315-1338.
79. C. Popa, 1995, "Preconditioning Conjugate Gradient Method for Nonsymmetric Systems", *Intl. J. Comput. Math.*, **58**, 117-133.
80. M. Papadrakakis and S. Bitzarakis, 1996, "Domain Decomposition PCG Methods for Serial and Parallel Processing", *Adv. Engr. Software*, **25**, 291-307.
81. C.N. Chen, 1995, "A Global Secant Relaxation (GSR) Method-based Predictor-corrector Procedure for the Iterative Solution of Finite Element Systems", *Comput. Struct.*, **54**, 199-205.
82. J.M. Ortega, *Introduction to Parallel and Vector Solution of Linear Systems*, Plenum Press, 1988.
83. C.C. Ashcraft, R.G. Grimes, J.G. Lewis, B.W. peyton and H.D. Simon, 1987, "Progress in Sparse Matrix Methods for Large Linear Systems on Vector Supercomputers", *Intl. J. Supercomputer Appls.*, **1**, 10-30.
84. T.K. Agarwal, O.O. Storaasli and D.T. Nguyen, 1994, "A Parallel-Vector Algorithm for Rapid Structural Analysis on High-Performance Computers", *Comput. Struct.*, **51**, 503-512.
85. A.R. Gourlay and G.A. Watson, *Computational Methods for Matrix Eigenproblems*, John Wiley & Sons, 1973.
86. C.B. Moler and G.W. Stewart, 1973, "An Algorithm for Generalized Matrix Eigenvalue Problems", *SIAM J. Numer. Anal.*, **10**, 241-256
87. K.J. Bathe and E.L. Wilson, 1973, "Solution Methods for Eigenvalue Problems in Structural Mechanics", *Intl. J. Numer. Methods Engr.*, **6**, 213-226.
88. A. Jennings and W.J. Stewart, 1975, "Simultaneous Iteration for Partial Eigensolution of Real Matrices", *J. Inst Math. Appls.*, **15**, 351-361.
89. M. Papadrakakis, 1984, "Solution of the Partial Eigenproblem by Iterative Methods", *Intl. J. Numer. Methods Engr.*, **20**, 2283-2301.
90. B. Nour-Omid, B.N. Parlett and R.L. Taylor, 1983, "Lanczos Versus Subspace Iteration for Solution of Eigenvalue Problems", *Intl. J. Numer. Methods Engr.*, **19**, 859-871.
91. P. Robert, 1984, "The Accelerated Power Method", *Intl. J. Numer. Methods Engr.*, **20**, 1179-1191.
92. F.A. Akl, W.H. Dilger and B.M. Irons, 1982, "Acceleration of Subspace Iteration", *Intl. J. Numer. Methods Engr.*, **18**, 583-589.
93. M. Papadrakakis and M. Yakoumidakis, 1987, "On the Preconditioned Conjugate Gradient Method for Solving $(\mathbf{A} - \lambda\mathbf{B})\mathbf{X} = \mathbf{0}$", *Intl. J. Numer. Methods Engr.*, **24**, 1355-1366.
94. S.P. Timoshenko, *Vibration Problems in Engineering, Third Edition*, D. Van Nostrand Company, Inc., 1955.
95. H. Reismann and P.S. Pawlik, *Elasticity Theory and Applications*, John Wiley & Sons, 1980.
96. J.M. Gere and S.P. Timoshenko, *Mechanics of Materials, Second SI Edition*, Wadsworth International, 1985.

97. C.N. Chen, 2002, "DQEM Vibration Analyses of Nonprismatic Beams Resting on Elastic Foundations", *Intl. J. Struct. Stab. Dyns.*, **1**, 99-115.
98. Lau, J. H., 1984, "Vibration Frequencies of Tapered Bars with End Mass", *J. Applied Mechanics*, **51**, 179-181.
99. C.N. Chen, 2003, "DQEM Analysis of Buckling of Nonprismatic Columns with and without Elastic Foundation", *Intl. J. Struct. Stab. Dyns.* , **3**, 183-193.
100. S.P. Timoshenko and J.M. Gere, *Theory of Elastic Stability, Second Edition*, McGraw-Hill, 1961.
101. J.D. Faires and R. Burden, *Numerical Methods, Second Edition*, Brooks/Cole Publishing Company, 1998.
102. C.N. Chen, 2001, "Analysis of 3-D Frame Problems by DQEM Using EDQ", *Adv. Engr. Software*, **32**, 395-407.
103. C.N. Chen, 2001, "The DQEM Analysis of Vibration of Frame Structure Having Nonprismatic Members Considering Warping Torsion", *Intl. J. Nonl. Sci. Numer, Simul.*, **2**, 235-255.
104. J.B. Carr, 1970, "The Effects of Shear Flexibility and Rotary Inertia on the Natural Frequencies of Uniform Beams", *Aeron. Quart.*, **21**, 79-90.
105. A.H. Gibson and E.G. Ritchie, *A Study of the Circular Arc Bow Girder*, Constable & Co. Ltd., 1914.
106. C.N. Chen, 2003, "Out-of-Plane Deflection of Nonprismatic Curved Beam Structures Solved by DQEM", *Adv. Engr. Software*, **34**, 297-306.
107. E. Volterra, "Deflection of a Circular Beam out of Its Initial Plane", *Trans. Am. Soc. Civil Engrs.*, Paper No. 2727, **120**, 65-91.
108. F. Kikuchi, 1982, "Accuracy of Some Finite Element Models for Arch Problems", *J. Commu. Numer. Methods Engr.*, **35**, 315-345.
109. B.K. Lee and J.F. Wilson, 1989, "Free Vibrations of Arches with Variable Curvature", *J. Sound Vib.*, **136**, 75-89.
110. C.N. Chen, 2005, "Differential Quadrature Element Method Analysis of In-Plane Deflection of Vertically Curved Beam Structures", *CAD Civ. Infr. Engr.*, **20**, 132-141.
111. T. Tsumura, et al., *Strength Design Data Book*, Shokabo Publishing Co., LtD., Tokyo, 1956.
112. A.I. Borisenko and I.E. Tarapov, *Vector and Tensor Analysis with Applications*, Prentice-Hall, Inc., 1968.
113. W.L. Oberkampf, 1976, "Domain Mappings for the Numerical Solution of Partial Differential Equations", *Intl. J. Numer. Methods Engr.*, **10**, 211-223.
114. A. El-Zafrany and R.A. Cookson, 1986, "Derivation of Lagrangian and Hermitian Shape Functions for Quadrilateral Elements", *Intl. J. Numer. Methods Engr.*, **23**, 1939-1958.
115. A. El-Zafrany and R.A. Cookson, 1986, "Derivation of Lagrangian and Hermitian Shape Functions for Triangular Elements", *Intl. J. Numer. Methods Engr.*, **23**, 275-285.
116. B. Bigdeli and D.W. Kelly, 1997, "C^*-convergence in the Finite Element Method", *Intl. J. Numer. Methods Engr.*, **40**, 4405-4425.
117. C.N. Chen, 2000, "Solution of Continuum Mechanics Problems by GDQEM Using EDQ", *Intl. J. Comput. Engr. Sci.*, **1**, 207-233.
118. C. Hadlock, *Field Theory and Its Classical Problems*, Mathematical Association of America, 2000.
119. I.N. Sneddon, *Mixed Boundary Value Problems in Potential Theory*, North-Halland Pub. Co., 1966.

120. S.C. Hunter, *Mechanics of Continuous Media*, Ellis Horwood Ltd., 1976.
121. P.R. Garabedian, *Partial Differential Equations*, John Wiley & Sons, 1964.
122. N.N. Lebedev, I.P. Skalskaya and Y.S. Uflyand, *Worked Problems in Applied Mathematics*, Dover, 1965.
123. C.N. Chen, 2004, "DQEM and DQFDM Irregular Elements for Analyses of 2-D Heat Conduction in Orthotropic Media", *Appl. Math. Model.*, **28**, 617-638.
124. J.R. Vinson and R.L. Sierakowski, *The Behavior of Structures Composed of Composite Materials*, Martinus Nijhoff Publishers, 1986.
125. C.N. Chen, 2003, "DQEM AND DQFDM for the Analysis of Composite Two-dimensional Elasticity Problems", *Comp. Struc.*, **59**, 3-13.
126. R. Szilard, *Theory and Analysis of Plates - Classical and Numerical Methods*, Prentice-Hall, Inc., 1974.
127. C.N. Chen, 2003, "Vibration of plate structures solved by DQEM using EDQ", ASME, Proc. 22th Intl. Conf. Offshore Mechs. Arctic Engr., Paper No. OMAE2003-37332, in CD-ROM, Cancun, Mexico.
128. C.N. Chen, 1998, "A Generalized Coordinate Differential Quadrature Element Method", Proc. 4th Intl. Conf. Numer. Methods Appls., 775-783, Sofia, BULGARIA.

I.S. Sokolnikoff, *Mathematical Theory of Elasticity*, second edition, McGraw-Hill, 1956.

Index

MIX
Papier aus verantwortungsvollen Quellen
Paper from responsible sources
FSC® C105338

If you have any concerns about our products,
you can contact us on
ProductSafety@springernature.com

In case Publisher is established outside the EU,
the EU authorized representative is:
Springer Nature Customer Service Center GmbH
Europaplatz 3, 69115 Heidelberg, Germany

Printed by Libri Plureos GmbH
in Hamburg, Germany